T0093077

Mathematics of the Big Four Casino Table Games

AK Peters/CRC Recreational Mathematics Series

The Untold Story of Everything Digital
Bright Boys, Revisited
Tom Green

Wondrous One Sheet Origami
Meenakshi Mukerji

The Geometry of Musical Rhythm
What Makes a "Good" Rhythm Good?, Second Edition
Godfried T. Toussaint

A New Year's Present from a Mathematician
Snezana Lawrence

Six Simple Twists
The Pleat Pattern Approach to Origami Tessellation Design, Second Edition
Benjamin DiLeonardo-Parker

Tessellations
Mathematics, Art, and Recreation
Robert Fathauer

Mathematics of Casino Carnival Games
Mark Bollman

Mathematical Puzzles
Peter Winkler

X Marks the Spot
The Lost Inheritance of Mathematics
Richard Garfinkle, David Garfinkle

Luck, Logic, and White Lies
The Mathematics of Games, Second Edition
Jörg Bewersdorff

Mathematics of The Big Four Casino Table Games
Blackjack, Baccarat, Craps, & Roulette
Mark Bollman

Star Origami
The Starrygami™ Galaxy of Modular Origami Stars, Rings and Wreaths
Tung Ken Lam

For more information about this series please visit: https://www.routledge.com/AK-PetersCRC-Recreational-Mathematics-Series/book-series/RECMATH?pd=published,for thcoming&pg=2&pp=12&so=pub&view=list

Mathematics of the Big Four Casino Table Games

Blackjack, Baccarat, Craps, & Roulette

Mark Bollman

CRC Press
Taylor & Francis Group
Boca Raton London New York

CRC Press is an imprint of the
Taylor & Francis Group, an **informa** business

A CHAPMAN & HALL BOOK

First edition published 2022
by CRC Press
6000 Broken Sound Parkway NW, Suite 300, Boca Raton, FL 33487-2742

and by CRC Press
2 Park Square, Milton Park, Abingdon, Oxon, OX14 4RN

CRC Press is an imprint of Taylor & Francis Group, LLC

Library of Congress Cataloging-in-Publication Data

Names: Bollman, Mark, author.
Title: Mathematics of the big four casino table games : blackjack,
   baccarat, craps, & roulette / Mark Bollman.
Description: First edition. | Boca Raton : C&H\CRC Press, 2021. | Includes
   bibliographical references and index.
Identifiers: LCCN 2021019434 (print) | LCCN 2021019435 (ebook) | ISBN
   9780367742294 (hardback) | ISBN 9780367740900 (paperback) | ISBN
   9781003156680 (ebook)
Subjects: LCSH: Games of chance (Mathematics) | Gambling--Mathematics. |
   Probabilities.
Classification: LCC QA271 .B66 2021 (print) | LCC QA271 (ebook) | DDC
   519.2/7--dc23
LC record available at https://lccn.loc.gov/2021019434
LC ebook record available at https://lccn.loc.gov/2021019435

ISBN: 978-0-367-74229-4 (hbk)
ISBN: 978-0-367-74090-0 (pbk)
ISBN: 978-1-003-15668-0 (ebk)

DOI: 10.1201/9781003156680

Typeset in Lartin Modern font
by KnowledgeWorks Global Ltd.

For my nephices:
Haley, Joe, Patrick, Kendyl, Kate, Jessica,
Emily, Sarah, Mathew, Megan, Natalie, & Henry.

Contents

Preface   ix

Acknowledgments   xi

1 Essential Probability   1
   1.1 Elementary Ideas   1
   1.2 Addition and Multiplication Rules   6
   1.3 Combinatorics   11
   1.4 Random Variables and Expected Value   16

2 Roulette   27
   2.1 Roulette Basics   27
   2.2 History of Roulette   32
   2.3 New Wheels   36
   2.4 Electronic Roulette Games   45
   2.5 Roulette with Cards   54
   2.6 Side Bets   64
   2.7 Roulette Betting Systems   82
   2.8 Exercises   94

3 Craps   105
   3.1 Craps Basics   106
   3.2 Hazard   116
   3.3 Street Craps   119
   3.4 Crooked Dice   124
   3.5 Controlled Shooting   136
   3.6 Variations   142
   3.7 Card Craps   155
   3.8 Side Bets   167
   3.9 Exercises   184

4 Baccarat   191
   4.1 Baccarat Basics   191
   4.2 Card Counting in Baccarat   197
   4.3 Variations   204
   4.4 Side Bets   217
   4.5 Exercises   232

5   Blackjack   **239**
    5.1   Blackjack Basics   239
    5.2   Basic Strategy   246
    5.3   Card Counting   257
    5.4   Variations   266
    5.5   California Games   284
    5.6   Side Bets   301
    5.7   Exercises   320

Answers to Selected Exercises   **325**

References   **333**

Index   **345**

Preface

In *Mathematics of Casino Carnival Games* [9], I wrote about the mathematics behind a wide variety of unusual casino games, but made a deliberate choice to minimize work on variations to the "Big Four" games of baccarat, blackjack, craps, and roulette. The numerous variations and modifications of these four games are the focus of this book. These games are the foundation of any casino's table games pit and have basic rules that have become standardized, with good mathematics behind the basic games. In addition, there are many rule variations and proposed alterations that are less standard, enough to fill this separate volume on their own.

In addition to exploring the mathematics behind the base games, I am interested in preserving some of the history of changes, both successful and unsuccessful. Often in mathematics and science, we lose track of the failed attempts at solving problems, and this extends to casino game development. Additionally, some of these examples raise interesting questions for game designers, such as how to design pay tables to achieve a desirable—for both players and casinos—house advantage.

The four games considered here are described and analyzed in increasing order of their mathematical complexity.

- Many probabilities that we encounter in **roulette** are a simple matter of dividing by 37 or 38 (or, with the recent growth in popularity of Sands or triple-zero roulette, 39). Some electronic configurations and alternate wheels lead to more complicated calculations, but very few game elements or variations require advanced arithmetic. This chapter includes an examination of betting systems for roulette—while many of these, including the martingale and Fibonacci progression, are equally applicable to other games with bets paying 1–1, they are most frequently marketed to roulette players.

- **Craps** mathematics calls for some means of handling the low-probability event that a single Pass line decision might take a very large number of rolls to be resolved. Since we're dealing with convergent geometric series, this is easily done.

- **Baccarat** is not subject to the assumption of independence, since cards dealt in one hand cannot be dealt again until the shoe is shuffled. However, many good approximations are possible using the infinite deck assumption to simplify calculations, and we shall do so here when appropriate.

- Like baccarat, **blackjack** deals dependent hands, and some calculations may be possible using the infinite deck approximation. When dealing with the mathematics behind card counting, though, it is necessary to account for the removal of cards as the dealer moves through a shoe, and so the simple assumptions must be discarded.

Acknowledgments

I am immensely grateful for the comments of two anonymous reviewers who saw an early rough version of this book and made numerous helpful suggestions that greatly improved the text and the mathematics.

It has always been a pleasure to work with the editorial team at Taylor & Francis in bringing this project forward. Callum Fraser continues to be a great editor and a great colleague, and Mansi Kabra has done excellent work shepherding this project through to publication. On the production side, I'd like to thank Paul Boyd, Production Editor, and Riya Bhattacharya, Associate Project Manager, for their guidance in the final days of manuscript preparation.

Spider craps (page 149) was developed by Albion College student Jacob Engel during a summer research program in 2011. Funding for this project was provided by Albion's Foundation for Undergraduate Research, Scholarship, and Creative Activity (FURSCA).

Chapter 1

Essential Probability

The mathematics behind roulette, craps, baccarat, and blackjack is drawn from *probability theory* at various levels of complexity. This is appropriate, because the origins of this branch of mathematics lie with the analysis of simple games of chance. This began with a series of letters between Blaise Pascal and Pierre de Fermat in 1654 that raised and answered several questions at the foundation of probability theory while addressing questions that arose in resolving gambling games [35]. Later mathematicians have developed and expanded this topic into a rigorous field of mathematics with many important applications unrelated to gambling. In this chapter, we shall outline the fundamental ideas of probability that are essential for analyzing casino games.

1.1 Elementary Ideas

We begin our study of probability with the careful definition of some important terms.

Definition 1.1. An *experiment* is a process whose outcome is determined by chance.

This may not seem like a useful definition. We illustrate the concept with several examples.

Example 1.1. Roll a standard six-sided die and record the number that results. ■

Example 1.2. Roll two standard six-sided dice (abbreviated as 2d6) and record the sum. ■

Example 1.3. Roll 2d6 and record the larger of the two numbers rolled (or the number rolled, if both dice show the same number). ■

Example 1.4. Deal a five-card video poker hand and record the number of aces it contains. ■

One trait of an experiment is that it results in a single definite outcome. While we will eventually concern ourselves with individual outcomes, we begin by looking at all of the possible results of an experiment.

Definition 1.2. The *sample space* **S** of an experiment is the set of all possible outcomes of the experiment.

Example 1.5. In Example 1.1, the sample space is **S** = {1, 2, 3, 4, 5, 6}. The same sample space applies to the experiment described in Example 1.3. ∎

Example 1.6. In Example 1.2, the sample space is **S** = {2, 3, 4, ..., 12}.

It is important to note that the 11 elements of **S** are not equally likely, as this will play an important part in our explorations of probability. Rolling a 7 is more likely than rolling any other sum on two dice; 2 and 12 are the least likely sums. ∎

When we're only interested in some of the possible outcomes of an experiment, we are looking at subsets of **S**. These are called *events*.

Definition 1.3. An *event* A is any subset of the sample space **S**. An event is called *simple* if it contains only one element.

Example 1.7. In Example 1.2, we rolled 2d6 and considered the sum as the outcome. The event "the roll is an even number" is the subset {2, 4, 6, 8, 10, 12}.

The event "The roll is a 7" is a simple event. Although there are 6 different ways to roll a sum of 7 on 2d6, we are only looking at the sum in this experiment, and not the two numbers appearing on the dice. ∎

Definition 1.4. Two events A and B are *disjoint* if they have no elements in common. In this case, we say that the event consisting of all outcomes common to A and B is the *empty set* ∅, which contains zero elements.

Example 1.8. The standard 52-card deck contains cards 13 cards in each of 4 suits: spades (♠), hearts (♡), diamonds (♢), and clubs (♣). Spades and clubs are traditionally black; hearts and diamonds are red. Each suit contains 13 cards: ace (1) through 10, jack, queen, and king. Some games add one or more jokers to a deck, which can have different values depending on the game.

Suppose that we draw one card from a standard deck and note its suit. The events A = {The card is a heart} and B = {The card is a club} are disjoint events, since no card has more than one suit. ∎

A particularly important example of disjoint sets that speeds some probability calculations is the *complement* of a given event.

Definition 1.5. The *complement* of an event A, denoted A', is the set of all elements of the sample space that do not belong to A.

Example 1.9. In Example 1.8, the complement of the event {The card is a heart} is simply {The card is not a heart}. A more complete description would be {The card is a spade, diamond, or club}. ∎

The following result, called the *Complement Rule*, connects the probability of an event A to the probability of its complement A'.

Theorem 1.1. *(Complement Rule): For any event A,*

$$P(A') = 1 - P(A).$$

The Complement Rule frequently turns out to be useful in simplifying probability calculations.

Example 1.10. The probability of winning at least once in 10 spins of a roulette wheel can be written as

$$P(1) + P(2) + \cdots + P(10).$$

where $P(n)$ represents the probability of winning n times, or much more easily as

$$1 - P(0),$$

the complement of the event "Lose on every spin". ∎

Informally, the *probability* of an event A is an attempt to measure how likely that event is to occur, which is a number $P(A)$ between 0 and 1—or between 0 and 100%. Simple probability is often a matter of little more than careful counting. For our work in probability, we will frequently be interested in the size of a set—that is, how many elements it has. For convenience, we introduce the following notation:

Definition 1.6. The expression $\#(A)$ denotes the number of elements in a set A.

Example 1.11. If A is a standard deck of playing cards, then $\#(A) = 52$. ∎

The challenge here is that a set of interest in gambling mathematics can be very large. If we are interested in the set A of all possible 7-card stud poker hands, then $\#(A) = 133,784,560$, and we'd like to have a way to come up with that number without having to list all of the hands and count them. Techniques for finding the size of such large sets will be discussed in Section 1.3.

To progress further, we need to develop procedures for assigning numerical probabilities to events.

Definition 1.7. Let $\mathbf{S}$ be a sample space in which all of the outcomes are equally likely, and suppose A is an event within $\mathbf{S}$. The *probability* of the event A, denoted $P(A)$, is

$$P(A) = \frac{\text{Number of elements in } A}{\text{Number of elements in } \mathbf{S}} = \frac{\#(A)}{\#(\mathbf{S})}.$$

Since the size of any event is necessarily less than or equal to the size of the sample space under consideration, the following theorem is a consequence of Definition 1.7.

Theorem 1.2. *For any event A,* $0 \leqslant P(A) \leqslant 1$.

There are several ways by which we might determine the value of $P(A)$. These methods vary in their mathematical complexity as well as in their level of precision. Each of them corresponds to a question we might ask or try to answer about a given probabilistic situation.

1. **Theoretical Probability**

   If we are asking the question *"What's supposed to happen?"* and relying on pure mathematical reasoning rather than on accumulated data, then we are computing the *theoretical probability* of an event.

   Example 1.12. For a single Pick 4 lottery ticket, there are 10,000 possible outcomes, from 0000–9999. Only one number is a winner. The theoretical probability of winning is then $\dfrac{1}{10,000}$. ∎

   Example 1.13. If we roll 2d6, what is the probability of getting a sum of 7?

   An incorrect approach to this problem is to note that the sample space is $\mathbf{S} = \{2, 3, 4, 5, 6, 7, 8, 9, 10, 11, 12\}$, and since one of those 11 outcomes is 7, the probability must be $\frac{1}{11}$. This fails to take into account the fact that some rolls occur more frequently than others—for example, while there is only one way to roll a 2, there are 6 ways to roll a 7: 1-6, 2-5, 3-4, 4-3, 5-1, and 6-1. (It may be useful to think of the dice as being different colors, so that 3-4 is a different roll from 4-3, even though the numbers showing are the same.) Counting up all of the possibilities shows that there are $6 \cdot 6 = 36$ ways for two dice to land. Since six of those yield a sum of 7, the correct answer is $P(7) = \frac{6}{36} = \frac{1}{6}$. ∎

2. **Experimental Probability**

   When our probability calculations are based on actual data drawn from repeated observations, the resulting value is the *experimental probability* of A. Here, we are answering the question *"What really did happen?"*

   Example 1.14. Suppose that you toss a coin 100 times and that the result of this experiment is 48 heads and 52 tails. The experimental probability of heads in this experiment is $48/100 = .48$, and the experimental probability of tails is $52/100 = .52$. ∎

   This experimental probability is different from the theoretical probability of getting heads on a single toss, which is ½. This is not unusual.

   Experimental probability can be a useful tool in its own right, as when processing a quantity of empirical data, and also as an approximation to theoretical probability when the numbers involved are too large for easy handling.

The connection between theoretical and experimental probability is described in a mathematical result called the *Law of Large Numbers*.

Theorem 1.3. *(Law of Large Numbers) Suppose an event has theoretical probability p. If x is the number of times that the event occurs in a sequence of n trials, then as the number of trials n increases, the experimental probability x/n approaches p.*

Using the language of limits, this becomes

$$\lim_{n \to \infty} \frac{x}{n} = p.$$

Informally, the Law of Large Numbers states that, in the long run, things happen in an experiment the way that theory says that they do. What is meant by "in the long run" is not a fixed number of trials, but will vary depending on the experiment. For some experiments, $n = 500$ may be a large number, but for others—particularly if the probability of success or failure is small—it may take far more trials before the experimental probabilities get acceptably close to the theoretical probabilities.

Example 1.15. When tossing a fair coin, the theoretical probability of heads is ½. In Example 1.14, the experimental probability of heads, based on 100 tosses, was $P(H) = .48$. While this is not the .5 we expect, 100 is a small number of trials. As the number of computer-simulated coin tosses increases as shown in Table 1.1, we see that the experimental probability approaches the expected theoretical probability of ½.

TABLE 1.1: Results from repeated tosses of a fair coin

Number of tosses	Number of heads	$P(H)$	$\left\lvert P(H) - \frac{1}{2} \right\rvert$
100	48	.4800	.02000
1000	532	.5320	.03200
10,000	5009	.5009	.00090
100,000	49,975	.49975	.00025
1,000,000	500,681	.50068	.00032
10,000,000	4,999,674	.49997	.00003
100,000,000	49,999,478	.49999	.00001

■

Note that the *number* of heads and tails do not get closer together as the number of tosses increases—in the run of 100 million tosses, there were 1044 more tails than heads—but the *ratio* of heads to tails gets closer to 1, which is what the Law of Large Numbers tells us to expect.

1.2   Addition and Multiplication Rules

Our next challenge will be to extend our understanding of probability to *compound* events: events that can be broken down into several simple events. We can find the probability of these simple events using techniques from Section 1.1. The results of this section allows us to combine those probabilities correctly to find probabilities of more complicated events.

Mutually Exclusive Events

Definition 1.8. Two events A and B are *mutually exclusive* if they have no elements in common—that is, if they cannot occur together.

Example 1.16. When rolling 2d6, the events "The sum is odd" and "The sum is even" are mutually exclusive. ■

Example 1.17. It is not necessary that mutually exclusive events exhaust the sample space, as they did in Example 1.16. In the same experiment, the events "The sum is even" and "The sum is a 9" are mutually exclusive. There are several outcomes that belong to neither event. ■

In computing probabilities, we may be in a situation where we know $P(A)$ and $P(B)$ and want to know the probability that either A or B occurs: $P(A$ or $B)$. The addition rules described next allow us to compute this new probability in terms of the known ones.

Theorem 1.4. *(First Addition Rule) If A and B are mutually exclusive events, then*

$$P(A \text{ or } B) = P(A) + P(B).$$

If A and B are not mutually exclusive, a slightly more complicated formula can be used to calculate $P(A$ or $B)$.

Theorem 1.5. *(Second Addition Rule) If A and B are any two events, then*

$$P(A \text{ or } B) = P(A) + P(B) - P(A \text{ and } B).$$

Proof. By definition,

$$P(A \text{ or } B) = \frac{\#(A \text{ or } B)}{\#(\mathbf{S})}.$$

What we need to do is compute $\#(A$ or $B)$. Elements of A or B can be counted by adding together the number of elements of A and of B, but if any elements belong to both, they have just been counted twice. In order that each element is only counted once, we must subtract out the number of elements that belong to both A and B. This gives

$$\#(A \text{ or } B) = \#(A) + \#(B) - \#(A \text{ and } B).$$

Dividing by $\#(\mathbf{S})$ completes the proof. $\qquad\square$

We can see that the First Addition Rule is a special case of the Second, for if A and B are mutually exclusive, then they cannot occur together; hence $P(A \text{ and } B) = 0$.

Example 1.18. Roll 2d6. If A is the event "The sum is odd" and B is the event "The roll is doubles", then A and B are mutually exclusive, since the sum of a doubles roll must be an even number. The First Addition Rule gives

$$P(A \text{ or } B) = P(A) + P(B) = \frac{1}{2} + \frac{6}{36} = \frac{2}{3}.$$

$\blacksquare$

Example 1.19. If we draw one card from a standard deck, the events $A =$ "The card is a face card" and $B =$ "The card is a diamond" are not mutually exclusive, since there are 3 diamond face cards: the $J\Diamond, Q\Diamond$, and $K\Diamond$ that are part of both events. As a result, the probability that A and B occur together is $\frac{3}{52}$. By the Second Addition Rule, we have

$$P(A \text{ or } B) = P(A) + P(B) - P(A \text{ and } B) = \frac{12}{52} + \frac{1}{4} - \frac{3}{52} = \frac{11}{26}.$$

$\blacksquare$

Independent Events

Definition 1.9. Two events A and B are *independent* if the occurrence of one has no effect on the occurrence of the other one.

From the perspective of probability, this definition states that the occurrence of one of two (or more) independent events does not affect the probability of the others occurring.

Two events that are mutually exclusive are explicitly *not* independent, since the occurrence of one eliminates the chance of the other occurring. Moreover, two events that are independent cannot be mutually exclusive.

Example 1.20. In drawing a single card from a deck, the events $A =$ "The card is a spade" and $B =$ "The card is a face card" are independent. We can confirm this by computing some probabilities. $P(A) = \frac{1}{4}$. If we are told that the card is a face card before its suit is revealed, we have $P(A) = \frac{3}{12} = \frac{1}{4}$, since there are 12 face cards in a deck and 3 of them are spades. Since $P(A)$ has not changed, knowing that B has occurred does not affect the chance of A occurring. Similar reasoning shows that knowing that a drawn card is a spade does not affect the probability that it's a face card: $P(\text{Face card}) = \frac{3}{13}$ either way. $\blacksquare$

It is a fundamental principle of gambling mathematics that *successive trials of random experiments are independent*. This includes successive die rolls at craps, wheel spins at roulette, and weekly drawings of six Powerball numbers, but *not* dealt hands in baccarat or blackjack. In these games, a card played in one hand is a card that cannot be played in the next hand. Since the composition of the deck has changed, we are not considering successive trials of the same random experiment.

This principle is not always well-understood by gamblers, and the inability or unwillingness to understand the doctrine of independent trials is sometimes called the *Gambler's Fallacy*. This fallacy is commonly committed by roulette players who have too strong a belief in the Law of Large Numbers, although it can crop up in any game where successive trials are independent.

Example 1.21. When betting on the number 28 for two consecutive spins of a roulette wheel, the outcome on the first spin has no effect on the results of the second. Winning (or losing) on the first spin neither raises nor lowers the probability of winning the second wager, since the two wheel spins are independent. ∎

Example 1.22. Many casinos provide a scoreboard of sorts near their roulette wheels which displays the results of the last 10–20 spins. If 14 of the last 15 spins have been red numbers, it would nonetheless be incorrect to assume that the wheel is biased to favor red numbers and bet big on red. Successive spins are independent, with no influence felt from previous spins or exerted on subsequent spins. ∎

Gambling devices don't understand the laws of probability. They have no knowledge of the mathematics we humans have devised to describe their actions, and they certainly don't understand what the long-term distribution of results is supposed to be. For the same reason, in Example 1.22, it would be equally erroneous to bet on the black numbers on the grounds that they're somehow "due".

If A and B are independent events, it is a simple matter to compute the probability that they occur together, with the use of a theorem called the *Multiplication Rule*.

Theorem 1.6. *(Multiplication Rule)* *If A and B are independent events, then*

$$P(A \text{ and } B) = P(A) \cdot P(B).$$

Informally, the Multiplication Rule states that we can find the probability that two successive independent events occur by multiplying the probability of the first by the probability of the second. The Multiplication Rule can be extended to any finite number of independent events: the probability of a sequence of n independent events is simply the product of the n probabilities of the individual events.

Example 1.23. The probability of a sum of 7 when 2d6 are rolled is $\frac{1}{6}$. Since successive rolls are independent, the probability of two 7s in consecutive rolls is

$$\frac{1}{6} \cdot \frac{1}{6} = \frac{1}{36}.$$

∎

Example 1.24. Return to Example 1.22. The probability of 14 straight American roulette spins turning up red is

$$\left(\frac{18}{38}\right)^{14} \approx \frac{1}{34,927}.$$

While this is very unlikely, looking at all roulette wheels in all the casinos of the world over time gives a vast pool to draw from and makes it almost certain that this will happen somewhere in the world at some time. The record for the longest run of one color in American roulette history is a run of 32 consecutive red numbers [149]. ∎

Example 1.25. For quick calculations, it may be useful to assume that the probability of drawing a card of a particular rank in baccarat or blackjack is constant, regardless of how many cards have been previously dealt. This is called the *infinite deck approximation*, because it is equivalent to assuming that the game is dealt with infinitely many decks of cards shuffled together. With this assumption in force, successive cards dealt in these games are independent, and if 6 or more decks or a continuous shuffling machine are used, the results computed are usually very good estimates of the actual probabilities from a finite deck.

Using the infinite deck approximation in a hand of blackjack, the probability of a pair of aces is

$$\left(\frac{1}{13}\right)^{2} = \frac{1}{169}.$$

∎

Conditional Probability

If the events A and B are not independent, we will need to generalize Theorem 1.6 to handle the new situation. This generalization requires the idea of *conditional probability*. We begin with an example.

Example 1.26. If we draw one card from a standard deck, the probability that it is a queen is $\frac{4}{52} = \frac{1}{13}$. If, however, we are told that the card is a face card, the probability that it's a queen is $\frac{4}{12} = \frac{1}{3}$—that is, additional information has changed the probability of our event by allowing us to restrict the sample space. If we denote the events "The card is a queen" by Q and "The card is a face card" by F, this last result is written $P(Q \mid F) = \frac{1}{3}$ and read as "the (conditional) probability of Q given F is $\frac{1}{3}$". ∎

Example 1.27. In Example 1.25, we saw that the probability of a pair of aces in the infinite deck approximation is $\frac{1}{169}$. If the hand is dealt from a fresh 4-deck shoe and the infinite deck approximation is not used, this probability is

$$\frac{16}{208} \cdot \frac{15}{207} = \frac{5}{897}.$$

The approximation differs from this true value by less than 4×10^{-4}. ∎

The fundamental idea here is that more information can change probabilities. If we know that the event A has occurred and we're interested in the event B, we are now not looking for $P(B)$, but $P(B$ and $A)$, because only the part of B that overlaps with A is possible. With that in mind, we have the following formula for conditional probability:

Definition 1.10. The *conditional probability* of B given A is

$$P(B \,|\, A) = \frac{P(B \text{ and } A)}{P(A)}.$$

This formula divides the probability of the intersection of the two events by the probability of the event that we know has already occurred. Note that if A and B are independent, we immediately have $P(B \,|\, A) = P(B)$, since then $P(B$ and $A) = P(A) \cdot P(B)$. This is one case where more information—in this case, the knowledge that A has occurred—does not change the probability of B occurring.

As with the addition rules, we can state a second, more general, version of the Multiplication Rule that applies to any two events—independent or not—and reduces to the first rule when the events are independent. This more general rule simply incorporates the conditional probability of B given A, since we are looking for the probability that both occur.

Theorem 1.7. (*General Multiplication Rule*) *For any two events A and B, we have*

$$P(A \text{ and } B) = P(A) \cdot P(B \,|\, A).$$

Proof. This result follows from the fact that $P(A$ and $B) = P(B$ and $A)$ and from Definition 1.10. □

Example 1.28. When rolling 2d6, let A be "The sum is an even number". and B be "The sum is greater than 9". We have

$$P(A) = \frac{1}{2}$$

and

$$P(B \,|\, A) = \frac{4}{18},$$

from which the General Multiplication Rule gives

$$P(A \text{ and } B) = \frac{1}{2} \cdot \frac{4}{18} = \frac{4}{36}.$$

Direct counting among the 36 possible rolls of 2d6 confirms this result. ∎

1.3 Combinatorics

Combinatorics is the branch of mathematics that studies counting techniques. In many applications of Definition 1.7, $P(A) = \#(A)/\#(S)$, the sheer number of elements comprising an event or the sample space is far too large to count them one by one. When computing probabilities, we seldom have need to consider each of these simple events individually; we are usually only interested in how many there are.

Fundamental Counting Principle

Frequently in gambling mathematics, we find ourselves considering the number of ways in which several events can happen in sequence, such as successive spins of a roulette wheel or the draw of several cards to a poker hand. If we know the number of ways that each individual event can happen, elementary combinatorics tells us that simple multiplication can be used to find the answer.

Theorem 1.8. (*Fundamental Counting Principle*) *If there are n independent tasks to be performed, such that task T_1 can be performed in m_1 ways, task T_2 can be performed in m_2 ways, and so on, then the number of ways in which all n tasks can be performed successively is*

$$N = m_1 \cdot m_2 \cdot \ldots \cdot m_n.$$

That the Fundamental Counting Principle is a reasonable result can be easily seen by testing out some examples with small numbers and listing all possibilities—for example, when rolling 2d6, one red and one green, each die is independent of the other and can land in any of six ways. By the Fundamental Counting Principle, there are $6 \cdot 6 = 36$ ways for the two dice to fall, and this may be confirmed by writing out all of the possibilities.

As an example of how this principle is used, we can show that the number of possible events associated with a probability experiment is exponentially greater than the number of elements in the sample space. Specifically, we have the following result.

Theorem 1.9. *If $\#(S) = n$, then there are 2^n events that may be chosen from S.*

Another way to state Theorem 1.9 is to say that a set with n elements has 2^n subsets. For example, if we toss a coin, we have $\mathbf{S} = \{H, T\}$. Since we have $\#(\mathbf{S}) = 2$, we expect $2^2 = 4$ events. The complete list of events contained within $\mathbf{S}$ is $\varnothing, \{H\}, \{T\}, \{H, T\}$, which is indeed a 4-element set.

Proof. Let A be an event contained in a sample space **S**. For every element x of the sample space, either x belongs to A or x does not belong to A. That is, there are two possibilities for each element: "Yes, it is in A" or "No, it is not in A." Every arrangement of the n possible yes and no answers corresponds to a different subset of **S**, and thus to a different event. For example, choosing "no" for every element gives $\varnothing$, and at the other extreme, choosing "yes" each time gives **S** itself.

There are n choices to be made, and two options for each of the n choices. Therefore, by the Fundamental Counting Principle, there are

$$\underbrace{2 \cdot 2 \cdot \ldots \cdot 2}_{n \text{ terms}} = 2^n$$

possible events. $\qquad\qquad\qquad\qquad\qquad\qquad\qquad\qquad\qquad\qquad\qquad\square$

A special case of the Fundamental Counting Principle arises when we consider the number of ways N to arrange a set of n elements, with no repetition allowed, in different orders. The first element may be chosen in n ways, the second in $n - 1$, and so on, down to the last item, which may be chosen in only 1 way. The total number of orders for a set of n elements is thus $N = n \cdot (n - 1) \cdot (n - 2) \cdot \ldots \cdot 3 \cdot 2 \cdot 1$. This number is given a special name, n *factorial*.

Definition 1.11. If n is a natural number, the *factorial* of n, denoted $n!$, is the product of all of the positive integers up to and including n:

$$n! = 1 \cdot 2 \cdot 3 \cdot \ldots \cdot (n - 1) \cdot n.$$

$0! = 1$, by definition.

It is an immediate consequence of the definition that $n! = n \cdot (n - 1)!$. Factorials get very big very fast. $4! = 24$, but then $5! = 120$ and $6! = 720$. $10!$ is greater than 3 million, and $52!$, which is the number of different ways to arrange a standard deck of cards, is approximately 8.0658×10^{67}.

Permutations: Counting When the Order Matters

Definition 1.12. A *permutation* of r items from a set of n items is a selection of r items chosen so that the order matters.

For example, ABC is a different choice of three alphabet letters than CBA. It should be noted that "order" may appear in several forms. One way to determine whether or not order matters in making a selection is to ask if different elements of the selection are being treated differently once they are chosen.

Example 1.29. One option when betting on a horse race is a *trifecta* wager, where the bettor must pick the first 3 horses to finish in order. Order matters: it's not enough simply to select the first three finishers. $\qquad\qquad\blacksquare$

Example 1.30. *Rouleno* is a casino game played with 6 standard pool balls, numbered from 1–6. The balls are scrambled in a box and all 6 are drawn out one by one [9]. Players may wager on the number of balls which are drawn into their "correct" place—as when the #3 ball is drawn third, for example—or whether a given ball will be drawn into its correct place. Since the order in which the balls are drawn matters, we use permutations to count the number of ways that the balls may be drawn. ∎

We are usually not as interested in a list of all of the permutations of a set as in how many permutations there are. The following theorem allows easy calculation of that number, which is denoted $_nP_r$.

Theorem 1.10. *The number of permutations of r items chosen from a set of n items is*

$$_nP_r = \frac{n!}{(n-r)!}.$$

Proof. There are n ways to select the first item. Once an item is chosen, it cannot be chosen again, so the second item may be chosen in $n-1$ ways. There are then $n-2$ items remaining for the third choice, and so on until there are $n-r+1$ numbers remaining from which to choose the rth and final term. By the Fundamental Counting Principle, we have

$$_nP_r = n \cdot (n-1) \cdot \ldots \cdot (n-r+1).$$

Multiplying the right-hand expression by $1 = (n-r)!/(n-r)!$ gives

$$_nP_r = n \cdot (n-1) \cdot \ldots \cdot (n-r+1) \cdot \frac{(n-r)!}{(n-r)!}$$
$$= \frac{n \cdot \ldots \cdot (n-r+1) \cdot (n-r) \cdot \ldots \cdot 3 \cdot 2 \cdot 1}{(n-r)!}$$
$$= \frac{n!}{(n-r)!}.$$

□

Example 1.31. In Example 1.29, suppose that the race had 12 horses starting. The number of possible ways to fill out a trifecta ticket is then

$$_{12}P_3 = \frac{12!}{9!} = 12 \cdot 11 \cdot 10 = 1320.$$

∎

Example 1.32. Rouleno (Example 1.30) draws 6 balls out of 6, and the order matters. The number of possible rouleno draws is

$$_6P_6 = \frac{6!}{0!} = 720.$$

∎

Example 1.33. The first column of a bingo card, under the letter B, contains 5 numbers selected from the range 1–15. See Figure 1.1.

FIGURE 1.1: A bingo card. Each column contains numbers drawn from an interval of 15 numbers.

Since order matters in laying out a bingo card, the number of ways to fill in the first column is

$$_{15}P_5 = \frac{15!}{10!} = 360,360.$$

The same count applies to columns I, G, and O. Due to the center Free square, column N contains only 4 numbers, drawn from the interval 31–45. Accordingly, the number of choices for column N is

$$_{15}P_4 = 15 \cdot 14 \cdot 13 \cdot 12 = 32,760.$$

By the Fundamental Counting Principle, the number of possible bingo cards is

$$(360,360)^4 \cdot 32,760 \approx 5.524 \times 10^{26}.$$

■

Combinations: Counting Without Regard to Order

Most of the time when gambling, we are not so concerned about the order of events, as when a hand of cards is dealt or a set of Powerball numbers is drawn. For counting these arrangements, we are interested in *combinations* rather than permutations.

Definition 1.13. A *combination* of r items from a set of n items is a subset of r items chosen without regard to order. The number of such combinations is denoted $\binom{n}{r}$, which is read as "n choose r." This value is sometimes denoted $_nC_r$.

Here, ABC and CBA are interchangeable combinations, as they are subsets of the alphabet consisting of the same three letters. The different order is not a concern here. If the elements of a selected subset are receiving the same treatment once selected, then the choice is a combination, not a permutation.

Theorem 1.11. *The number of combinations of r items chosen from a set of n items is*

$$\binom{n}{r} = \frac{n!}{(n-r)! \cdot r!} = \frac{_nP_r}{r!}.$$

Proof. We begin with the formula for the number of permutations:

$$_nP_r = \frac{n!}{(n-r)!}.$$

Since we are looking for combinations, two permutations that differ only in the order of the elements are identical to us. Any combination of r elements from a set of n can be rearranged into $r!$ different orders that are interchangeable, by the Fundamental Counting Principle. We then have

$$\binom{n}{r} = \frac{_nP_r}{r!} = \frac{n!}{(n-r)! \cdot r!},$$

as desired. □

Example 1.34. The number of 5-card poker hands that can be dealt from a standard 52-card deck is

$$\binom{52}{5} = 2,598,960.$$

Combinations are used here rather than permutations since the order in which the cards are dealt does not matter. A poker player may arrange the 5 cards in some particular order as he or she may prefer, but changing the order of the cards does not change the rank of the hand. ∎

Example 1.35. Contract bridge deals 13 cards to each of 4 players, using the full deck. The number of possible 13-card bridge hands is then

$$\binom{52}{13} = \frac{52!}{13! \cdot 39!} = 635,013,559,600.$$

∎

The following theorem collects several simple facts about combinations.

Theorem 1.12. *For all $n \geqslant 0$:*

1. $\binom{n}{0} = \binom{n}{n} = 1$ *and* $\binom{n}{1} = n.$

2. For all k, $0 \leqslant k \leqslant n$, $\binom{n}{k} = \binom{n}{n-k}$.

3. $\sum_{r=0}^{n} \binom{n}{r} = 2^n$.

Proof.

1. Given a set of n elements, there is only one way to select none of them— that is, there is only one way to do nothing, so $\binom{n}{0} = 1$. Similarly, since the order does not matter, there is only one way to choose all of the items: $\binom{n}{n} = 1$.

   If we are choosing only one item, we may select any element from among the n, and there are thus n choices possible.

2. We note that every selection of k items from a set of n partitions the set into two disjoint subsets: one of size k and the other of size $n - k$, and so choosing k items to take is equivalent to choosing $n - k$ items to leave behind. The conclusion follows immediately.

   Alternately, direct application of the formula for combinations gives the following:

   $$\binom{n}{k} = \frac{n!}{(n-k)! \cdot k!} = \frac{n!}{k! \cdot (n-k)!} = \frac{n!}{[n-(n-k)]! \cdot (n-k)!} = \binom{n}{n-k}.$$

3. If we think of a combination of r items from a set A with $\#(A) = n$ as choosing a subset of A with r elements, then the left side of this equation is simply the total number of subsets of A of all sizes.

   From Theorem 1.9, we know that the number of subsets of a set with n elements is 2^n. Since we have counted the set of all subsets of A in two different ways, those two expressions must be the same, completing the proof.

   $\square$

1.4   Random Variables and Expected Value

Definition 1.14. A *random variable* is an unknown quantity X whose value is determined by a chance process.

This is another definition that, on its face, isn't terribly useful—indeed, this phrasing comes perilously close to using the words "random" and "variable" in the definition of random variable. Once again, a sequence of examples will illustrate this important idea far better than a formal definition.

Example 1.36. Roll 2d6 and let X denote their sum. X then takes on a value in the set $\{2, 3, 4, 5, 6, 7, 8, 9, 10, 11, 12\}$. We see that X is simply the outcome of the experiment described in Example 1.2. ■

Example 1.37. In a five-card poker hand, let X count the number of aces it contains. $X = 0, 1, 2, 3,$ or 4. ■

Example 1.38. In a hand of blackjack, let X denote the sum of the first two cards, counting the first ace as 11. Here, X lies in the set $\{3, 4, 5, 6, \ldots, 19, 20, 21\}$. (A hand containing two aces would be counted here as 12, not 2 or 22.) ■

Definition 1.15. A *probability distribution* for a random variable X is a list of the possible values of X, together with their associated probabilities.

Probability distributions are most commonly presented in a table of values or as an algebraic formula, which is called a *probability distribution function* or *PDF*.

Example 1.39. Let X denote the number of times a fair d6 must be rolled until the first 4 comes up. The event "$X = k$" corresponds to $k-1$ consecutive non-4s and then a single roll of 4. Since successive rolls of a fair die are independent, the probability distribution function is

$$P(X = k) = \left(\frac{5}{6}\right)^{k-1} \cdot \frac{1}{6}.$$

■

Example 1.40. If X denotes the number that appears when a fair d6 is rolled, then the probability distribution for X is shown in Table 1.2.

TABLE 1.2: Probability distribution when a d6 is rolled

x	$P(X = x)$
1	1/6
2	1/6
3	1/6
4	1/6
5	1/6
6	1/6

Since the probabilities in the table are all the same, the simple events listed are equally likely. ■

Expected Value and House Advantage

The notion of *expected value* is fundamental to any discussion of random variables and is especially important when those random variables arise from a game of chance. The expected value of a random variable X is, in some sense, an "average" value, or what we might expect in the long run if we were to sample many values of X.

The common notion of "average" corresponds to what statisticians call the *mean* of a set of numbers: add up all of the numbers and divide by how many numbers there are. For a random variable X, this approach requires some fine-tuning, as there is no guarantee that a small sample of values of X will be representative of the range of possible values. Our interpretation of average will incorporate each possible value of X together with its probability, computing what is in some sense a long-term average over a very large hypothetical sample.

Definition 1.16. Let X be a random variable with a given probability distribution function $P(X = x)$. The *expected value* or *expectation* $E(X)$ of X is computed by multiplying each possible value for X by its corresponding probability and then adding the resulting products:

$$E(X) = \sum_x x \cdot P(X = x).$$

This expression may be interpreted as a standard mathematical mean drawn from an infinitely large random sample. If we were to draw such a sample, we would expect that the *proportion* of sample elements with the value x would be $P(X = x)$; adding up over all values of x gives this formula for $E(X)$.

We may abbreviate $E(X)$ to E when the random variable is clearly understood. The notation $\mu = E(X)$, where μ is the Greek letter mu, is also common, particularly when the expected value appears as a term in another expression, as in Definition 1.21 (page 22).

Example 1.41. If X denotes the outcome when a single d6 is rolled, the PDF from Example 1.40 shows that the expected value of X is

$$E(X) = \sum_{k=1}^{6} k \cdot \frac{1}{6} = \frac{21}{6} = 3\frac{1}{2}.$$

■

The notation used in Definition 1.16 does not indicate the limits of the indexing variable x, as is customary with sums; this is because those values may not be a simple list running from 1 to some n. When written this way, we should take the sum over *all* possible values of the random variable X, as in the next example.

Example 1.42. Let X denote the outcome of a $1 bet on any one single number in roulette. X has 2 possible values:

- If the bet wins, $X = 35$.
- If the bet loses, $X = -1$.

The expected value of X is then

$$E(X) = (35) \cdot P(X = 35) + (-1) \cdot P(X = -1).$$

The values of $P(X = 35)$ and $P(X = -1)$ depend on the wheel in use; we shall compute them in Chapter 2. ■

A concept related to expected value is the *house advantage* associated with a game of chance.

Definition 1.17. The *house advantage (HA)* of a game with a wager of N and payoffs given by the random variable X is $-E(X)/N$.

If the expectation is negative, as it is in virtually every casino game, the HA will be positive. The HA measures the proportion of the total amount wagered that can be reliably expected to be retained by the casino or lottery agent, in the long run.

Example 1.43. The Michigan Lottery's Daily 3 drawing calls on players to select a 3-digit number, from 000–999. If the player's number matches the 3-digit number selected in the daily drawing, the ticket pays 500–1, but since the player's wager is not returned with the winnings, the net win is 499–1. The possible values of a random variable X measuring the outcome are therefore 499 and –1. The expected value of a $1 wager is

$$E = (499) \cdot \frac{1}{1000} + (-1) \cdot \frac{999}{1000} = -\frac{1}{2}.$$

The house advantage is 50%—in the long run, the state keeps half of the money wagered. ■

Definition 1.18. If X is a random variable measuring the payoffs from a game, we say that the game is *fair* if $E(X) = 0$.

Example 1.44. Suppose you gamble with a friend on the toss of a coin. If heads is tossed, you win $1; if tails is tossed, you pay $1. Since a fair coin can be expected to land heads and tails equally often, the expected value for this game is $E = (1)(\frac{1}{2}) + (-1)(\frac{1}{2}) = 0$. The game is fair. ■

If a game is fair, then in the long run, we expect to win exactly as much money as we lose, and thus, aside from any possible entertainment derived from playing, we expect no gain. This is often summarized in the following maxim:

> *If a game is fair, don't bother to play.*
> *If a game is unfair, make sure it's unfair in your favor.*

Failure to heed this advice, of course, is responsible for the ongoing success of the gambling industry, for games which are unfair and favor the gambler are rare.

An important principle of expected value is contained in the following theorem.

Theorem 1.13. *If* $X_1, X_2, \ldots, X_n$ *are random variables, then*

$$E(X_1 + X_2 + \cdots + X_n) = E(X_1) + E(X_2) + \cdots + E(X_n).$$

Another way to state this result is to say that *expectation is additive*.

Example 1.45. Example 1.43 showed that the expected value of a $1 Daily 3 lottery ticket is –50¢: half the cost of the ticket. Buying n tickets on different numbers, while it will increase your chance of winning, does not improve the expected value of your investment. Each ticket contributes –50¢ to the expected value of the total wager, which is $-\$\dfrac{n}{2}$: half the amount spent on tickets. ∎

Binomial Distribution

We begin by considering an example:

> *Suppose you buy one Florida Pick 2 lottery ticket every day for 150 straight days. What is the probability that you will win 5 times?*

Solving this problem is facilitated by introducing the concept of a *binomial experiment*.

Definition 1.19. A *binomial* experiment has the following four characteristics:

1. The experiment consists of a fixed number of successive identical trials, denoted by n.

2. Each trial has exactly two outcomes, denoted *success* and *failure*.

   In practice, it is often possible to amalgamate multiple outcomes into a single category to get down to two. For example, in the Daily 2 lottery question above, with 100 different possible numbers, we can collect all 99 losing numbers into a single outcome—if you lose your bet, it matters little what the winning number was.

3. The probabilities of success and failure are constant from trial to trial. We denote the probability of success by p and the probability of failure by q, where $q = 1 - p$.

   The Complement Rule is easily applied here, since we have defined the problem so that there are only 2 outcomes.

4. The trials are independent.

Definition 1.20. A random variable X that counts the number of successes of a binomial experiment is called a *binomial* random variable. The values n and p are called the *parameters* of X.

The experiment described in the example above meets the four listed criteria and is therefore a binomial experiment. If we change the experiment to "Start buying one lottery ticket per day, and let the random variable X be the number of tickets required to win exactly 5 times," then the new experiment is not binomial. Since the number of trials is no longer fixed at the outset, criterion 1 is no longer true.

If X is a binomial random variable with parameters n and p, the formula for $P(X = r)$ can be derived through the following three-step process:

1. Select which r of the n trials are to be successes. This can be done in $\binom{n}{r}$ ways, as the order in which we select the successes does not matter.

   If we think of the trials as a row of n boxes, each to be designated "success" or "failure," what we're doing here is determining which r of the n boxes are successes.

2. Compute the probability of these r trials resulting in successes. Since the trials are independent, this probability is p^r.

3. We must now ensure that there are *only* p successes. This is done by assigning the outcome "failure" to the remaining $n - r$ trials. The probability of this many failures is $(1 - p)^{n-r} = q^{n-r}$.

Multiplying these three factors together gives the following result, called the *binomial formula*:

Theorem 1.14. *If X is a binomial random variable with parameters n and p, then*

$$P(X = r) = \binom{n}{r} \cdot p^r \cdot q^{n-r} = \binom{n}{r} \cdot p^r \cdot (1 - p)^{n-r}.$$

We can now revisit the question that started this section with this new insight.

Example 1.46. Since the pool of two-digit numbers for Florida's Daily 2 game includes numbers with a leading zero such as 00 or 09, the probability of winning on a \$1 straight bet is $\frac{1}{100}$. We have $n = 150$ and $p = \frac{1}{100}$. The probability of winning on r tickets is then

$$P(r) = \binom{150}{r} \cdot \left(\frac{1}{100}\right)^r \cdot \left(\frac{99}{100}\right)^{150-r}.$$

If $r = 5$, then $P(5) \approx .0138$, so the answer to the original question is "Slightly less than 1.4%". ∎

If a random variable is binomial, computing its expected value is simple.

Theorem 1.15. *If X is a binomial random variable with parameters n and p, then $E(X) = np$.*

Put simply, the average number of successes is the number of trials multiplied by the probability of success on a single trial.

Example 1.47. In Example 1.46, the expected number of wins in 150 days of ticket-buying is

$$np = 150 \cdot \frac{1}{100} = 1.5.$$

Another way to reach this conclusion is to imagine 150 tickets purchased for the same drawing: one on every even number and two on every odd number. Half of the time, the winning number will be even, and there will be 1 winner. The other half of the time, the winning number will be odd, and thus there will be 2 winners. Averaging these two results, since odd and even numbers are equally likely, gives an expected value of 1.5 winning tickets. ∎

It follows from this result that your average winnings after making 150 of these bets would be

$$(1.5) \cdot 50 - 150 = -\$75.$$

On the average, you should expect to lose half of the money you wagered.

Aside from the general rule that "gambling games always favor the house," it might be reasonable to ask if there is any way to identify particularly bad bets such as this one before risking money. One tool that may be useful is the *standard deviation* of a random variable, which is denoted by the Greek letter sigma: σ.

Definition 1.21. The *standard deviation* σ of a random variable X with mean μ is

$$\sigma = \sqrt{\sum [x^2 \cdot P(X = x)] - \mu^2}$$

where the sum is taken over all possible values of X.

Informally, σ is a measure of how far a typical value of X lies from the mean. Computing σ using the formula above is an arithmetically intense process:

- Compute the mean of the random variable.

- For each value of x that X can attain, multiply x^2 by $P(X = x)$ and add up the products.

- Subtract the square of the mean from this sum.

- Take the square root of the difference. Notice that the use of the square root guarantees that $\sigma \geqslant 0$.

Fortunately, it is seldom necessary to perform these calculations by hand, as calculators and computer software will readily compute σ. In the special case where the random variable is binomial, we have the following simple result:

Theorem 1.16. *If X is a binomial random variable with parameters n and p, then the standard deviation of X is given by*

$$\sigma = \sqrt{np(1 - p)} = \sqrt{npq}.$$

While the expected value of a random variable tells us where a "typical" value of X lies, the standard deviation gives us information about the "spread" of the values of X. The nature of the random variable X allows us to use the mean and standard deviation to derive useful information about the distribution of the data set.

For example, in considering a suitably large collection of rolls of 2d6, the distribution of sums is approximately bell-shaped (many values near the mean, and fewer values as we move away from the mean of 7 in either direction) and symmetrically distributed about the mean. See Figure 1.2, which shows a plot of the sums of 100,000 simulated rolls, for an example.

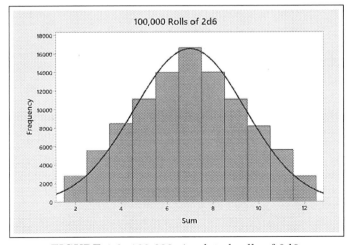

FIGURE 1.2: 100,000 simulated rolls of 2d6.

The histogram bars suggest a distribution that is roughly symmetric about the mean of 7, and the approximating curve overlaid on the graph is appropriately bell-shaped. Data sets that are distributed this way are called *normal*. In this circumstance, a result called the *Empirical Rule* is a good description of the data.

Theorem 1.17. *(Empirical Rule) If the values of many samples of a random variable with mean μ and standard deviation σ are bell-shaped and symmetric, then we have the following result:*

1. *Approximately 68% of the data points lie within 1 standard deviation of the mean: between $\mu - \sigma$ and $\mu + \sigma$. (This is often stated as "about 2/3.")*

2. *Approximately 95% of the data points lie within 2 standard deviations of the mean: between $\mu - 2\sigma$ and $\mu + 2\sigma$.*

3. *Approximately 99.7% of the data points lie within three standard deviations of the mean: between $\mu - 3\sigma$ and $\mu + 3\sigma$. (This is often stated as "almost all.")*

For many practical purposes, an experimental result is deemed to be statistically significant, meaning that the result is likely to be the result of something other than random variation, if it is at least 2 standard deviations (SDs) away from the mean. This means that its probability—under the assumption that there is no unusual effect and all of the deviation from the mean is due to random chance—is less than 5%. Since the random variable is symmetric, this 5% is evenly distributed between results greater than 2 SDs above μ and results less than 2 SDs below μ. This 2 SD standard is a convention agreed upon by the statistics community; it does not fall out of any equation as a rigorously derived standard.

By using 95% as a minimum, what we are saying is that 19 out of 20 times that we identify a result as due to something other than random variation, we will be correct, and this level of confidence is acceptable in many lines of inquiry. Some fields may have more exacting standards: in experimental particle physics, for example, the standard for confirming a discovery is "5 sigma," or at least 5 standard deviations away from the expected value, corresponding to P (Chance event) < 1 in 3.5 million.

Example 1.48. For the 100,000 simulated rolls shown in Figure 1.2, the standard deviation of the sample data can be computed directly, and is found to be approximately 2.4155. Using this number as an approximation to the population SD σ, we find that the Empirical Rule gives the following conclusions:

- Approximately 68% of all sums should lie between 4.58 and 9.42. Since the sum of 2d6 must be an integer, we round this result inward (up or down, as needed, to the nearest integer within the interval) to $[5, 9]$.

- Approximately 95% of all sums should lie between 3 and 11, rounded inward.

- Approximately 99.7% of all sums should lie between 0 and 14.

The actual data show 66.7% within 1 SD of 7, 94.4% within 2 SDs, and ˜100% within 3 SDs, and so the Empirical Rule is confirmed. ■

Returning to the Florida Daily 2 lottery, repeated trials purchasing 150 tickets would yield a mean of 1.5 wins and a standard deviation of 1.2186 wins. The Empirical Rule tells us that 68% of the time, we would expect between .2814 and 2.7186 wins. We round this inward to [1, 2]. 95% of the time, the number of wins will fall between −.9372 and 3.9372, which rounds inward to [0, 3]. Scoring more than 5 wins in 150 tickets (the 3-sigma level) is highly improbable, and might be cause for an investigation into lottery operations.

Chapter 2

Roulette

Roulette is a glamorous casino table game in which a small ball is spun on a rotating wheel and comes to rest in a numbered pocket on the wheel. *European* roulette uses a wheel divided into 37 pockets, numbered 1 to 36 and 0. The 0 is colored green, and the remaining 36 slots are divided evenly between red and black. In *American* roulette, a 38th pocket, numbered 00 and also colored green, is added—see Figure 2.1.

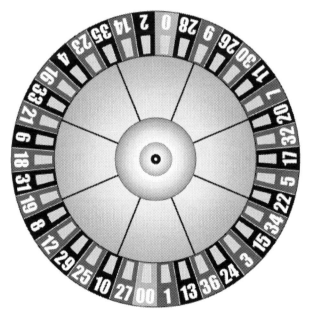

FIGURE 2.1: American roulette wheel [46]

2.1   Roulette Basics

Roulette players can make a variety of wagers, at a range of payoffs, on which pocket the ball lands in. Roulette bets are made by placing the wager on the layout on a single number or at the intersection of several numbered spaces. Figure 2.2 shows an illustration of the American roulette betting layout.

FIGURE 2.2: American roulette layout [45].

Table 2.1 lists the standard bets and their corresponding payoffs. Wagers made on the grid of numbers, covering 6 or fewer numbers, are called *inside* bets, while bets on 12 or more numbers, placed at the edges of the layout, are known as *outside* bets.

- *Straight* bets cover a single number.
- *Split* bets are made on two adjacent numbers on the layout by placing a chip on their common line segment. Except for 0 and 2, numbers that are adjacent on the layout are not adjacent on the wheel.
- A *street* bet is made on a row of three consecutive numbers, for example 28, 29, and 30, by laying that chip at the edge of the row.
- *Corner* bets may be made on any block of 4 numbers forming a 2 × 2 square, by placing the chip at the center of that block.
- The *basket* bet is available only on American roulette wheels and can only be made on the five-number combination 0, 00, 1, 2, and 3. A player makes this bet by putting chips down at the edge of that block of five

TABLE 2.1: Roulette bets and payoffs

Bet	# of numbers	Payoff
Straight	1	35 to 1
Split	2	17 to 1
Street	3	11 to 1
Corner	4	8 to 1
Basket	5	6 to 1
Double street	6	5 to 1
Dozen	12	2 to 1
Even-money	18	1 to 1
Double column	24	½ to 1

numbers, either at the corner where 0 and 1 meet or the corner where 00 meets 3.

- *Double street* bets cover 6 numbers, 2 adjacent rows, and are made by placing chips on the edge of the grid of numbers where two numbers meet, as where 24 and 27 come together. This double street bet would cover the numbers 22 through 27.

- A *dozen* bet may be made on the numbers from 1 to 12, 13 to 24, or 25 to 36, or on any of the three columns on the betting layout.

- An *even-money* bet may be made on odd, even, red, black, low (1–18), or high (19–36) numbers. Every even-money bet covers exactly 18 of the 38 numbers; 0 and 00 are neither even nor low.

- *Double column* bets, which are not offered at many casinos, cover two adjacent columns of 12 numbers each.

A challenge to roulette players is keeping track of their bets on a layout where many other people may be making very similar bets. Many casinos assist their patrons by providing special roulette chips—called *checks*—in a variety of colors, each one specific to an individual player. Gamblers exchange their cash for checks at the table, and when purchasing checks, may designate their value. A player buying into a game for $50 has the option of buying 50 dollar chips, 100 chips valued at 50¢ each, 10 $5 chips, or another configuration that the casino accepts. These checks must be redeemed for standard casino chips before leaving the table, as there is no record off the table of a particular color's value.

Figure 2.3 shows roulette checks from several Nevada casinos. While ordinary casino chips may be played at roulette tables, using color-coded checks makes the game easier for everyone involved.

Casinos derive their advantage at roulette bets by paying off bets at less than correct odds. For example, a bet on a single number, which carries odds against of 37–1 or 36–1, is paid only 35–1 if it wins. European and American

FIGURE 2.3: Roulette checks. The value of a check is determined by the player at the time of purchase. Letters or numbers on the checks indicate the table where they are used.

roulette wagers have the same payoff structure: a bet on n numbers pays off at $\dfrac{36-n}{n}$ to 1, except for the basket bet. That bet pays only 6 to 1 instead of the $6\frac{1}{5}$ to 1 that this formula would produce. The expectation of any other $1 bet on n numbers, on a wheel with N pockets, is then

$$E = \left(\frac{36-n}{n}\right) \cdot \frac{n}{N} + (-1) \cdot \frac{N-n}{N} = \frac{36-N}{N}.$$

For the basket bet on an N-pocket wheel, the expected value of a $1 bet is slightly lower:

$$E = (6) \cdot \frac{5}{N} + (-1) \cdot \frac{N-5}{N} = \frac{35-N}{N}.$$

Note that the expected value is independent of n and depends only on the number of pockets on the wheel.

In European roulette, the *en prison* rule is frequently applied to even-money wagers. If the casino follows this rule, even-money bets are not immediately lost when the ball lands in the 0 pocket. Rather they are held "in prison" for the next spin. If that spin turns up a winning number, the bet is returned to the player, though without any additional winnings. If the bet loses a second time, including if the second spin is also a 0, it is taken by the casino. There are 3 outcomes to an en prison bet:

- **Win 1 unit**: This can occur on the first spin only. Its probability is

$$P(\text{Win } 1) = \frac{18}{37}.$$

- **Win 0 units**: This happens when the first spin is a 0 and the second spin wins. The probability of breaking even is

$$P(\text{Win } 0) = \frac{1}{37} \cdot \frac{18}{37} = \frac{18}{37^2}.$$

- **Lose 1 unit**: There are two ways to lose the all of initial bet: when the first spin loses outright or when the first spin is 0 and the second spin loses, including a second spin of 0, the bet is lost in full. The probability of losing is

$$P(\text{Lose } 1) = \frac{18}{37} + \frac{1}{37} \cdot \frac{19}{37} = \frac{685}{1369},$$

slightly more than ½.

Taken together, these probabilities show that the house edge on an even-money bet with en prison in effect is 1.39%.

A variant of en prison, the *partage* rule collects half of an even-money wager when the wheel shows 0. This rule effectively cuts the HA on even-money bets in half, to 1.35%.

In 2017, the Venetian Casino in Las Vegas introduced *Sands Roulette*. This game added a 39th pocket to an American roulette wheel, which was colored green and bore an "S" for Sands, the parent company of the Venetian. The S space acts like a third zero for betting and payoff purposes. It is represented on the layout by a rectangle above the 0 and 00 spaces, so it also belongs to no column. Basket bets are available on this layout.

The following year, 000 spaces appeared on some roulette wheels along the Strip, at New York–New York and Planet Hollywood, and the Luxor added a 39th pocket bearing its pyramid logo. The payoffs remain the same: $\frac{36 - n}{n}$ to 1 except for the basket bet, but the probability of winning drops and so the HA rises.

Table 2.2 shows the house advantage for each of the three wheel configurations and the separate basket bets. Increasing the number of green pockets

TABLE 2.2: Roulette house advantages

Wheel	**N**	**HA**
European *en prison* bet	37	1.35%
European	37	2.70%
American	38	5.26%
American basket bet	38	7.89%
Sands	39	7.69%
Sands basket bet	39	10.26%

while retaining the payoffs in Table 2.1 is easily seen to raise the HA, making the game somewhat less favorable to gamblers.

Example 2.1. Adding a quadruple-zero space on the wheel, but retaining the standard payoff odds, would give a expected return on an n-number bet of

$$E = \left(\frac{36-n}{n}\right) \cdot \frac{n}{40} + (-1) \cdot \frac{40-n}{40} = -\frac{4}{40} = -\$.10.$$

The HA would be an even 10%.   ∎

Example 2.2. For a brief while in 1991, the Four Queens Casino in downtown Las Vegas offered a 36–1 payoff on single-number American roulette wagers instead of the usual 35–1 [30]. Other payoffs were unchanged. This raised the expected value of a $1 single-number bet from –5.26¢ to

$$E = (36) \cdot \frac{1}{38} + (-1) \cdot \frac{37}{38} = -\frac{1}{38} \approx -\$.0263,$$

and halved the HA to 2.63%.   ∎

2.2   History of Roulette

Biribi

There is some evidence that roulette finds its roots in *biribi*: an Italian game of chance with 70 numbers in use and Figure 2.4 as its layout.

1	9	17	25		39	47	55	63
2	10	18	26	33	40	48	56	64
3	11	19	27	34	41	49	57	65
4	12	20	28	35	42	50	58	66
5	13	21	29	36	43	51	59	67
6	14	22	30	37	44	52	60	68
7	15	23	31	38	45	53	61	69
8	16	24	32		46	54	62	70

FIGURE 2.4: Biribi layout.

Biribi did not use a wheel; the 70 numbers were printed on tickets that were mixed together in a bag. One number was drawn from the bag by the game operator. In form, then, biribi was akin to a modern lottery. Players could bet on the number of their choice. Only single-number bets were available, and a winning bet paid 64–1. The expected value of a biribi wager was

$$E = (64) \cdot \frac{1}{70} + (-1) \cdot \frac{69}{70} = -\frac{5}{70} = -\frac{1}{14} \approx -\$.0714,$$

and so the game operator held a 7.14% edge.

Example 2.3. Devise a reasonable payoff for a 2-number biribi bet.
Let the payoff be x to 1. The expected value as a function of x is

$$E(x) = (x) \cdot \frac{2}{70} + (-1) \cdot \frac{68}{70} = \frac{2x - 68}{70}.$$

To get the same 7.14% HA as for the single-number bet would require

$$\frac{2x - 68}{70} = -\frac{1}{14},$$

or $x = \frac{63}{2} = 31.5$. If we wish to avoid fractional payoffs, a payoff of 31–1 gives
a 8.57% HA, and 32–1 reduces the edge to 5.71%. ∎

This result suggests a general formula for the payoff to a winning biribi
bet on n numbers:

$$\frac{65 - n}{n} \text{ to } 1,$$

rounded as necessary to give an easy-to-implement integer payoff. While this
would not yield a constant HA, it would assure that the HAs are reasonably
close to one another.

Rounding is inevitable when using this formula because 65 has few divisors:
only 1, 5, 13 and 65 divide 65. By contrast, the 36 that is used in the payoff
formula for standard American and European roulette is divisible by 1, 2, 3,
4, 6, 9, 12, 18, and 36, so fewer fractions are encountered when calculating
payoff odds.

Example 2.4. There are 4 places on the biribi layout where a 3-number
bet may be placed easily: 25–26–33, 33–39–40, 31–32–38, and 38–45–46. The
formula above returns

$$\frac{65 - 3}{3} = \frac{62}{3} = 20\frac{2}{3},$$

as the payoff, which could be rounded to either 20–1 or 21–1. At 20–1, the
HA is 10%, 21–1 gives the house an edge of 5.71%. ∎

Biribi was soon combined with a spinning wheel and evolved into the
familiar game of roulette.

The First American Wheel

Before American roulette evolved into the 38-pocket format with 0 and 00
pockets, an earlier wheel had pockets numbered 1–28 together with 0, 00, and
a picture of an eagle which functioned as a third zero. The layout is shown in
Figure 2.5.

0	Eagle.		00
1	2	15	16
3	4	17	18
5	6	19	20
7	8	21	22
9	10	23	24
11	12	25	26
13	14	27	28
3 for 1	3 for 1	3 for 1	3 for 1

FIGURE 2.5: Early American roulette layout, with 0, 00, and Eagle spaces [114].

There were 3 bets available [114]:

- Single-number bets paid 27 for 1.

- Column bets covered 7 numbers and paid 3 for 1. As Figure 2.5 shows, the zeros and eagle were not part of any column.

- Bets on Red or Black, which paid 2 for 1 (even money). These bets were placed to the right or left, respectively, of the layout pictured.

A single-number bet had expected value

$$(27) \cdot \frac{1}{31} - 1 = -\frac{4}{31} \approx -.1290,$$

giving the operator a 12.90% advantage. Advantages on both the column bet and the color bet were 9.68%.

Example 2.5. A 4-number bet on this wheel could take the form of a row bet, for example, 5–6–19–20, or a bet at the point common to 4 spaces on the layout, as with 9–10–11–12. If such a bet were allowed and paid X for 1, the expectation would be

$$E(X) = (X) \cdot \frac{4}{31} - 1 = \frac{4X - 31}{31}.$$

This is negative if X is an integer less than 8; taking $X = 7$ gives a house advantage of 9.68%. ∎

This is the same HA that the existing even money and column bets carry, which suggests that a general payoff formula of $\frac{28 - n}{n}$ to 1, for a bet on n numbers, might cover wagers that were possible on the layout but not part of the game as played in 19th-century America at the same 9.68% HA. An exception is a hypothetical 3-number bet that would have to include 0, 00, or the eagle. Like its analog the basket bet, the formula would require rounding down to produce an integer payout.

These high HAs were sometimes enhanced by cheating. One clever trick involved the *frets*: the small metal barriers that separate adjacent numbers on a wheel. A carefully rigged wheel had every other fret slightly taller than the ones adjacent to it. With this fret pattern and the numbers alternating in color on the wheel, the croupier could increase the chance of winning money for the house by the direction in which he spun the ball. A survey of the layout would reveal which color had attracted more action, and the ball could then be spun to increase the chance of the other color appearing [1].

Advantages this high made the game untenable in the 19th century, but the idea of a 000 or equivalent pocket was ripe for revival in the early 21st century. Increasing the number of numbered pockets from 28 to 36, with payoffs increasing accordingly, proved to be the key to success for a triple-zero wheel.

Polish Roulette

If more zeros raises the HA, what would be the effect of eliminating the zeros entirely? That was the gimmick for a roulette variation launched at Vegas World in the 1980s. Casino owner Bob Stupak drew on his ethnic heritage with the zero-free Polish Roulette. Advertising for this game offered the reader a logical paradox [130]:

INTRODUCING POLISH ROULETTE!
WE LOSE MONEY ON EVERY SPIN,
BUT WE HOPE TO MAKE IT UP IN VOLUME.
NO ZEROS

A roulette wheel without the zeros, and with the payoffs unchanged from the standard $\dfrac{36 - n}{n}$ to 1 on an n-number bet (as was the case at Vegas World), would be a fair game, with no advantage to the casino or the player. Accordingly, it would not be completely accurate to say that the casino lost money on every spin, but one cannot help but wonder where the catch was in this generous offer. While the player held no long-term edge either, it was certainly plausible that a lucky gambler, or one with a good system, might be able to ride a hot streak to short-term profit. While this is possible in any casino game, it is more likely when the player is bucking a 0% HA.

For example, a common but ultimately unsuccessful betting strategy is the *martingale* (page 85), where a gambler begins with a 1-unit bet on an even-money proposition. If the bet wins, the gambler pockets the profit and begins a new sequence. If the bet loses, the next bet is doubled to 2 units; if it wins, the gambler is 1 unit ahead and can restart the sequence. If it loses, the next bet is 4 units, and so on. Once a bet wins, the player collects a 1-unit profit. However, if the player hits an unlucky streak and the next doubling would put the wager over the table limit, he or she takes a significant loss.

Example 2.6. Beginning with a \$1 bet, subsequent martingale wagers would be \$2, \$4, \$8, and so on until the player wins or the table limit is reached. If that limit is \$100, a gambler who loses 7 straight bets has lost \$127 and can only wager \$100 on the next decision. Even if that spin of the wheel is a winner, he or she is still down \$27. ■

The martingale is used for bets with an even-money payoff; these typically have $P(\text{Win})$ just less than ½. The scheme would have a better chance of succeeding on a sequence of successive even-money wagers that were truly 50/50 bets before hitting the table maximum.

Table maximum?

Ah...there was the catch. The maximum wager at Polish Roulette was \$5, which rendered the game essentially immune to quick profit-taking on the part of system players. Any player wagering more than \$5 was subject to the 0 and 00 spaces and their effect on gameplay; this meant that Polish Roulette could be played on a standard American roulette wheel without modification. Wagers of \$5 or less were simply carried over to the next spin when the ball landed in the 0 or 00 slots. An even-money martingale starting at a \$1 minimum could only lose three times before reaching the bet limit, winning \$7 for the casino. The probability of this happening was $\frac{1}{8}$, whereas a martingale bettor who won three straight even-money wagers—all at the \$1 minimum, and with probability of $\frac{1}{8}$ again—would win only \$3.

2.3   New Wheels

A challenge facing the casino game designer who wishes to develop a variation on roulette is the increased cost of a nonstandard wheel. Professional-grade roulette wheels in European or American configurations cost thousands of dollars; there would be considerable additional expense in manufacturing a specialized wheel with a different number of pockets. While a large number of alternate casino games have been marketed using standard cards and dice in new ways, new roulette games—as opposed to new wagers using a standard wheel—often require a modified wheel, and so are considerably rarer.

Riverboat Roulette

One simple and comparatively inexpensive way to modify a roulette wheel was exhibited in *Riverboat Roulette*, which debuted in 2012. This game variation expanded the roulette spectrum beyond red and black, and added a new wagering option. In addition to retaining their red or black colors for the purposes of standard color wagering, 32 of the 36 numbers from 1 to 36 are also tagged with a second color, which appears in the pocket instead of behind the number on the wheel. Table 2.3 shows these additional colors. The numbers 25 and 36 remain red, and 26 and 35 remain black, without a second color.

TABLE 2.3: Riverboat Roulette: Secondary colors

Color	Numbers
White	1, 2, 9, 10, 13, 14, 27, 28
Blue	3, 15, 22, 24, 34
Orange	4, 16, 21, 23, 33
Teal	5, 17, 32
Yellow	6, 18, 31
Purple	7, 11, 20, 30
Pink	8, 12, 19, 29

By reference to Figure 2.1 (page 27), one can see that each new color group except for white is comprised of a block of adjacent numbers on the wheel, and so these color groups are called *neighborhoods*. Each set of two numbers immediately adjacent to one of the zeroes, on either side, is colored white.

Riverboat Roulette allows for all of the bets available on a standard roulette wheel and adds a new type of wager on six of the seven new neighborhoods. Bets on the white neighborhood are one-spin bets, like bets on red or black. In ordinary roulette and many of its variants, each spin of the wheel resolves all bets, and the layout is cleared before the next spin. In Riverboat Roulette, bets on the neighborhoods other than white function differently. A bet on one of the other six colors is a bet that a number in that neighborhood will be spun before a white number. If a number of any other color—including red, black, and green—is spun, the result is a push, and the bet remains active. If a white number is spun, this is called a "White Out", and all neighborhood bets lose; this is similar to the fate of place bets when a 7 is rolled at the craps table. Neighborhood bets may be "taken down" (withdrawn) or increased following a push.

Example 2.7. Are any of these new neighborhood bets especially better or worse than the others?

For each color, we can focus on the wheel numbers that resolve the bet: the eight white numbers and the three, four, or five numbers in the neighborhood. Denote this latter number by x. The neighborhood bets have the following values for x and corresponding payoffs:

Color	x	Payoff
Teal & Yellow	3	2 to 1
Pink & Purple	4	8 to 5
Orange & Blue	5	7 to 5

For the teal and yellow neighborhoods, there are 11 numbers that resolve the bet. The expectation on a $1 bet is

$$E = (2) \cdot \left(\frac{3}{11}\right) + (-1) \cdot \left(\frac{8}{11}\right) = -\frac{2}{11} \approx -.1818,$$

and the HA is 18.18%.

In the remaining two cases, we assume a $5 wager to eliminate fractions in the payoff. For pink and purple neighborhoods, 12 numbers resolve the bet.

$$E = (8) \cdot \left(\frac{4}{12}\right) + (-5) \cdot \left(\frac{8}{12}\right) = -\frac{8}{12} \approx -.6667.$$

Dividing by the $5 wager gives an HA of 13.33%.

Orange and blue wagers involve 13 numbers. The expectation on a $5 bet is

$$E = (7) \cdot \left(\frac{5}{13}\right) + (-5) \cdot \left(\frac{8}{13}\right) = -\frac{5}{13} \approx -.3846,$$

so the HA is 7.69%. ∎

While these bets may increase gamblers' interest in roulette through the different betting options and the novelty of multi-spin wagers, it seems unlikely that wagers with such a high house edge will attract much action if that edge becomes apparent to gamblers. The option to take down a neighborhood bet after a push may make these bets slightly less unattractive.

It is worth noting that these percentages are the house edge *per resolved bet*. An alternate interpretation of the phrase "house advantage" is the HA *per spin of the wheel*. This calculation considers all 38 numbers on the wheel, including those for which the wager is a push and the payoff is 0. Denote the expectation per spin by E', to distinguish it from the expected values already computed. For the teal and yellow neighborhoods, the expectation per spin is

$$E' = (2) \cdot \left(\frac{3}{38}\right) + (-1) \cdot \left(\frac{8}{38}\right) + (0) \cdot \left(\frac{27}{38}\right) = -\frac{2}{38} \approx -\$.0526,$$

and so the recalculated HA is the 5.26% we're accustomed to in roulette. A bet on the teal or yellow neighborhoods will stay on the layout for an average of

$$\sum_{k=1}^{\infty} k \cdot \frac{11}{38} \cdot \left(\frac{27}{38}\right)^{k-1} = 3\frac{5}{11}$$

spins before being resolved; multiplying this number by 5.26% returns the 18.18% HA found originally.

Similar calculations will show that for the pink and purple neighborhoods, the expected return per spin on a $5 bet is $E' = -\$.2105$, for a new HA of 4.21%—better than 5.26%—and for a $5 bet on orange or blue, $E' = -\$.1315$, so the HA there is 2.63%, half the edge of a standard roulette wager.

Example 2.8. Eight numbers comprise the white neighborhood. According to our standard American roulette formula, this bet should pay off at

$$\frac{36 - 8}{8} = \frac{28}{8} = 3\frac{1}{2}$$

to 1. To avoid fractions, the payoff on this one-spin bet is written as 7–2. Assuming a $2 bet for convenience, find the HA.

The expected value of a $2 bet on white is

$$E = (7) \cdot \frac{8}{38} + (-2) \cdot \frac{30}{38} = -\frac{4}{38}.$$

The HA is—once again—5.26%.

However, this bet may be susceptible to casino-friendly rounding if it's made for an odd amount. If a roulette system player makes a $5 bet on white with checks valued at $1 and the casino does not stock 50¢ chips, the payoff will be $17 rather than $17.50, and so the HA will rise to 7.37%. ■

Super 62 Roulette

Another easy way to modify the 38-pocket American roulette wheel is simply to add more pockets. This is the thinking behind *Super 62 Roulette*, marketed by Empire Global Gaming, Inc. As the name suggests, Super 62 Roulette involves a wheel with 62 numbers: 1–60 plus 0 and 00. The layout contains 12 rows of 5 numbers each, which can easily accommodate bets on 1, 2, 4, 5, 10, or 12 numbers in addition to 20-number bets (high/middle/low, similar to the dozens bets on a standard wheel) and 30-number even-money bets (red/black, high/low, odd/even). Three-number bets must involve one or both zeroes [109].

The standard bets on n numbers pay off at $\frac{60 - n}{n}$ to 1, so a single-number bet at Super 62 roulette pays off at 59–1, a split bet pays 29–1, and so on. Once again, the expectation of a $1 bet can be computed in a single calculation:

$$E = \left(\frac{60 - n}{n}\right) \cdot \left(\frac{n}{62}\right) + (-1) \cdot \left(\frac{62 - n}{62}\right) = -\$\frac{1}{31} \approx -\$.0323.$$

The house edge on any such bet is thus a fixed 3.23%—better than American roulette: a lower probability of winning has been more than balanced by the higher payoff odds.

A different wagering option in Super 62 Roulette is a *digits bet*: a simple bet that a selected digit from 0–9 will appear in the number spun on the wheel. Since the digits from 1–5 are substantially more likely to appear than the others, the payoffs for digit bets depend on the digit selected.

Digit	Winning numbers	Payoff
1–5	15	3–1
6	7	7–1
7–9, 0	6	9–1

Each digit except 6 has payoff odds that fit the $\dfrac{60-n}{n}$ to 1 pattern and thus carry a 3.23% house edge.

For the digit bet on 6, the expectation is

$$E = (7) \cdot \left(\frac{7}{62}\right) + (-1) \cdot \left(\frac{55}{62}\right) = -\frac{6}{62} \approx -\$.0968,$$

for a higher HA of 9.68%—three times higher than the other wagers.

Alphabetic Roulette

One roulette game incorporating a new wheel that reached the casino floor, at Fitzgerald's in downtown Las Vegas (now The D), was Alphabetic Roulette, which used letters to label pockets. The Alphabetic Roulette wheel was divided into 25 pockets: one for each letter of the alphabet from A to X and one labeled YZ. The house advantage, as one might expect, came from the YZ pocket. Bets were made on letters, singly or in combinations; players could bet on their names or on other significant words.

Standard Alphabetic Roulette bets could be made on 1, 2, 3, 4, 6, or 12 letters. Blocks of four consecutive letters were colored the same color, making for six four-letter color bets. Additionally, the layout provided for two customizable spaces that could be labeled and defined as the casino chose, perhaps with the casino's name. A casino such as the Tropicana in Las Vegas could use its name to create an 8-letter bet, paying off at 2–1, on the letters {T, R, O, P, I, C, A, N}. A default option was to label these spaces "Party Pit" and "Roulette", which corresponded respectively to bets on the six letters {P, A, R, T, YZ, I} or {R, O, U, L, E, T}.

The payoff odds at Alphabetic Roulette are very simple; they are compiled in Table 2.4.

A bet on n letters pays off at $\dfrac{24-n}{n}$ to 1; these would be fair bets if the wheel had only 24 pockets. The expectation on any standard Alphabetic Roulette wager is then

$$E = \left(\frac{24-n}{n}\right) \cdot \frac{n}{25} + (-1) \cdot \frac{25-n}{25} = -\$\frac{1}{25} = -\$.04,$$

and the corresponding house advantage is 4%.

An optional Alphabetic Roulette wager is the Bonus Box bet, which is a bet that the letter just spun will be repeated up to three more times. If the letter appears again, a Bonus Box bettor may either take down $23 in

TABLE 2.4: Alphabetic Roulette payoffs [2]

Number of letters	Payoff
1	23–1
2	11–1
3	7–1
4	5–1
6	3–1
12	1–1

winnings or let the bet ride for another spin. If the same letter comes up a third time, the payoff is $550 (a hypothetical payout of just under 22–1 on the $24 being risked), which again may be taken down as profit or left to ride for a final spin. If that letter hits a fourth straight time, the Bonus Box bet pays off $14,400, just over 25–1 on the money at risk.

Example 2.9. The house edge for a Bonus Box bettor who cashes out after the number comes up once is the same 4% for a single-letter bet. What are the expectations for the player choosing to cash in after two repetitions of a letter or the one who stays in hoping for a third repeat?

The probability of two repeats of a given letter is $\left(\frac{1}{25}\right)^2 = \frac{1}{625}$. The corresponding expectation is then

$$E = (550) \cdot \frac{1}{625} + (-1) \cdot \frac{624}{625} = -\frac{74}{625} = -\$.1184,$$

for a house edge of 11.84%: nearly triple the edge on a standard bet.

For the player who rides a streak of repeated letters to the limit, the expectation is

$$E = (14,400) \cdot \left(\frac{1}{25}\right)^3 + (-1) \cdot \left(1 - \left(\frac{1}{25}\right)^3\right) \approx -\$.0783.$$

The HA against this bettor is less than for the bettor who stops after two repetitions, only 7.83%. ∎

We may safely conclude that the best Bonus Box strategy is to avoid this bet altogether.

Fast Action Roulette

In Example 2.1, we considered the 10% house advantage of a hypothetical 4-zero roulette wheel with standard payoffs. *Fast Action Roulette*, found at the Marina Bay Sands, made that wheel a reality in 2020, but with a change to the rules that takes the HA below 10%.

The Marina Bay Sands is owned by the parent company of the Venetian Casino where Sands Roulette premiered, so it is not surprising to see an S on the wheel that functions as a 000 pocket. Fast Action Roulette also adds a Marina Bay Sands logo, represented in Figure 2.6 as an $\mathfrak{M}$ and the equivalent of a 0000 space.

$\mathfrak{M}$	0	00	S
1	2	3	4
5	6	7	8
9	10	11	12
13	14	15	16
Super Bonus Win		Scattered Bonus	
17	18	19	20
21	22	23	24
25	26	27	28
29	30	31	32
33	34	35	36
Bonus Win			

FIGURE 2.6: Fast Action Roulette layout.

The twist in Fast Action Roulette is that, on every spin, 10 Bonus spaces, chosen at random, are illuminated on the layout and qualify the bettor for enhanced payoffs. Only 4 bets are permitted at the Fast Action Roulette table:

- **Single-number bets**. No other inside bets are permitted. These bets pay the standard 35–1 if the number is not one of the 10 Bonus numbers, 40–1 if the number is a Bonus number, and 80–1 if all 4 numbers in the same row as the selected number are Bonus numbers. This last condition is called the *Super Bonus*.

  Since the minimum payoff is 35–1 and there is a possibility of higher payoffs, the HA on this bet will be less than the 10% that we saw in Example 2.1 for a 4-zero wheel with standard payoffs. The probability that the wining number is not a Bonus number and thus pays off at 35–1 is ¾. The chance that all 4 numbers in the winning number's row are selected as Bonus numbers is

  $$p = \frac{\binom{36}{6}}{\binom{40}{10}} = \frac{21}{9139} \approx .0023,$$

  since we simply need to count the number of ways to choose the other 6 Bonus numbers from the 36 numbers that are not in the row of interest.

The expected value of a \$1 bet on a single number at Fast Action Roulette is then

$$E = \left[(35) \cdot \frac{3}{4} + (40) \cdot \left(\frac{1}{4} - p \right) + (80) \cdot p \right] \cdot \frac{1}{40} + (-1) \cdot \frac{39}{40} \approx -.0665,$$

giving the casino a 6.65% advantage. This falls in between the 5.26% HA on an American roulette wheel and 7.69% from Sands Roulette.

- **Bonus Win.** The Bonus Win wager pays off if the winning number—whatever it is—is also a Bonus number. The payoff is 2–1.

  The expectation here is easily computed. As noted above, the probability that the winning number is a Bonus number is ¼. We have

  $$E = (2) \cdot \frac{1}{4} + (-1) \cdot \frac{3}{4} = -\$0.25.$$

  This is effectively a bet on 10 numbers, although the gambler does not know which numbers are hers when she wagers. Its HA is 25%.

  If the payoff odds are changed to 5–2, the house edge drops by half, to 12.5%.

- **Super Bonus Win.** This wins if the winning number is part of a Super Bonus quartet, and pays 350–1.

  Using the probability p computed above, we find that the HA of of this wager is 19.35%.

- **Scattered Bonus.** This wager wins if the 10 bonus numbers are evenly distributed, with 1 in every row of the layout. It pays off at 700–1.

  Since there are 4 choices for each row's representative among the Bonus numbers, there are $4^{10} = 1{,}048{,}576$ winning sets of Bonus numbers. The probability of a win is then

  $$\frac{4^{10}}{\binom{40}{10}} \approx \frac{1}{808}.$$

With a 700–1 payoff, the house holds a 13.28% advantage.

Example 2.10. If the Marina Bay Sands expanded its betting menu to allow 2-number Split bets, with payoffs of 17–1 if neither number was drawn as a Bonus number and larger payoffs for 1 or 2 Bonus numbers, find payoffs that give a HA between 5.26% and 7.69%.

The probability that neither number is a Bonus number is

$$\frac{\binom{38}{10}}{\binom{40}{10}} = \frac{29}{52},$$

just more than ½. The probability that 1 of the 2 numbers is drawn for a bonus is

$$2 \cdot \frac{\binom{38}{9}}{\binom{40}{10}} = \frac{20}{52}.$$

Subtracting gives

$$P(2 \text{ Bonus numbers}) = \frac{3}{52}.$$

Let X and Y be the payoffs to 1 for landing 1 or 2 Bonus numbers, respectively. The expected value of this wager is

$$E = \left[(17) \cdot \frac{29}{52} + (X) \cdot \frac{20}{52} + (Y) \cdot \frac{3}{52} \right] \cdot \frac{2}{40} + (-1) \cdot \frac{38}{40}.$$

To hit the specified range requires that

$$-.0769 \leqslant \frac{40X + 6Y - 990}{2080} \leqslant -.0526$$

or

$$830.048 \leqslant 40X + 6Y \leqslant 880.592.$$

At this point, we have some choices. If $17 < X < Y$, which seems reasonable—more Bonus numbers should qualify for a higher payoff—then there are several integer solutions to this inequality. Table 2.5 shows the possibilities together with the HA of the Split bet made with the specified values of X and Y.

TABLE 2.5: Fast Action Roulette: House edge for several hypothetical Split bets

X	Y	$40X + 6Y$	HA
18	19	834	7.50%
18	20	840	7.21%
18	21	846	6.92%
18	22	852	6.63%
18	23	858	6.35%
18	24	864	6.06%
18	25	870	5.77%
18	26	876	5.48%
19	20	880	5.29%

These values may not attract players—the difference in payoffs is simply too small. An alternative would be to offer an enhanced payoff only if both numbers covered by the split bet are drawn as Bonus numbers. By setting $X = 17$, we get a new inequality in the single variable Y:

$$25.008 \leqslant Y \leqslant 33.432,$$

so integer values of Y from 26 through 33 result in a HA within the specified range. Paying this bet at 17–1, with a bump to 30–1 if both numbers are Bonus numbers, would give a house advantage of 6.25%. ■

2.4 Electronic Roulette Games

Some casino patrons are reluctant to play live table games because they're intimidated by the other players or by the need to interact with a dealer. To invite these customers to try the games, many gaming companies have developed and marketed electronic machines that allow gamblers to play table games. A prospective player should examine these games carefully, because the payoff odds may not be as favorable as the live game odds. An electronic American roulette game installed at the Rivers Casino in Des Plaines, Illinois used Table 2.6 as its pay table.

TABLE 2.6: Electronic roulette pay table

Wager	Payoff
Straight	32 for 1
Split	16 for 1
Street	11 for 1
Corner	9 for 1
Basket	7 for 1
Double street	6 for 1
Dozen	3 for 1
Even-money	2 for 1

For an electronic game, this pay table where bets pay off at x for 1 reflects the reality that the player's wager is deducted from his or her account balance before the bet is resolved. With this interpretation of Table 2.6, several of the payoffs are seen to be worse than their counterparts at a live table.

Example 2.11. The straight 1-number bet pays 32 for 1, or 31 to 1, while the standard payoff on a winning 1-number bet is 35–1. The expected value of a $1 straight bet is

$$E = (31) \cdot \frac{1}{38} + (-1) \cdot \frac{37}{38} = -\frac{6}{38} \approx -\$.1579$$

—giving the casino *triple* the edge of a live game.

Another way to calculate the expectation which is useful in games where the payoff is declared as "for 1" regards the $1 wager as lost under any circumstance and simply considers the probability of winning and the payoff listed in the odds:

$$E = (32) \cdot \frac{1}{38} - 1 = -\frac{6}{38} \approx -\$.1579.$$

Using this method makes it unnecessary to consider the probability of losing.

■

The split and street bets also carry higher HAs than their live equivalents; the other wagers, which offer lower payoff odds, have the same 5.26% HA of live American roulette.

Roulette and the Ohio Lottery

State, provincial, and national lottery games typically have much higher house advantages than casino games, even when lotteries offer variations of casino games such as keno or poker. This higher HA is achieved by changing the pay table, as was done for a roulette-like game introduced in Ohio in 2018, when the state lottery launched *The Lucky One* as an add-on to its keno game. Several roulette-like wagers were available in this game, which was drawn every 4 minutes, following each live keno game.

- Select a single number in the range 1–36.

- Bet Odd or Even.

- Bet Low or High.

Each Lucky One wager included an autopick option where the computer selected the player's bet [88]. The lottery computer drew one number for each game. With no zeros in play, this game looked fair until the pay table was examined.

- A single number bet paid off at 24 for 1 if the number was drawn.

- Odd, Even, Low, and High paid winners at 1.50 for 1; a winning bet paid off only 50¢ in profit for each dollar wagered.

The single numbers were paying 23–1 on a 35–1 shot. This generated a huge edge for the lottery: the expected value of a $1 ticket was

$$E = (24) \cdot \frac{1}{36} - 1 = -\frac{12}{36} = -\$\frac{1}{3},$$

or an expected loss of one-third of the wager. The four even-chance wagers held 25% of the wager in the long run; it seems inaccurate to call them "even-money" bets.

Mini Roulette

Online roulette games eliminate the need for a physical wheel, and a computer-animated roulette wheel allows for considerably more flexibility in wheel design. If an expanded physical wheel is not to your liking, the online game *Mini Roulette* might fit the bill. Mini Roulette features a 13-number wheel, 1–12 plus a single zero. The usual array of wagers is available: one may bet on 1, 2, 3, 4, or 6 numbers in various subsets [82]. Zero remains green, the numbers 1, 3, 5, 8, 10, and 12 are black while the others are red.

As with American and European roulette, the payoffs are consistent from wager to wager: a Mini Roulette bet on n numbers pays off at $\dfrac{12-n}{n}$ to 1. Additionally, a variation of the European *la partage* rule provides that if the 0 is spun, all losing bets lose only half the amount wagered. With this modification, a bet on n numbers has expected value

$$E = \left(\frac{12-n}{n}\right) \cdot \frac{n}{13} + \left(-\frac{1}{2}\right) \cdot \frac{1}{13} + (-1) \cdot \frac{12-n}{13} = -\frac{1}{26}.$$

The HA is then 3.85%, in between the values for American and European roulette.

The partial refund on a 0 may have been added late in the game design process. Absent that partial refund, the common expectation drops to

$$E = \left(\frac{12-n}{n}\right) \cdot \frac{n}{13} + (-1) \cdot \frac{13-n}{13} = -\frac{1}{13} \approx -.0769,$$

leading to a HA of 7.69%, which is probably too high for a successful casino game without a huge jackpot available. If a player makes a single-number bet on 0, the partial refund cannot be invoked, and so the HA on that bet remains 7.69%.

100 To 1 Roulette

At the other end of the scale from Mini Roulette lies *100 To 1 Roulette*, a game offered online at caesarscasino.com that uses an electronic wheel with 105 pockets: the numbers from 1–100 and 5 symbols, ○, □, △, ☆, and ◇, that stand in for zeros. The layout is shown in Figure 2.7. The numbers are divided between black and red in the checkerboard pattern shown; the symbols are in green.

The 100 To 1 Roulette pay table is given in Table 2.7. Several of these bets have clear roulette analogs; some exceptions are noted here.

- A 3-number bet must include one symbol and the 2 adjacent numbers from the top row of numbers. For example, a player may bet on 7, 8, and ☆.

- The only 5-number bet covers all of the green symbols.

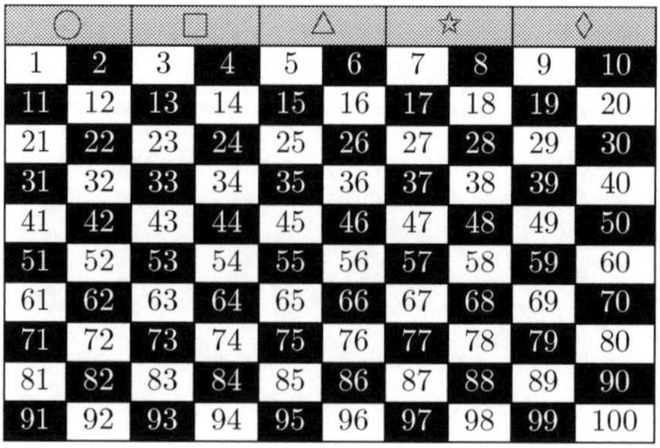

◯		□		△		☆		◇	
1	2	3	4	5	6	7	8	9	10
11	12	13	14	15	16	17	18	19	20
21	22	23	24	25	26	27	28	29	30
31	32	33	34	35	36	37	38	39	40
41	42	43	44	45	46	47	48	49	50
51	52	53	54	55	56	57	58	59	60
61	62	63	64	65	66	67	68	69	70
71	72	73	74	75	76	77	78	79	80
81	82	83	84	85	86	87	88	89	90
91	92	93	94	95	96	97	98	99	100

FIGURE 2.7: 100 To 1 Roulette layout.

TABLE 2.7: 100 To 1 Roulette bets and payoffs

# of numbers or symbols	Payoff
1	100–1
2	49–1
3	32–1
4	24–1
5	19–1
10	9–1
15	11–2
20	4–1
50	1–1

- The 10-number bets may be made on either a row or a column, and are called "Street" bets.
- There is one 15-number, or "Avenue", bet. It covers the 5 symbols and the numbers from 1–10.
- 20-number bets are made on two adjacent rows or columns.
- Even-money bets may be made on Even, Odd, Red, or Black.

Many of the payouts follow a common formula: in this case, a bet on n numbers pays $\dfrac{100 - n}{n}$ to 1. These bets have a common HA of 4.76%.

Example 2.12. The single-number bet is an exception to this rule. Following the formula would give a 99–1 payoff, but there is surely some promotional value in a bet paying 100–1. This bet has expected value of

$$E = (100) \cdot \frac{1}{105} + (-1) \cdot \frac{104}{105} = -\frac{4}{105} \approx .0381,$$

leading to a lower HA of 3.81%. ∎

Example 2.13. The Avenue bet on 15 numbers should pay

$$\frac{100 - 15}{15} = \frac{85}{15} = 5\frac{2}{3}$$

to 1. Rounding this to 11–2 or 5½–1 is convenient (though not necessary, as this game is online and all payoffs are handled by computer) and raises the HA from 4.76% to 7.14%. ∎

Chinese Roulette

In the payoff formula $\dfrac{36 - n}{n}$ to 1 for a bet on n numbers at a standard American or European roulette wheel, one sees that every proper divisor of 36 except 9 corresponds to a roulette wager. *Chinese roulette*, an electronic roulette game launched in 2018 and developed by Win Systems, fills in this gap among other new wagers. Chinese roulette comprises a video layout for use with a standard European wheel; the layout is shaped like a figure 8 in recognition of that number's significance as a harbinger of good luck in Chinese mysticism.

As the $\dfrac{36 - n}{n}$ formula indicates, a 9-number bet pays off at 3–1, and thus has the familiar 5.26% HA we find on American roulette wheels. Groups of 9 numbers in Chinese roulette comprise 3 groups of three numbers each. Each group of 3 (which can be bet separately, with the usual 11–1 payoff for a street bet) is linked to one of the 12 animals of the traditional Chinese calendar; the groups of 9 each correspond to one of the four elements of wood, fire, metal, and water, as in Table 2.8 [159].

Chinese roulette also offers a Zero 50/50 protection feature. When the ball lands on 0, a playing card is dealt. If the card is an 8 or higher, players retain their wagers rather than losing them. Counting an ace as high, this refund happens slightly more than half the time. The expectation of a \$1 n-number bet that does not include 0 is then

$$E = \left(\frac{36 - n}{n}\right) \cdot \frac{n}{37} + (-1) \cdot \left(\frac{36 - n}{37} + \frac{1}{37} \cdot \frac{6}{13}\right) = -\frac{6}{481} \approx -\$0.0125.$$

The HA drops from 2.70% to 1.25%.

Double Bonus Spin Roulette

Other electronic roulette games expand the standard wheel beyond merely adding numbers and make use of new possibilities afforded by computer gaming. In *Double Bonus Spin Roulette*, developed by gaming firm IGT, a yellow Bonus pocket is added to the standard American wheel. This space is 1½ times

TABLE 2.8: Chinese roulette numbers sorted by elements and zodiac animals

Element	Animal	Numbers
	Sheep	18, 22, 29
Wood	Monkey	7, 12, 28
	Rooster	3, 26, 35
	Tiger	15, 19, 32
Fire	Pig	2, 4, 21
	Rat	17, 25, 34
	Rabbit	8, 10, 23
Metal	Dog	11, 30, 36
	Ox	6, 13, 27
	Horse	9, 14, 31
Water	Snake	1, 20, 33
	Dragon	5, 16, 24

wider than the other 38 pockets on the wheel. Players may place a separate wager on the bonus space, which pays off at 12–1, or may combine it with a bet on the nearby 0 and 00 spaces on the layout. If the ball lands in the Bonus pocket, all players' wagers are continued for two free spins of the virtual wheel [38]. Since everything is done electronically, this is easily facilitated with a new graphic of a ball spinning on two concentric rings which rotate and then come to a stop. The virtual ball indicates the two new winning numbers.

Standard roulette wagers can be adapted to the Double Bonus wheel by accounting for 1.5 additional virtual spaces, which changes the denominator in probabilities from 38 to 39.5. Absent the free spins, this would change the expectation on an n-number bet (excepting the basket bet) from–\$.0526 to

$$E = \left(\frac{36 - n}{n}\right) \cdot \left(\frac{n}{39.5}\right) + (-1) \cdot \left(\frac{39.5 - n}{39.5}\right) = -\frac{3.5}{39.5} \approx -\$.0886$$

—but this assumes that the bet loses if the bonus space hits. To account for this, we define a new random variable $Z(n)$ to be the amount that a player wins in the two bonus spins when he or she bets on n numbers.

A full calculation of the expectation of a bet on n numbers is then

$$E = \left(\frac{36 - n}{n}\right) \cdot \left(\frac{n}{39.5}\right) + (-1) \cdot \left(\frac{38 - n}{39.5}\right) + E[Z(n)] \cdot \left(\frac{1.5}{39.5}\right),$$

where $E[Z(n)]$ denotes the expected value of $Z(n)$.

For convenience, let $p = \dfrac{n}{39.5}$, the probability that a bet wins; the probability that it loses is then $q = 1 - p = \dfrac{39.5 - n}{39.5}$. $Z(n)$ has the following probability distribution:

Number of wins	0	1	2
Payoff	-1	$\dfrac{36-n}{n}$	$2 \cdot \left(\dfrac{36-n}{n}\right)$
Probability	q^2	$2pq$	p^2

and the corresponding expectation is

$$E[Z(n)] = \left(2 \cdot \frac{36-n}{n}\right) p^2 + \left(\frac{36-n}{n}\right)(2pq) + (-1) \cdot q^2.$$

This simplifies to

$$E[Z(n)] = \frac{2844 + 453n - 89n^2}{6241},$$

which is positive for all $n \leqslant 8$ and negative thereafter. See Figure 2.8.

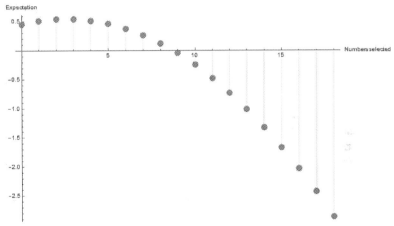

FIGURE 2.8: $E[Z(n)]$: Bonus round expectation for Double Bonus Spin Roulette.

Armed with this expression for $E[Z(n)]$, we can now assess Double Bonus Spin Roulette's house advantage. For a bet on n numbers, $n \neq 5$, we have

$$E = \left(\frac{36-n}{n}\right) \cdot \left(\frac{n}{39.5}\right) + (-1) \cdot \left(\frac{38-n}{39.5}\right) + \frac{2844 + 453n - 89n^2}{6241} \cdot \frac{1.5}{39.5},$$

which can once again be simplified as a function of n, yielding

$$E(n) = \frac{-16,432 + 1359n - 267n^2}{493{,}039}.$$

TABLE 2.9: Overall expectation in Double Bonus Spin Roulette

n	$E(n)$	HA
1	−.0311	3.11%
2	−.0300	3.00%
3	−.0299	2.99%
4	−.0310	3.10%
6	−.0363	3.63%
12	−.0782	7.82%
18	−.1592	15.92%

Table 2.9 records the expectation $E(n)$ and house advantage for values of n corresponding to standard roulette wagers. This table shows that smaller house advantages from the bonus spins are achieved when playing fewer numbers; the HA runs at around 3% for inside bets on up to 6 numbers, but then jumps quickly when column, dozen, or even-money bets are made.

Other wagers specific to Double Bonus Spin Roulette involve the yellow bonus pocket. On the layout, the bonus space is a rectangle placed atop the 0 and 00 betting spaces. This allows for three additional bets [37]:

- The *straight* yellow bet, on the bonus space alone.

- The *split* yellow bet, covering the bonus space and one of the two zeros.

- The *trio* yellow bet, covering the bonus space and both zeros.

On a non-bonus spin, these bets pay off at 12–1, 6–1, and 4–1, respectively. In a bonus round, the payoffs are multiplied by 10 if one wheel wins and by 100 if a winning space is hit on both wheels.

Example 2.14. Find the expected value of a $1 straight yellow bet.

The probability that the ball will land in the yellow space is $p = \frac{1.5}{39.5}$, regardless of whether or not the game is in a bonus spin situation. The chance that it will not land in the yellow space is $q = 1 - p = \frac{38}{39.5}$. The expectation is then

$$E = \left(12q^2 + 120(2pq) + 1200p^2\right) \cdot p - q \approx -\$.1416,$$

and so the house edge is about 14.16%.   ∎

The split and trio yellow bets require slightly different handling, because if 0 or 00 comes up, the bets pay off but do not trigger two bonus spins.

Example 2.15. For the split bet involving 00, define the following probabilities:

- $p = \dfrac{1.5}{39.5}$, the chance that the yellow pocket wins, and a bonus round follows.

- $q = \dfrac{37}{39.5}$, the probability that the bet loses.

- $r = \dfrac{1}{39.5}$, the probability that 00 comes up and the bet wins without starting the bonus round.

In the bonus round, the following outcomes, listed with their associated probabilities, are possible.

Result	Probability
6	q^2
60	$2(p+r)q$
600	$(p+r)^2$

Since the probability of entering the bonus round is p, we have

$$E = p \cdot \left[6q^2 + 60 \cdot 2(p+r)q + 600(p+r)^2 \right] + (-1) \cdot q + (6) \cdot r,$$

which gives $E \approx -\$.2235$—hardly an attractive wager. ∎

Lightning Launch Roulette

In the 2010s, several casino game developers sought to market some kind of "social" casino game that might appeal to millennial gamblers raised on video games. One such variation on an established game was *Lightning Launch Roulette*, developed by Scientific Games and introduced in 2018. This game was a completely electronic game, with ball and wheel in video form only. The game console could accommodate four players, each one assigned to a colored virtual ball: red, blue, yellow, and green [72].

Players had two wagering options: all of the standard roulette bets were available on their ball, and a Bonus bet was based on the outcome of all 4 balls. If fewer than 4 players were at the table at a time, the extra balls were released if any Bonus bets were made. The Bonus bet was restricted to $1 and to single-number wagers only. This bet paid off according to Table 2.10 on the number of virtual balls landing on the player's chosen number. One

TABLE 2.10: Lightning Launch Roulette: Bonus bet pay table

# of Balls	Payoff odds
1	5 for 1
2	75 for 1
3	1000 for 1
4	10,000 for 1

immediately notes that the payoff for catching 1 of the 4 balls is only 5 for

1, or 4–1—a far cry from the 35–1 paid on a non-bonus single-number wager. While the payoffs for catching more than 1 of the 4 balls are better than 35–1, one expects a high HA on the Bonus bet. What would fair odds on this bet look like?

The number of winning balls is a binomial random variable X with $n = 4$ and $p = \frac{1}{38}$. It follows that the probability of x winning balls is

$$P(X = x) = \binom{4}{x} \cdot \left(\frac{1}{38}\right)^x \cdot \left(\frac{37}{38}\right)^{4-x},$$

which is tabulated with the correct odds for 1 in Table 2.11. The probability

TABLE 2.11: Lightning Launch Roulette: Probabilities and correct odds

x	$P(X = x)$	Equivalent odds
0	.8988	N/A
1	.0972	10.3 for 1
2	.0039	253.9 for 1
3	7.0979×10^{-5}	14,088.8 for 1
4	4.7959×10^{-7}	2,085,136 for 1

of a player win on the Bonus bet is just over 10%.

Example 2.16. The casino is, of course, paying off winning bets at less than true odds—that's what casinos do—but at what disadvantage to the player?

Denote the binomial probability $P(X = x)$ with parameters n and p by $B(x; n, p)$. The expectation of the Bonus bet is

$$E = (10000) \cdot B(4; 4, \frac{1}{38}) + (1000) \cdot B(3; 4, \frac{1}{38})$$
$$+ (75) \cdot B(2; 4, \frac{1}{38}) + (5) \cdot B(1; 4, \frac{1}{38}) - 1 \approx -.1429,$$

from which we have a HA of 14.29%. ∎

2.5 Roulette with Cards

California state law forbids the use of dice or a ball as the sole device determining a gambling outcome. The state's Native American casinos are bound by this law as well. This effectively bans traditional roulette and craps games from California casinos, but game designers have demonstrated considerable ingenuity in modifying these games so that they meet Golden State standards while preserving the gameplay and the essential odds.

In 2004, Oklahoma voters approved the Tribal-State Gaming Compact, which specified permissible games in the state's Native American casinos. Tribes in Oklahoma were permitted to offer electronic amusement and bingo games and card games not banked by the house—but not craps or roulette [140]. Some of the same adaptations made to play craps and roulette with cards in California made their way to Oklahoma casinos. In 2018, as a way to generate more revenue for education in the state, the law was changed to allow dice and ball games in Oklahoma tribal casinos.

Some roulette games intended for California and Oklahoma casinos use cards in order to conform to those states' gambling laws and simply substitute cards for a ball on a wheel, without any distinctive additions.

- INAG International is a gaming company that specializes in card-based games, including versions of games such as roulette and craps that typically are played with other devices. These games can be attractive to casinos in states such as California and Oklahoma. One of INAG's forays into card-based table games is *Mystery Card Roulette*. This game replaces the numbers on a roulette wheel with a circular card rack that holds specially-marked cards. The deck of 38 cards is shuffled by the dealer and the cards are distributed among 38 slots on the wheel. The wheel functions something like a horizontal Big Six wheel or Wheel of Fortune when spun; instead of a roulette ball, a rubber marker moves through a sequence of pins on the wheel's circumference. As the wheel slows, the marker indicates the winning card, which is then removed from the wheel and exposed to the players.

- The local variation of roulette at the Firelake Grand Casino in Shawnee, Oklahoma, was called *Spinette* and was played with 38 standard playing cards on a wheel. The cards are divided among the four suits, 9 per suit, and two green jokers. Instead of betting on the numbers from 1–36, Spinette players bet on standard playing cards or groups of cards. Players may bet on any of the four suits, a 9-card option that pays 3–1, just like the element bets in Chinese roulette.

- A game simply called *Card-Based Roulette* was on offer at the Riverwind Casino in Norman, Oklahoma. A deck of 38 standard playing cards was used, with each spin of a hypothetical roulette wheel corresponding to three cards dealt to the table. The middle card was scanned by a computer that feeds a video display which shows a roulette wheel with a ball dropping into the winning pocket [101].

Richer Roulette

California's gaming laws can be compressed to "the outcome has to be determined by cards" as a first approximation without much distortion. *Richer Roulette*, the wheel game of choice at the Valley View Casino in Valley Center, manages to offer American roulette by extending the definition of what constitutes a "card". Noting that California's objection to roulette centers on the use of a ball to determine the outcome, Richer Roulette follows Mystery Card Roulette by using a pointer and a wheel with pins around its perimeter. The game uses standard roulette numbers, each printed on a tile that rests face down on the wheel. The wheel is spun, and when it comes to a stop, the pointer indicates a tile, which is removed and displayed to show the winning number. Each tile also bears the image of a playing card, keeping the game in line with California law although its rank and suit are not used in the play of the game.

The Richer Roulette wheel is described as a "Double Action Roulette Wheel" offering a more exciting game due to the presence of an additional high-payoff wager. This wheel consists of three concentric rings: an outermost wheel divided into 38 unlabeled sectors, a middle wheel which stores the 38 card tiles face down, and an innermost wheel bearing two hearts and four spades spaced around its surface. The three rings rotate independently and lock together as the spin slows down. The Richer Bonus bet, which may be made on any of the 38 numbers separately from the main layout, pays off if the winning number on the middle wheel is aligned with a suit on the inner wheel. This results in 6 possible Richer Bonus bet numbers on every spin. However, the Richer Bonus is only paid when the randomly-chosen winning number is locked on to a suit; this happens with probability $\frac{6}{38}$. If the suit is a spade, the payoff is 100–1; a heart pays 300–1.

We can develop a probability distribution function for a \$1 Richer Bonus bet. One way that the bet loses is if the number is not spun on the outer wheel; this contributes $\frac{37}{38}$ to the probability of losing. The bet also loses if the number is spun but does not line up with a heart or spade. Since there are only 6 suit symbols among the 38 spaces on the inner wheel, the probability of this further loss is

$$\frac{1}{38} \cdot \frac{32}{38} = \frac{32}{1444}.$$

Adding gives

$$\frac{37}{38} + \frac{32}{1444} = \frac{719}{722} \approx .9958$$

as the total probability of losing \$1.

The probability of a win is then seen to be

$$1 - \left(\frac{37}{38} + \frac{32}{1444} \right) = \frac{3}{722}.$$

Of this tiny fraction, the bet wins \$100 with probability $\frac{2}{3}$ and wins \$300 with probability $\frac{1}{3}$. Table 2.12 gives the complete probability distribution for the winnings X.

A \$1 Richer Bonus bet then has an expectation of

$$E = (100) \cdot \frac{4}{38} \cdot \frac{1}{38} + (300) \cdot \frac{2}{38} \cdot \frac{1}{38} + (-1) \cdot \left(\frac{37}{38} + \frac{1}{38} \cdot \frac{32}{38} \right) \approx -\$.3033,$$

TABLE 2.12: Probability distribution for Richer Roulette's Richer Bonus bet

x	$P(X = x)$
–1	$\dfrac{719}{722} \approx .9958$
100	$\dfrac{2}{722} \approx .0028$
300	$\dfrac{1}{722} \approx .0014$

making this an exceptionally bad roulette bet—among the worst bets in any casino table game.

Triple Flop Roulette

While it was well-suited to casinos in California and Oklahoma, *Triple Flop Roulette* (TFR) made an appearance at the Orchid Casino in Aruba in 2018. TFR uses a deck of 38 cards, 6s through aces from a standard deck plus two distinguishable jokers, to play the mathematical equivalent of American roulette without a ball or wheel. The game could also be installed in a 37-card European roulette version with a single joker.

A motivation for TFR's developers was to bring excitement back to the casino floor, which they did by claiming that the most exciting carnival games all had a poker component [143]. With that as a guiding principle, TFR began with a single drawn card to resolve roulette bets, but then went on to draw two additional cards and form a three-card poker hand, which invited a separate set of wagers. On the first card, the jokers function like the 0 or 00 on a roulette wheel; on the second or third card, a joker is wild. TFR's roulette bets were the standard bets, with traditional payouts, on 1, 2, 3, 4, 6, 12, or 18 cards; either layout-adjacent cards or cards of a common color. There was no basket bet at TFR, and although the 38 cards included 9 cards in each suit, there was no simple suit bet paying 3–1 on offer either. Each of these bets carried the 5.26% HA we have come to know well from American roulette.

Example 2.17. The TFR layout permits a gambler to bet on all 9 cards of one suit by making 3 street bets paying 11–1 each, one each on the 6 through 8, 9 through jack, and queen through ace. At $1 per wager, this bet combination pays a net profit of $9 if a card of the chosen suit is drawn and loses $3 otherwise. The expected value of this combination is

$$E = (9) \cdot \frac{9}{38} + (-3) \cdot \frac{29}{38} = -\frac{6}{38} \approx -\$.1579.$$

Dividing by the $3 wagered gives the familiar 5.26% HA associated with American roulette. ■

The optional poker hand wagers were based on the strength of the 3-card hand. Six bets were available; they are listed in Table 2.13. Jokers must be

TABLE 2.13: Triple Flop Roulette: Poker bets

Wager	Payoff
Pair:	6s through kings: 2–1
	Aces: 5–1
Straight	7–1
Flush	11–1
Straight flush	40–1
Three of a kind	50–1
Royal Jackpot:	Royal flush, AKQ of diamonds: 1000–1
	Royal flush, AKQ of other suits: 250–1
	Jokered royal flush: 100–1

valued as the card that makes the resulting 3-card hand as high as possible.

Example 2.18. A hand consisting of $A\heartsuit, 7\diamondsuit$, and a joker would be read as a pair of aces rather than a pair of 7s, and so would pay off Pair bettors at 5–1, not 2–1.

If the 3 cards are the $A\clubsuit, 8\clubsuit$, and a joker, then the joker is valued as the $K\clubsuit$, and the hand is a flush rather than a pair of aces or 8s. ∎

It is noteworthy that straight flush and royal flush hands also paid off on appropriate lower-level bets, so the Straight and Flush bets won if a straight flush was turned up, for example. This makes counting those bets easier, since there is no need to remove straight or royal flushes from the pool of winning hands. However, a Pair bet does not pay off if 3 of a kind is dealt.

Example 2.19. In assessing the HA of the poker bets at TFR, we are drawn to the 5.26% HA on the roulette bets as a standard. How does the HA of the Pair bet compare to this value?

A hand containing both jokers will be valued as a straight flush or royal flush, and so those hands need not be counted here. There are 4 cases to consider:

1. **A "natural" pair of 6s through kings, without a joker.** These are easy to count: there are 8 choices for the rank and then $\binom{4}{2} = 6$ ways to choose the cards comprising the pair. The third card must not be a joker or a third card of the rank of the pair, so there are 32 ways to pick the odd card. The probability of these natural pairs is

$$p_1 = \frac{8 \cdot 6 \cdot 32}{\binom{38}{3}} = \frac{1536}{8436}.$$

2. **A natural pair of aces.** There are $6 \cdot 32 = 192$ of these, so the probability of this hand is

$$p_2 = \frac{192}{8436}.$$

3. **A pair of 6s through kings that includes a joker.** There are 2 ways to choose the joker; we shall set that aside temporarily. The third card must not complete any higher hand; this is a possibility if the two non-jokers are suited or in sequence.

It is immediately clear that there are no pairs of 6s that include a joker, since the joker would instantly be paired with the higher card. It is perhaps less clear that there can also be no jokered pairs of 7s or 8s, even if the other card is lower in rank. For example, in the hand 8♠ 6♣ Joker, the joker would be valued as a 7 and the hand called a straight.

There are exactly 3 hands containing a joker and valued as a pair of 9s: the third card can be any 6 that is not of the same suit as the 9. Similarly, there are 6 ways to pick the third card to join with a 10 and a joker to create a jokered pair of 10s. Any 6 or 7 of a different suit will do that. Continuing up the list of card ranks, there are 9 ways to pick the 3rd card to form a pair of jacks, 12 ways to get a pair of queens, and 15 ways to form a jokered pair of kings. Adding everything up gives

$$p_3 = \frac{2 \cdot 4 \cdot (3 + 6 + 9 + 12 + 15)}{\binom{38}{3}} = \frac{360}{8436}.$$

4. **A jokered pair of aces.** We can choose the ace and joker in $4 \cdot 2 = 8$ ways. The third card must not be an ace, a joker, or any card that completes a higher-ranked hand. The pattern in case 3 can be extended: there are 18 ways to pick that 3rd card, which can be any 6 through jack of a different suit than the ace. This gives

$$p_4 = \frac{2 \cdot 4 \cdot 18}{\binom{38}{3}} = \frac{144}{8436}.$$

The probability of losing a Pair bet is then

$$1 - p_1 - p_2 - p_3 - p_4 = 1 - \frac{2232}{8436} = \frac{6204}{8436} \approx 73.5\%,$$

and so the expectation of a Pair bet is

$$E = (2) \cdot (p_1 + p_3) + (5) \cdot (p_2 + p_4) + (-1) \cdot \frac{6204}{8436} \approx -.0868,$$

so the house holds an 8.68% advantage on the Pair bet, rather higher than the edge on the roulette bets. ∎

The Flush bet can be considered similarly, by dividing the 3-card hands based on the number of jokers. The number of 3-card flushes without any jokers is simply $4 \cdot \binom{9}{3} = 336$. If the hand holds 1 joker, there are $4 \cdot \binom{9}{2} = 144$ ways to choose two other cards of the same suit, which must be multiplied by 2 to account for the choice of joker. Finally, there are 36 hands with 2 jokers—simply count the number of non-joker cards.

Adding gives

$$\frac{336 + 288 + 36}{8436} \approx \frac{1}{12.78} \approx .0782$$

as the probability of a 3-card flush, from which it follows that the casino holds a 6.12% edge on this bet.

For the Royal Jackpot bet, there are 4 natural royal flushes, one per suit, and 36 jokered royal flushes, divided as follows:

- If the hand includes 1 joker, there are $4 \cdot \binom{3}{2} = 12$ ways to pick the other 2 cards, thus we have 24 1-joker royal flushes since there are 2 ways to pick the joker.

- For a royal flush with 2 jokers, the third card must be a queen, king, or ace, and there are 12 of those.

Since the diamond royal flush pays 1000–1 and the other natural royal flushes pay 250–1, the resulting expectation is

$$E = (1000) \cdot \frac{1}{8436} + (250) \cdot \frac{3}{8436} + (100) \cdot \frac{36}{8436} + (-1) \cdot \frac{8396}{8436} \approx -\$.3611.$$

The house holds a 36.1% edge on this bet. This could be brought down to a more player-friendly value by changing the pay table. Simply paying 1000–1 on all natural royal flushes cuts the HA to 9.44% and simultaneously makes the game easier for dealers to administer. If there is marketing value in having one natural royal flush pay a bonus—perhaps in a "Suit of the Day" promotion where the bonus suit is selected daily—that could be preserved and the HA reduced to 3.51% by paying one suit at 1500–1 and all other suits at 1000–1.

An alternative would be to increase the payoff on a jokered royal flush, though there is little room to do so while preserving the other payoffs. Raising the payoff to 150–1 gives a new house edge of 14.77%, but going to 200–1 gives the players a 6.57% advantage.

Pala Roulette

In Pala, California, the Pala Casino deals a form of roulette where a modified wheel is used to select a winning card. The cards are a special deck of 38 bearing each of the numbers on an American roulette wheel in its traditional color. The wheel is changed by removing the numbers and coloring 12 pockets red, 12 white, 12 blue, and 2 green. Four cards, one designated for each color, are dealt facedown at the start of each spin of the wheel, and the color of the pocket where the ball comes to rest determines which card is turned over as the winner.

All of the probabilities and expectations in this variation match those of American roulette. The Pala adds an extra wager to the layout: the Super Green Bet. This bet is a winner if the ball lands in a green pocket and the card turned over is a green number: 0 or 00. The payoff on this bet is 275–1. Since the pocket and the card are independent events, the probability of winning is

$$\left(\frac{2}{38}\right)^2 = \frac{1}{361},$$

giving the casino an edge of 23.55%.

Example 2.20. Recall that every bet on an American roulette wheel except for the basket bet carries a 5.26% HA ($\frac{1}{19}$ rounded to two decimal places). What whole-number payoff, x to 1, on the Super Green Bet would come closest to this value?

We need to solve the linear equation

$$(x) \cdot \frac{1}{361} + (-1) \cdot \frac{360}{361} = \frac{x - 360}{361} = -\frac{1}{19}$$

for x, which results in $x = 341$. While this value might be inconvenient, there is certainly room to increase the payoff odds on the Super Green Bet while maintaining a healthy house advantage. ∎

2018: Changes in Oklahoma

In 2018, the Oklahoma Legislature legalized "ball and dice games" in the state's tribal casinos as part of an effort to raise revenue for Oklahoma teachers, who were among the lowest-paid teachers in America. It is common practice in Oklahoma to charge a per-hand fee at card games independent of a player's wager, and this practice was carried over to craps and roulette tables in Oklahoma.

In roulette, this took the form of a $1 ante charged per spin of the wheel. Combined with a $5 minimum bet, the effective payout on a winning even-money bet was 5–6. The expected value of such a bet was then

$$E = (5) \cdot \frac{18}{38} + (-5) \cdot \frac{20}{38} - 1 = -\frac{48}{38},$$

giving a house advantage of $\frac{48}{190}$, or 25.26%—nearly 5 times the 5.26% advantage without the fee.

Since the fee was fixed rather than set as a proportion of a player's wager, it was possible to decrease the casino's advantage by making a bigger bet. On an $N even-money bet, the expectation becomes

$$E(N) = (N) \cdot \frac{18}{38} + (-N) \cdot \frac{20}{38} - 1 = \frac{-2N - 38}{38}.$$

Dividing by N gives a HA of

$$-\frac{-2N - 38}{38N} = \frac{1}{19} + \frac{1}{N},$$

which decreases as N increases. For a \$100 wager, the player is fighting a house edge of 6.26%.

How does the ante affect other roulette bets? A \$5 single-number bet would have an expected value of

$$E = (175) \cdot \frac{1}{38} + (-5) \cdot \frac{37}{38} - 1 = -\frac{48}{38},$$

as with even-money bets. The HA remains 25.26%, and it decreases in inverse proportion to the amount of the wager as well.

The same mathematics applies to every bet on an American wheel, except for the basket bet. The house advantage on a \$5 basket bet in Oklahoma is 27.89%.

Joker Seven

Card-based games similar to roulette date back at least to the 1960s in England. Twenty years prior to the Indian Gaming Regulatory Act, *Joker Seven* was an English card game launched on cruise ships. Joker Seven's rules card promised "the variety and style of gaming associated with roulette, with the international fascination of playing cards". Each 7-card hand of Joker Seven was dealt from a 54-card deck that included 2 jokers and was shuffled between hands. The game was not based on poker hands beyond 3 of a kind; the available wagers focused on colors, pairs, and *prials*, or 3-of-a-kind hands.

Jokers carried neither color nor rank, although if both jokers were dealt to the table, they counted as one pair for betting purposes.

Example 2.21. Joker Seven provides for "No Pair" and "Two Pair" bets, each paying 3–1. With the understanding that 4 of a kind is counted as 2 pairs, which of these events has the higher probability?

The probability of a 7-card no-pair hand must include both joker-free hands and hands with 1 joker. There are

$$\binom{13}{7} \cdot 4^7$$

ways to choose one card each from 13 different ranks not including jokers, and

$$2 \cdot \binom{13}{6} \cdot 4^6$$

hands containing a joker and 6 cards of different ranks, Taken together, the probability of a no-pair hand is

$$\frac{\binom{13}{7} \cdot 4^7 + 2 \cdot \binom{13}{6} \cdot 4^6}{\binom{54}{7}} \approx .2381.$$

For hands with 2 pairs, there are 8 disjoint cases to consider. Each can be subdivided for easier counting by considering the number of jokers in the hand.

Consider the simplest case: two unlike pairs and 3 single cards, and denote it as $AABBCDE$. If the hand contains no jokers there are

$$\binom{13}{2}\binom{4}{2}^2\binom{11}{3} \cdot 4^3 = 29,652,480$$

ways to draw this hand. With a single joker, the hand may be drawn in

$$\binom{13}{2}\binom{4}{2}^2\binom{11}{2} \cdot 4^2 \cdot 2 = 4,942,080$$

ways, and if one of the pairs consists of the 2 jokers, the number of hands is

$$13 \cdot \binom{4}{2} \cdot \binom{12}{3} \cdot 4^3 = 1,098,240.$$

Adding these up and dividing by $\binom{54}{7}$ gives $P(AABBCDE) \approx .2015$.

$AABBCDE$, by itself, accounts for nearly 90% of all two-pair hands. Table 2.14 shows the probabilities of the remaining 7 hand configurations.

TABLE 2.14: Joker Seven: Two-pair hand probabilities

| Configuration | —— Number of jokers —— | | | Probability |
	0	1	2	
$AABBCDE$	29,652,480	4,942,080	1,098,240	.2015
$AABBCCD$	61,816	123,552	123,552	.0017
$AAABBCC$	123,552	0	3744	.0007
$AAABBCD$	3,294,720	329,472	45760	.0207
$AAABBBC$	54,912	2496	0	.0003
$AAAABCD$	183,040	27,456	0	.0012
$AAAABBC$	41,184	1872	624	.0002
$AAAABBB$	624	0	0	3.523×10^{-6}

Summing the last column gives a probability of .2265 for a two-pair hand, somewhat less than for No Pair. ∎

The respective HAs are 4.75% for No Pair and 9.40% for Two Pairs. A small difference in the probabilities masks a house advantage that is nearly doubled.

2.6   Side Bets

A *side bet* at a table game is a separate bet unrelated to the main game, as when the Lucky Ladies blackjack side bet pays off based solely on the player's and dealer's first two cards, before the hands are played out. Developing a lucrative side bet for baccarat, blackjack, craps, or roulette may be an attractive pursuit for a casino game designer because the basic game is in place; the side bet can be marketed to casinos as an add-on to a game already on offer. Without the need to develop and sell a table game from scratch, the potential for selling a good new small idea increases.

Side bets should be easy to understand for both players and dealers, without distracting from the main game. Many side bets have considerably higher house advantages than the primary wagers; this can be a way for casinos to make more money on an optional bet that draws less action or smaller wagers, or a way to build excitement over a large jackpot with a low win probability.

Example 2.22. A \$100 bet on a game with a .5% HA, such as blackjack using basic strategy, has an expected return for the casino of 50¢. This is the same average return the casino collects from a \$2 side bet bearing a 25% house edge.   ∎

Some of the biggest jackpots are *progressive*: these prizes are seeded with an initial payoff and are increased by a small percentage of each bet until they are won, after which the jackpot starts over with the initial seed.

Table 2.15 shows HAs for the main bets on the big 4 table games, together with the HA of a side bet covered in this book.

TABLE 2.15: Side bets

Game	Main bet	HA	Side bet	HA
Blackjack	Player hand	.5–5%	Lucky Ladies	24.94%
			WINSURE	$\geqslant 10.20\%$
Baccarat	Player	1.36%	Baccarat World	$\geqslant 2.32\%$
	Banker	1.17%	WINSUIT	2.89%
Craps	Pass	1.41%	Field	5.56%
	Don't Pass	1.34%	Tall, Small	7.76%
American	Any except basket	5.26%	Colors	4.34%
roulette	Basket bet	7.89%	WINGO	3.78%

It may seem that most of the good ideas for side bets have been taken. Craps, blackjack, baccarat, and roulette have long histories and have been studied extensively, and the list of proposed side bets is long and varied, with widely ranging degrees of success. Nonetheless, new ideas designed to enliven traditional table games continue to appear on casino floors.

Newar Bets

The Golden Nugget Casino operated in London from 1965–2014, and included a number of European roulette wheels [123]. From time to time, a pair of extra bets called the *Newar bets* was available. One Newar bet paid 3–1 on red even numbers; the second paid the same 3–1 on black odd numbers. Both bets paid off at 2–1 if the single zero was spun.

The expectation of a Newar bet can be found by referring to Figure 2.9. For either bet, we count 8 winning numbers plus the one zero. We have

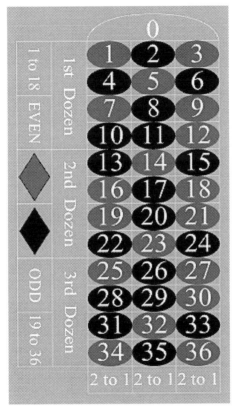

FIGURE 2.9: European roulette layout

$$E = (3) \cdot \frac{8}{37} + (2) \cdot \frac{1}{37} + (-1) \cdot \frac{28}{37} = -\$\frac{2}{37} \approx -\$0.0541.$$

The HA at European roulette is 2.70% on every bet except the even money bets with the *en prison* rule in force, which offer the player a low 1.35% HA. The Newar bets, with their 5.41% HA, are at least twice as bad for the player.

Prime Time

A *prime* number is an integer $p > 1$ which is only divisible by itself and 1. The prime numbers on a roulette wheel are 2, 3, 5, 7, 11, 13, 17, 19, 23, 29, and 31. The *Prime Time* side bet for roulette is a wager on the length of a consecutive string of spins that produce prime numbers. Several pay tables were devised for Prime Time; one paying off on 1–7 prime spins is shown in Table 2.16.

TABLE 2.16: Prime Time roulette side bet pay table [93]

Number of primes	Payoff
1	1–1
2	3–1
3	6–1
4	10–1
5	20–1
6	50–1
7	299–1

Note that the payoff on a single prime is 1–1; since there are 11 prime numbers, this is substantially worse than a dozen or column bet, both of which pay 2–1 and cover one more number. Any allure that Prime Time might hold comes in the potential for a long string of prime spins.

The probability of spinning a prime number on an American roulette wheel is $\frac{11}{38}$, just under 30%. It follows from the independence of successive roulette spins that the probability of spinning exactly k straight primes, $0 \leqslant k \leqslant 6$, is

$$P(k) = \left(\frac{11}{38}\right)^k \cdot \frac{27}{38}.$$

The probability of 7 straight primes is simply

$$P(7) = \left(\frac{11}{38}\right)^7 \approx \frac{1}{5871},$$

without the need for a factor of $\frac{27}{38}$ since the wager ends at this point, regardless of what the 8th spin is.

Example 2.23. The HA on a column or dozen bet is, as we know, 5.26%. Define the payoff (to 1) for a string of k prime spins shown in Table 2.16 as $A(k)$. Prime Time carries an expected value of

$$E = \sum_{k=1}^{7} [A(k) \cdot P(k)] + (-1) \cdot \frac{27}{38} \approx -\$.0722,$$

giving a higher HA of 7.22%. The potential for a higher payoff than can ordinarily be attained by a $1 bet somewhat mitigates the downside of a higher casino edge. ∎

WINGO

Gaming writer Stanley Roberts proposed several new side bets, for roulette, baccarat, and blackjack, in 1990. His roulette bet was called *WINGO*, after *WIN* magazine. WINGO was a multispin wager for even-money bets. Like the odds bet in craps, WINGO paid off at true odds, with no house advantage, once the initial bet was resolved.

To place a WINGO bet, a player first had to place and win an even-money roulette bet: High, Low, Red, Black, Odd, or Even. Instead of accepting the even-money payoff, the player called out "WINGO!", which was a request for the bet to be moved to the lowest rung of a 5-bet ladder. Effectively, this amounted to repeating the previous winning bet by letting the winnings ride. The bet ascended the ladder as long as it kept winning; if 5 straight spins (for a total of 6) then came up showing the indicated even-money option, the bet was paid off at 64–1. While the bet was on the ladder, 0 and 00 were not losing spins, but pushes, which kept the bet alive [104].

Example 2.24. Suppose a player makes a $1 bet on Odd, and the first spin is 9. The player has the option of taking her bet down and collecting $2. If she declares "WINGO!" instead, the wager moves to the bottom of the ladder. If the next three spins are 19, 7, and 19, the wager moves up 3 times and is 2 steps from the top. A 5th spin turns up 0, and the bet does not move. When the 6th spin reveals a 2, the bet loses. Unlike No Lose Free Roll, there is no payoff for the 4 odd numbers spun. ∎

WINGO bets, once entered onto the ladder, could not be withdrawn and had to ride to the top of the ladder or be lost. The probability of winning $64 on a $1 bet was

$$p = \left(\frac{18}{38}\right) \cdot \left(\frac{1}{2}\right)^5 = \frac{18}{1216} \approx .0148.$$

The expected value was then

$$(64) \cdot p + (-1) \cdot (1-p) = 65p - 1 \approx -\$.0378,$$

slightly less than the −$.0526 for the even-money wager without the WINGO option. This shows that WINGO tilted the game's chances slightly toward the player. The HA decreases when spins of 0 and 00 function as standoffs on even-money bets rather than casino wins.

Colors

The *Colors* bet is a descendant of WINGO that was introduced in 2008 at the Orleans Casino in Las Vegas [13]. To make this bet, a gambler selects either

Red or Black, and if that color then turns up on three consecutive spins, the payoff is 8 to 1. No matter which color is chosen, we have $P(\text{Win}) = \frac{18}{38}$ for a single spin on an American roulette wheel. Since the spins are independent, the probability of winning the Colors bet is

$$\left(\frac{18}{38}\right)^3 = \frac{729}{6859} \approx .1063,$$

and so we have

$$E = (8)\cdot\left(\frac{729}{6859}\right) + (-1)\left(\frac{6130}{6859}\right) = -\frac{298}{6859} \approx -\$.0434.$$

The house advantage is about 4.34%—somewhat less than the HA for any other bet on the board.

The same calculation for European roulette wheels, where $P(\text{Win}) = \frac{18}{37}$, shows that the expected value of a \$1 Colors bet is

$$E = (8)\cdot\left(\frac{5832}{50,653}\right) + (-1)\left(\frac{44,821}{50,653}\right) = \frac{1835}{50,653} \approx \$.0362,$$

about 3.6¢. Since this is positive, the player has a 3.6% advantage over the casino—which means that you won't see the Colors bet with an 8–1 payoff on a European wheel. If you do, someone has made a serious calculation or judgment error. The bet is unfair in your favor, and you should settle in and prepare to win money.

Roulette Rage

Roulette Rage is a side bet that extends Colors; it is approved for use in British casinos as well as in Mississippi [53]. A Roulette Rage bet is a wager that a single color will be spun at least 5 times in a row. The bet is available immediately following any change of colors between red and black in a sequence of spins—a spin landing on 0 terminates the previous color but does not open the board to new bets. For example, if a red number is spun after 6 black numbers, the croupier announces that the Roulette Rage bet on red is available. This first red spin merely establishes the color of the wager; it does not count in the subsequent sequence.

Following the establishing spin, the length of the sequence of subsequent spins of the same color determines the payout. Payoffs are made for 4–14 straight spins; the minimum approved payoff odds are shown in Table 2.17.

Example 2.25. Using a European wheel with a single 0, the probability of extending a sequence is $\frac{18}{37}$ on each spin. The probability of a winning sequence of 4 or more spins following the establishment of a color is then

$$\left(\frac{18}{37}\right)^4 \approx .0560,$$

slightly more than 1 chance in 18. ∎

TABLE 2.17: Roulette Rage minimum pay table [53]

Consecutive spins	Minimum payoff
4	5–1
5	8–1
6	15–1
7	20–1
8	40–1
9	60–1
10	100–1
11	200–1
12	400–1
13	1000–1
14	2000–1

Table 2.17 yields a 17.00% house advantage, hence its designation as a table of minimum permitted payoffs [53]. Since Roulette Rage has a low probability of winning, balancing the game would call for larger payoffs, preferably toward the beginning of the pay table where wins are more attainable. An even-money payoff on a run of 3 spins would bring the HA down to 5.18%.

On a larger scale, one approved alternate pay table is shown in Table 2.18, with the improved payoffs in bold. The expected value of a $1 bet against

TABLE 2.18: Alternate Roulette Rage pay table [53]

Consecutive spins	Payoff
4	**6–1**
5	**10–1**
6	**20–1**
7	**30–1**
8	40–1
9	60–1
10	100–1
11	200–1
12	400–1
13	1000–1
14	2000–1

Table 2.18 is 4.61%.

Pro-Aggressive Roulette

When we hear the word "progressive" in regard to slot machines, we suspect immediately that the probability of winning on that machine is lower than on a non-progressive version of that machine, due to the need to fund the jackpot. *Pro-Aggressive Roulette*, a new American roulette wager patented in 2004 [94], is a progressive side bet that raises that same concern.

The standard roulette wagers are all available in this game; to that array of choices is added a progressive jackpot that pays off if three consecutive spins of the roulette wheel generate a line of three numbers (for example, 24, 27, and 30) on the layout (see Figure 2.10) in any direction—horizontally, vertically, or diagonally—and any order. A second, larger, progressive amount is paid if three straight spins hit the numbers 0, 00, and 2, which form a "V" on the layout.

FIGURE 2.10: American roulette layout [45]

To be eligible for this payoff, a gambler need not correctly pick and bet on the actual numbers, but instead must make a separate progressive bet on every one of the three spins involved; in practice, a player pursuing this jackpot would be likely to invest an extra unit on every spin. This progressive wager qualifies the player for the jackpot if a required line of three numbers is spun anywhere on the layout. Once a sequence is established, if the next spin does not fall on a winning line, a new sequence begins with that number as the first.

Example 2.26. Suppose that a player is betting the progressive wager on every spin and the following sequence of spins occurs:

TABLE 2.19: Sample of roulette spins with Pro-Aggressive Roulette side bet active

24	Sequence starts. Eligible next numbers are 16, 18, 20, 21, 22, 23, 26, 27, 28, and 30.
34	Cannot form a line of 3 numbers with 24, so the first sequence ends. A new sequence starts, with eligible numbers 28, 30, 31, 32, 35, and 36.
23	The wager ends, and a new sequence starts. Eligible next numbers are 17, 19, 20, 21, 22, 24, 25, 26, 27, and 29.
7	The wager ends, and a new sequence starts. Eligible next numbers are 1, 3, 4, 5, 8, 9, 10, 11, 13, and 15.
15	On a line with 7, so the sequence continues. If the next number is 11, players win on a diagonal.
28	The wager ends, and a new sequence starts. Eligible next numbers are 22, 24, 25, 26, 29, 30, 31, 32, 34, and 36.
30	The wager continues. If the next number is 29, a row is complete, and players win.
2	The wager ends, but players are now eligible for both progressive jackpots. Eligible next numbers are 1, 3, 5, and 8 for the small jackpot, and 0 and 00 for the large jackpot.
6	Both wagers end, and a new sequence starts. Eligible next numbers are 3, 4, 5, 8, 9, 10, and 12.
9	Sequence continues. Players win if the next number is 3 or 12.
12	Sequence complete; progressive jackpot awarded.

The total amount wagered in pursuit of this jackpot was 11 units. ∎

The probability of winning the large progressive jackpot in three spins is simply the probability that three straight spins will result in 0, 00, and 2 in any order, and is easily found to be

$$\frac{3}{38} \cdot \frac{2}{38} \cdot \frac{1}{38} = \frac{6}{54,872} \approx \frac{1}{9145}.$$

The probability of winning the small progressive jackpot is somewhat trickier to compute, since it depends on the numbers in the sequence. The first number can be anything, of course, but where that number lies on the layout affects the probabilities for the second and third numbers.

Example 2.27. Suppose that the first number spun is 34, in the lower left corner of the layout. The only numbers that will contribute to completing a line of 3 are 28 and 31, 30 and 32, or 35 and 36. Once the second number is spun, only one number remains to complete the triple. The probability of winning the small progressive jackpot on a line starting with 34 is then

$$\frac{1}{38} \cdot \frac{6}{38} \cdot \frac{1}{38} = \frac{6}{54,872},$$

the same as the probability for winning the big progressive jackpot on 0, 00, and 2.   ■

Example 2.28. Suppose that the first number spun is 17. Since this number is in the middle of the layout, there are six different winning triples passing through 17:

$$11, 14, 17$$
$$13, 17, 21$$
$$14, 17, 20$$
$$15, 17, 19$$
$$16, 17, 18$$
$$17, 20, 23$$

—consequently, the probability of continuing the sequence following a spin of 17 is $\frac{12}{38} \approx .3158$.

The probability of winning on the third number depends on which number was spun second: 14 admits two possibilities in 11 and 20, whereas 17 followed by 21 only completes a winning triple if the next number is 13. The possibilities are tabulated below:

Number	11	13	14	15	16	18	19	20	21	23
Possibilities	1	1	2	1	1	1	1	2	1	1

Denoting the number of options by x gives the probability of a winning sequence starting at 17 and with a second number that gives x options for the third number as

$$P(x) = \frac{1}{38} \cdot \frac{1}{38} \cdot \frac{x}{38} = \frac{x}{54,872},$$

and summing across all possible second numbers results in

$$P(\text{Win starting at } 17) = 8 \cdot P(1) + 2 \cdot P(2) = \frac{8+4}{54,872} = \frac{12}{54,872},$$

twice as great as the chance of winning starting with 34.   ■

Every number outside of the first two and last two rows on the layout has eight adjoining numbers for which $x = 1$ and two numbers for which $x = 2$, as was the case for 17. By counting the individual possibilities for each of the 36 nonzero numbers and adding up the resulting probabilities, we find that the probability of winning the small progressive jackpot is approximately .0067, or about 1 in 149.

Rainbow Bet

In early 2015, Riverboat Roulette with its spectrum of additional roulette wheel colors was replaced by *Rainbow Bet Roulette*, which is marketed by INAG [100]. An optional side bet, the Rainbow Bet, relies on tagging each number from 1–36 with a secondary color. In addition to the standard red or black, each row of three numbers receives another color, which is shown on the layout as a ring around the number:

TABLE 2.20: Rainbow Roulette secondary colors [95]

Numbers	Color
1–3	Blue
4–6	Pink
7–9	*White*
10-12	Purple
13–15	Teal
16–18	*White*
19–21	Lime
22–24	Orange
25–27	*White*
28–30	Yellow
31–33	Brown
34–36	*White*

In form, the Rainbow bet is analogous to several craps wagers that track a list of outcomes on successive spins or rolls. A Rainbow wager is a bet that 3 or more secondary colors will come up consecutively on the wheel before either a white or green (0 and 00) number is spun or a color repeats. This bet is paid off depending on the number of colors that hit. Note that in the payoff table for an American Roulette wheel (Table 2.21), all payoffs are quoted as "for 1" rather than "to 1". This is so stated because player wagers are collected prior to the start of a Rainbow sequence and placed in a separate betting rack where the colors of successive spins are also recorded. When the wager is resolved, it is convenient for the casino simply to pay off any winners and retain their original chips.

Inspection of Table 2.20 shows that the probability that a color streak will end on a white or green number is $\frac{14}{38}$. The streak-ending probability increases as more colors are hit; in the end, the probability of a streak extending from 7 to 8 colors is only $\frac{3}{38}$.

Example 2.29. Find the probability that a color streak runs for exactly 3 colors.

TABLE 2.21: Rainbow bet payoffs [95]

Colors	Payoff
3	2 for 1
4	5 for 1
5	10 for 1
6	25 for 1
7	200 for 1
8	1000 for 1

Four events must happen in succession:

1. One of the eight secondary colors must come up on the first spin. This happens with probability $p_1 = \dfrac{24}{38}$.

2. A different secondary color must come up on the second spin. The probability of this event is smaller than p_1 due to the need not to duplicate the first color: $p_2 = \dfrac{21}{38}$.

3. A third secondary color must come up on the third spin, $p_3 = \dfrac{18}{38}$.

4. The streak must then end with either a repeated secondary color, which happens with probability $\dfrac{9}{38}$, or a white or green number. Using the Addition Rule, we have

$$p_4 = \frac{9 + 14}{38} = \frac{23}{38}.$$

Since successive spins are independent, the Multiplication Rule can be used to compute this probability:

$$p = p_1 \cdot p_2 \cdot p_3 \cdot p_4 = \frac{24}{38} \cdot \frac{21}{38} \cdot \frac{18}{38} \cdot \frac{23}{38}.$$

Evaluating this product gives

$$p = \frac{208{,}656}{38^4} \approx .1001.$$

∎

Similar calculations can be made for streaks of other lengths, resulting in the probabilities in Table 2.22. These probabilities can then be multiplied by the corresponding payoffs in Table 2.21 to find the house edge on the Rainbow bet. After subtracting \$1 from the sum of these products to account for the "for 1" payoffs, we find that the expected value of a \$1 Rainbow bet is –\$.1137. This bet is more than twice as bad for the bettor as any of the standard roulette bets. Any appeal that the Rainbow bet has derives from the potential for high payoffs on the 7- and 8-color streaks.

TABLE 2.22: Rainbow bet streak probabilities

Colors	Probability
0	.3684
1	.2825
2	.1837
3	.1001
4	.0447
5	.0157
6	.0041
7	7.099×10^{-4}
8	6.084×10^{-5}

Example 2.30. Pay tables, of course, can be changed, and a skilled game designer will prepare several payoff options that a casino adopting the game might choose from. An alternate payoff for the Rainbow bet doesn't start paying off until 6 colors have been spun in a row, but increases the payoffs for 6, 7, and 8 numbers in a partial effort to make up for this. This pay table is as follows:

Colors	Payoff
6	50 for 1
7	500 for 1
8	2000 for 1

The probabilities in Table 2.22 do not change. The corresponding expectation for a \$1 bet with this new pay table is

$$E = (50) \cdot .0041 + (500) \cdot .0007099 + (5000) \cdot .00006084 - 1 = -\$.1353.$$

Here, subtracting the initial wager at the end eliminates the need to incorporate the probabilities for the losing streaks of fewer than 6 colors. ■

Roulette Royale

Roulette Royale is an online variation with something unusual: a mandatory side bet. Typically, side bets at table games are entirely optional; Roulette Royale requires that the player make the progressive side bet in addition to any other wagers he or she makes at this European roulette derivative [108].

The progressive wager is a simple bet that a number will be spun in two or more consecutive spins. The jackpot is paid out when any number appears five straight times, an event with probability

$$\left(\frac{1}{37}\right)^4 = \frac{1}{1,874,161}.$$

Lesser amounts are paid for shorter runs of a single number:

Number of repeats	2	3	4	5
Payoff	$15	$200	$3000	Progressive

A $1 jackpot bet rides as long as the same number is spun; when a different number comes up, all jackpot bets are collected, appropriate winnings (if any) are paid, and the bets must be replaced.

Example 2.31. Assume that the progressive jackpot starts at $40,000. What is the expected return of a $1 jackpot wager? In light of the fact that this is a mandatory wager, how does it affect the expectation of a $1 wager on a single number?

The first calculation is a simple one:

$$E = (15) \cdot \frac{1}{37} + (200) \cdot \frac{1}{37^2} + (3000) \cdot \frac{1}{37^3} + (40{,}000) \cdot \frac{1}{37^4} + (-1) \cdot \frac{36}{37},$$

or $E = -34.1$¢.

We can then use the Addition Rule to combine this bet with a $1 even-money bet.

$$E = (1) \cdot \frac{18}{37} + (-1) \cdot \frac{19}{37} + -.341 \approx -\$.3679.$$

Dividing this value by the $2 wager gives a house advantage of 18.4%—a far cry from the 2.70% HA on the original wager without the progressive side bet.   ∎

Since Roulette Royale offers a progressive jackpot for a run of 5, it is possible that the advantage may turn to the gambler if the jackpot is sufficiently high. If the jackpot is denoted by J, the player has the edge if

$$E = (15) \cdot \frac{1}{37} + (200) \cdot \frac{1}{37^2} + (3000) \cdot \frac{1}{37^3} + (J) \cdot \frac{1}{37^4} + (-1) \cdot \frac{36}{37} > 0.$$

Simplifying this inequality gives

$$\frac{J}{37^4} - .3622 > 0,$$

and so the expectation is positive when J is at least $678,913.

This is, of course, per gambler. If multiple players ride a string of repeated numbers to a progressive jackpot, the jackpot must rise in proportion to their number to turn the expectation positive.

Lucky Ball

Lucky Ball is another roulette side bet that makes use of an electronic display. Four betting squares: red, green, blue, and yellow, are added to the top of a

European roulette layout. Once all bets are placed, four numbers are randomly chosen and one is assigned to each of the colors. Additionally, one color is chosen for an enhanced bonus payoff that exceeds the standard 35–1 on a straight number bet. If one of the four numbers is spun, Lucky Ball wagers on that number pay off as shown in Table 2.23.

TABLE 2.23: Lucky Ball payoffs [152]

Color	Payoff	Bonus payoff
Red	6–1	120–1
Green	12–1	100–1
Blue	20–1	75–1
Yellow	25–1	60–1

Since the non-enhanced color payoffs are each less than 35–1, the expected value of a Lucky Ball bet is a balance between the depressed ordinary payoff and the higher bonus that occurs ¼ of the time. The general formula, with an ordinary payoff of A to 1 and an enhanced payoff of B to 1, is

$$E = \left(\frac{3}{4}A + \frac{1}{4}B\right) \cdot \frac{1}{37} + (-1) \cdot \frac{36}{37} = \frac{3A + B - 144}{148}.$$

Each color has $3A + B < 144$, and so the casino always has the edge. The best choice for the player is the Red bet, with the lowest HA of 4.05%; this is still worse than the 2.70% edge the house holds on a single-number bet.

Touchdown Roulette

Touchdown Roulette is a side bet incorporated into electronic roulette games developed by Interblock USA. The game, which debuted at the Borgata Casino in Atlantic City in 2013, uses an American football field as the arena on which the bet is played [150].

Players may make a Touchdown bet from $1–5 along with any of the standard roulette bets. If the number spun is 0 or 00, the next 4 spins are interpreted as football plays (while standard roulette action continues). Each spin of the wheel moves the virtual football several yards toward either the red or black end zone. Players with an active Touchdown bet must choose the red or black team before the game starts; a random choice is made for players who do not make their own selection. No new Touchdown bets are accepted while a game is active.

Touchdown bets lose if the wheel turns up any nonzero number; this probability is $\frac{18}{19} \approx .9474$. Touchdown Roulette can also be configured for a European wheel. In that event, the football game is triggered if the wheel turns up 0 or the last number spun, which changes the probability of losing to $\frac{35}{37} \approx .9459$, slightly but perhaps not significantly smaller.

The virtual football starts at the 50-yard line and moves with the spins in accordance with Table 2.24.

TABLE 2.24: Touchdown Roulette: Yardage moved with the 4 game spins. The ball moves in the direction indicated by the color of the number spun.

Number	Yardage
0, 00	0
1–10	3
11–20	5
21–30	10
31–36	15

Example 2.32. Suppose that the 4 spins of the wheel are 36 (red), 4 (black), 26 (black), and 23 (red). The football would move 15 yards toward Red, 3 toward Black, 10 toward Black, and 10 toward Red. Its final position would be the 38-yard line on the Red side of the field. Players choosing the Red team would win.   ∎

The bet pays off according to the final position of the virtual football. The Borgata's original pay table is shown in Table 2.25. Players who selected

TABLE 2.25: Touchdown Roulette payoffs

Yard line	Payoff
Touchdown	500–1
1–10	250–1
11–20	75–1
21–30	50–1
31–40	25–1
41–49	10–1
50	5–1

the losing team receive an even-money payoff. If the ball ends up at midfield, both teams receive the 5–1 payoff. In Example 2.32, the payoff to Red bettors would be 25–1.

Example 2.33. What is the probability of a touchdown?

Any touchdown sequence must use all 4 spins. This may have been a deliberate game design choice: there is no way to score a touchdown in 3 or fewer spins, which would raise the question of whether or not a post-touchdown spin would be permitted to back the ball up and cancel out a touchdown.

There are three ways to score a touchdown:

1. Move the ball 15 yards, 4 times.

   This requires 4 straight spins, of the same color, in the range 31–36. There are 3 red and 3 black numbers in this range, so the probability of a 60-yard touchdown of one color is

   $$p_1 = \left(\frac{3}{38}\right)^4 = \frac{81}{2,085,136} \approx 3.885 \times 10^{-5}.$$

2. Move the ball 45 yards in any 3 spins of 15 yards apiece and then move 5 or 10 yards on the remaining spin. This odd spin need not occur at any particular spot in the sequence of 4 spins.

   The second case allows for 1 spin to be in the range 11–30, again of the same color as the 3 spins from 31–36. There are 4 places in the sequence for this odd spin to occur. Each color covers 10 of the 20 numbers between 11 and 30, so this probability is

   $$p_2 = 4 \cdot \left(\frac{3}{38}\right)^3 \cdot \frac{10}{38} = \frac{1080}{2,085,136} \approx 5.180 \times 10^{-4}.$$

3. Move 15 yards twice and 10 yards twice.

   There are $\binom{4}{2} = 6$ ways to distribute the four spins among two 15-yard gains and two 10-yard gains. In the interval from 21–30 on the wheel, there are 5 red numbers and 5 black numbers. We have

   $$p_3 = \binom{4}{2} \cdot \left(\frac{3}{38}\right)^2 \cdot \left(\frac{5}{38}\right)^2 = \frac{1350}{2,085,136} \approx 6.474 \times 10^{-4}.$$

Adding gives a touchdown probability of

$$p_1 + p_2 + p_3 = \frac{2511}{2,085,136} \approx .0012 \approx \frac{1}{830}$$

for each color. Touchdowns thus contribute

$$(500) \cdot \frac{2}{38} \cdot \frac{2511}{2,085,136} \approx \$.032,$$

just more than 3 cents, to the expected value of a $1 Touchdown bet. ■

Computing the probabilities of other outcomes often requires consideration of far more than 3 cases. Some results are impossible—for example, the game cannot end with the ball on either 6-yard line. A computer analysis of the game shows that the house advantage of the Touchdown bet is 9.00%, making this the worst bet available on the Touchdown Roulette menu. This high HA is largely attributable to the $\frac{18}{19}$ chance of an immediate loss; the average win on a bet that proceeds to a football game is $16.29 per dollar wagered [127].

Roulette Bets Including Dice

Roul 8

Roul 8 combined a roulette wheel with three custom six-sided dice, and was offered at the Grand Casino in Brussels, Belgium. The three dice were blank on 5 faces and had an ∞ symbol on the sixth face. With each spin of the wheel, the dice were tossed within an enclosed plastic dome. Payoffs were based on the number of ∞s on the dice and whether or not the roulette ball landed on the number 8 [126]. Table 2.26 shows the payoffs, which favor the digit 8, as this is thought to be lucky in several Asian cultures.

TABLE 2.26: Roul 8 pay table [126]

Roulette spin	Number of ∞s	Payoff
8	3	888
8	2	88
8	1	8
8	0	4
Not an 8	3	88
Not an 8	2	2
Not an 8	1	0
Not an 8	0	0

Since the roulette wheel and the dice are independent, assessing this side bet is a simple application of the Multiplication Rule.

Example 2.34. What is the probability of winning nothing on a Roul 8 bet?

The bet loses if the ball lands on a number other than 8 and 0 or 1 ∞s are rolled on the dice. We assume a European roulette wheel, so this probability is

$$\left(\frac{36}{37}\right) \cdot \left[\left(\frac{5}{6}\right)^3 + 3 \cdot \left(\frac{5}{6}\right)^2 \cdot \left(\frac{1}{6}\right)\right] = \frac{100}{111} \approx .9009.$$

∎

With a chance of losing that just tops 90%, Roul 8's payoffs need to be high to achieve a reasonable house advantage. Is the top payoff of 888–1 enough to do that?

The probability of the top prize on a European wheel is

$$\frac{1}{37} \cdot \left(\frac{1}{6}\right)^3 = \frac{1}{7992} \approx 1.251 \times 10^{-4}.$$

This contributes 11.11¢ to the expected return on a $1 bet.

Roul 8 actually fares quite well on the reasonability standard; the HA on a European wheel is 5.46%—this is about double the HA on most European roulette bets, but very close to the edge at almost every American roulette wager. On an American wheel, the HA rises, but only to 6.51%. Side bets with HAs under 10% are uncommon, and while the standard roulette bets offer better odds, Roul 8 is among the more player-friendly options.

Jackpot Streets Double Six

Another roulette wager that combines a wheel spin with a die roll is *Jackpot Streets Double Six*. As the name suggests, this new bet involves the street, or three-number, bets. The roulette result is combined with a roll of 2d6, and ⚅s enhance street bet payoffs.

- If one of the bettor's 3 numbers is spun but no 6s are rolled on the dice, the payoff is 5–1. This is a sharp reduction from the 11–1 payoff on a street bet with no dice involved.

- If a number is spun and one die shows a 6, the payoff is 17–1.

- If a number is spun and both dice show 6s, the payoff is 90–1.

The game was initially marketed for a European roulette wheel, where the HA on a street bet is 2.70%. The probabilities of the three possible wins on a Jackpot Streets Double Six bet are shown in Table 2.27.

TABLE 2.27: Jackpot Streets Double Six probability distribution

Payoff	Probability
5–1	$\frac{3}{37} \cdot \frac{25}{36} \approx .0563$
17–1	$\frac{3}{37} \cdot \frac{10}{36} \approx .0225$
90–1	$\frac{3}{37} \cdot \frac{1}{36} \approx .0023$

From the figures in this table, we can show that the HA on this bet is 5.18%—nearly double the edge on the standard street bet. On an American wheel, the HA on a street bet is 5.26% and the edge on Jackpot Streets Double Six is 7.68%, almost 50% higher.

Flush Roulette

Flush Roulette, a game addition invented by Stephen Au-Yeung, adds an extra random element to a European roulette wheel. The standard roulette bets remain available, but as the wheel is spun, two standard card suits are chosen by rolling special 12-sided dice bearing each suit on three faces, dealing cards, or choosing from ping-pong balls marked with suits [50]. Since the suits are chosen independently, it is possible for both chosen suits to be the same.

These suits are combined with single-number bets in an innovation called the Flush Bet. Players making this wager choose a suit along with their number, and are paid off if their number is spun, with the payoff odds depending on how many suits are matched, as in Table 2.28.

TABLE 2.28: Flush Roulette: Flush Bet payoffs [50]

Suit matches	Payoff odds
0	17–1
1	50–1
2	100–1

The table shows that matching zero suits comes at a reduction of about 50% in the single-number payoff odds, to 17–1. Do the increased payoffs at the higher suit match levels compensate for this?

Recall that a single-number bet at European roulette pays 35–1 and carries a HA of 2.70%. The average winning payoff on a $1 Flush Bet is

$$(17) \cdot \frac{9}{16} + (50) \cdot \frac{6}{16} + (100) \cdot \frac{1}{16},$$

where each payoff is weighted by the probability of matching the indicated number of suits. This number is $34\frac{9}{16}$—very close to 35, but lower. The house edge rises slightly: to 3.89%.

A 2.70% house edge could be preserved by raising the payoffs for matching 1 or 2 suits, to 51–1 and 101–1 respectively.

2.7   Roulette Betting Systems

No casino game has attracted more system sellers—merchants promoting combinations of wagers that are purported to overcome the house advantage—than roulette. In part this is because roulette is easy to understand and to play, and in part because the numbers, colors, and betting options provide ample opportunity for invoking mysterious spurious correlations that some gamblers find attractive.

The Gambler's Fallacy In Action

The Gambler's Fallacy (page 8) denies the independence of successive spins of a roulette wheel, tosses of a coin, or throws of dice. Informally, it is the belief that a short-term deficiency of one random result must soon be balanced by a short-term abundance of that result, so that things "even out".

Example 2.35. If a fair coin is tossed 10 times, yielding 7 heads and 3 tails, a subscriber to the Gambler's Fallacy would declare that there is an increased probability of tails on the next toss, so that the heads/tails ratio will get closer to 1. The ratio of heads to tails at this point is 2.333.

In reality, of course, successive tosses are independent, so the probability of heads and tails remains fixed at ½ regardless of past results. Moreover, the *ratio* between heads and tails can approach 1 even as the gap between the number of heads and the number of tails continues to grow. If, after 1000 flips, there have been 517 heads and 483 tails, the ratio of heads to tails is 1.070 even as there have been 34 more heads than tails. ∎

In the absence of evidence for an unbalanced or otherwise defective wheel, denying independence is flawed reasoning. Since casino games are profitable because of how the rules and payoffs are set, the casino has a strong interest in completely random behavior, and would swiftly move to repair or replace any equipment that was found to be delivering nonrandom results.

However, many casinos encourage the Gambler's Fallacy by providing lighted signs at the roulette tables that show the results of the last 20 or so spins. Armed with this information, many system sellers have claimed that the path to easy riches lies in watching the wheel trends and betting against the pattern shown in the last few spins, or possibly with it.

Example 2.36. A simple system for American roulette that ignores independence is called *Watching for 0s*. This system amounts to waiting for a stretch of at least 40 spins where 0 and 00 do not appear, then launching an increasing sequence of single-number wagers on both 0 and 00, starting with 1 unit on each and increasing by 1 after every loss [83]. Once a number hits, stop betting on it, but you may continue betting on the other number if you wish. ∎

Watching for 0s (which could, of course, be modified to cover any number or pair of numbers not seen in many spins) illustrates the "due number" fallacy: that a result which has not been seen for some time is somehow "due to appear" and thus has an increased probability.

The probability that any 2 selected numbers will not appear in 40 spins of an American roulette wheel is

$$\left(\frac{36}{38}\right)^{40} \approx .1150.$$

One immediate advantage of the Watching for 0s system is that you will lose less money simply by not betting on 40 or more spins while waiting for this shortage of 0s and 00s—5.26% of 0 units bet is 0 units.

"Lose money more slowly," however, is not a compelling advertising slogan for roulette systems.

Once the betting has begun, the increasing bets can quickly use up all of the gambler's bankroll. If neither 0 nor 00 comes up in the next 10 spins, the gambler has lost 110 units. While a win on the next spin would yield a net profit of $(11 \cdot 35 - 11) - 110 = 264$ units, independence of successive spins means that the probability of a win remains fixed at $\frac{2}{38}$.

Example 2.37. A system similar to Watching for 0s, one which requires a much shorter wait before placing the first bet, is a system called *Phase One*. A Phase One bettor watches the wheel until one color, red or black, appears on 5 successive spins. He or she then bets on the opposite color, which is judged to be due, if not overdue [52]. If red has come up 5 times, the flawed logic espoused here is that black is heavily favored, since the probability of 6 straight red numbers is about 1 in 88.5.

While it is true that this is the probability of 6 straight red numbers, that assumes that we start counting before the first spin. The relevant probability is conditional: the probability of 6 straight red spins, *given* that the first 5 spins were all red. That probability is the same $\frac{18}{38}$ of Red on any single spin. ∎

The betting progression for Phase One was 5 units on the first spin after the color is established, which moves up with each successive loss through the sequence 25–70–200–500–1000. The probability of losing all 6 of these bets, and thus dropping 1800 betting units, is the same 1 in 88.5 chance of 6 straight numbers of the same color that we saw earlier.

Gamblers who purchased Phase One for $25 were invited to send the seller 10% of their winnings until they had sent $500, at which time they could advance to Phase Two. If they continued to tithe up to $5000, they were invited to join the Inner Circle. Given the mathematics working against them, it is likely that few people qualified [52].

An alternative to betting on due numbers is betting on "hot numbers": numbers which have recently come up more frequently than expected. One such system is the *Favored 5*. This system calls for watching a wheel until 5 numbers repeat and then betting on those numbers. Effectively, this system is trying to bet into a short-term streak or streaks, which happen in any string of spins. Under the assumption that the wheel is operating properly without bias, there is no sound reason to believe that any observed pattern is going to continue. If the wheel is indeed so defective that certain numbers are favored, then you want to be betting on those numbers, but again, it is highly unlikely that a modern casino would fail to detect a flawed wheel and remove it from service.

Example 2.38. A system called *Chinese roulette* is not connected to the electronic game of the same name described beginning on page 49. This system is based on even-money bets, and divides the 6 bets into 2 subsets: *ROL*, including Red, Odd, and Low; and *BEH*, covering Black, Even, and High. The directions call for the player to start with one set and bet the options in order: say BEHBEHBEH... until the first loss. At that point, switch to the corresponding spot in the other sequence and continue betting that sequence until it loses, and so on. For example, if the first sequence loses when Even is bet, the next bet is Low [83]. The wagers are constant.

Essentially, this version of Chinese roulette is a sequence of independent even-money bets, each with win probability less than ½. Switching bets on every spin may make long losing streaks seem like they are less likely, but the house will still extract its 5.26% on every spin in the long run. ∎

Chinese roulette is an excellent illustration of a fundamental idea of gaming mathematics that we examined as Theorem 1.13:

> *The expected value of a sequence of wagers is the sum of the expected values of the individual wagers.*

It doesn't matter if that sequence is composed of the same bet repeated many times or different bets changing on a whim: the sum of negative expectations will never be positive.

Martingale

One of the most plausible-sounding gambling systems is the *martingale*, which is promoted as a system that guarantees a profit to the committed gambler. The idea is simple and tempting:

- Bet 1 unit on an even-money wager: red, black, high, low, odd, or even.
- If you win, collect your 1-unit winnings and leave the original bet out there for another spin.
- If you lose, bet 2 units on the next spin. If you win that, you have a 1-unit profit, and revert to 1 unit for the next spin.
- If you lose the second spin (total loss = 3 units), bet 4 units on the next spin.
- Continue, doubling your bet after each loss and reverting to a 1-unit bet after each win. After each win, your net profit will be 1 unit.
- Since the probability of never winning is 0, you will eventually win, cash out 1 unit ahead, and walk away a winner.

The flaw in this system is not that "the probability of never winning is 0"—it is. Over an infinitely long string of spins, the chance of them all being losers is indeed 0. Bucking a long string of losses is costly, but a gambler can be confident of eventual victory, even if it only comes after 1000 losses. The reason why the martingale is doomed to failure is that casinos have maximum bet limits on their roulette tables. Successive doubling and redoubling will, in the event of a long string of losses, require a bet that exceeds that maximum— and then the martingale player is stuck with a huge loss.

Example 2.39. Suppose that the bet limits on a roulette table run from $1–500. If a gambler plays a martingale with a $1 unit and loses 9 times in a row, her total loss is $511. The martingale now calls for a $512 bet, but she can only put down $500. If this bet finally wins, her loss is cut to $11, but if she loses a tenth decision, her total loss is \$1011, and her next bet is still restricted to \$500.

The probability of losing 9 consecutive decisions is

$$\left(\frac{18}{38}\right)^9 \approx \frac{1}{833}.$$

While this is a rare event, its effect would be financially ruinous for a small-stakes bettor at a \$500 maximum roulette table, as it wipes out the accumulated winnings from over 500 successful martingales.   ∎

Fibonacci Progression

The *Fibonacci sequence* is a well-known sequence of integers $\{F_n\}_{n \geqslant 0}$ that begins with $F_0 = 0$ and $F_1 = 1$. Each successive term in the sequence is defined recursively as the sum of the two previous terms: $F_n = F_{n-1} + F_{n-2}$, so the sequence proceeds

$$0, 1, 1, 2, 3, 5, 8, 13, 21, 34, 55, 89, 144, 233, 377, 610, \ldots$$

The *Fibonacci progression* is a variant on the martingale where successive bets after losing spins are sized by the next Fibonacci number rather than by doubling the previous bet [42]. If the sequence moves past its first few numbers, this has the effect of multiplying each bet by approximately

$$\phi = \frac{1 + \sqrt{5}}{2} \approx 1.6180,$$

a number popularly known as the *golden ratio*, to determine the next bet. Since $\phi < 2$, the effect of choosing the Fibonacci progression instead of the martingale is that bets rise more slowly, which both lowers losses and might allow a gambler to make more wagers before hitting the unavoidable maximum bet limit—which stands as an immovable barrier against this system just as it ultimately defeats the martingale.

The Fibonacci progression starts with $F_2 = 1$, the second occurrence of 1 in the sequence, so the second bet in the sequence would be 2 units. In Table 2.29, we see a comparison of the martingale and Fibonacci progressions through 9 straight losses. While martingales hit the table limit on the tenth spin, the Fibonacci gambler can continue through spin 13 (377 units) before bumping up against the limit, as the next bet would be for 610 units.

The Fibonacci progression doesn't usually wipe out all accumulated losses with a single win—for example, if the first win is on spin 5, the 8-unit win doesn't quite make up for 11 units lost on the first 4 spins. In this way, it can be seen as a moderate version of a martingale: while the wins aren't as big, neither are the losses. Both phenomena are due to the slower rise in wager amounts.

The Fibonacci and martingale systems differ in how they handle a loss.

TABLE 2.29: Martingale and Fibonacci betting systems compared through 9 consecutive losing spins

Loss #	Martingale Wager	Total loss	Fibonacci Wager	Total loss
1	1	1	1	1
2	2	3	2	3
3	4	7	3	6
4	8	15	5	11
5	16	31	8	19
6	32	63	13	32
7	64	127	21	53
8	128	255	34	87
9	256	511	55	142

After a win, a gambler using the Fibonacci progression backs up 2 numbers in the sequence rather than going all the way back to 1.

Example 2.40. If the first 5 spins lose, for a cumulative loss of 19, a 13-unit win on the next spin takes the gambler to a net loss of 6 units. The next bet would then be for 5 units. ∎

The cumulative loss after n straight losing spins with a martingale is

$$1 + 2 + \cdots + 2^n = 2^{n+1} - 1$$

units. We can express the losses with the Fibonacci progression using the following theorem.

Theorem 2.1. *The total amount lost in the Fibonacci progression after n consecutive losses is $F_{n+3} - 2$ units.*

Proof. After n losses, the total amount lost is

$$F_2 + F_3 + \cdots + F_{n+1}.$$

We shall show that this sum equals $F_{n+3} - 2$ by induction.
If $n = 1$, we have

$$F_2 + \cdots + F_{n+1} = F_2 = 1$$

and

$$F_{n+3} - 2 = F_4 - 2 = 3 - 2 = 1,$$

establishing the base case.
Assume that the statement $F_2 + F_3 + \cdots + F_{n+1} = F_{n+3} - 2$ is true for $n = k$. For $n = k + 1$, we seek to show that

$$F_2 + F_3 + \cdots + F_{k+1} + F_{k+2} = F_{k+4} - 2.$$

We have

$$F_2 + F_3 + \cdots + F_{k+2} = (F_2 + \cdots + F_{k+1}) + F_{k+2}$$
$$= (F_{k+3} - 2) + F_{k+2}$$

by the induction hypothesis. By the definition of the Fibonacci sequence, $F_{k+3} + F_{k+2} = F_{k+4}$, completing the proof.   □

One way to overcome this slow increase in the amount wagered is to play the progression, omitting the 2-bet regression after a win, until two consecutive wins occur. If the first 4 spins are losers, the total loss is 11 units. A win on spin #5 gets 8 units back and sets up a bet of 13 units on the 6th spin. If that one wins, the player can quit with a 10-unit profit. If it loses, then the total loss is 24 units, with 21- and 34-unit bets up next as the quest for a positive return continues. Table 2.30 shows the profit when the Fibonacci progression endures n losses before scoring 2 straight wins. We note that the ultimate win

TABLE 2.30: Fibonacci progression: Net profit after n losses followed by 2 wins

n	Total win
0	3
1	4
2	5
3	7
4	10
5	15
6	23
7	36

after n losses and 2 wins is $F_{n+2} + 2$ units. The proof (Exercise 2.26) follows readily from Theorem 2.1.

Example 2.41. Wins and losses do not always come in nice unbroken streaks. Consider the outcome WLLLWLWLLW. This sequence would allow a gambler to take a 7-unit profit without waiting for a second straight win, or she could place the next bet, for 144 units, and try for a 151-unit win—at considerable risk.   ∎

With this sequence of wins and losses, the player can retire a winner after a single win because the string of 3 losses occurs early, so the terminal win comes when a large bet is made. Two consecutive early wins will necessarily occur when the bets are still small.

An alternative for a roulette player dismayed by the low returns from the Fibonacci progression when applied to even-money bets is to use it on a 2–1 wager: one of the columns, Low, Middle, or High. The first bet is for \$1; if it wins, the profit is \$2. On a loss, the second bet is for \$2; if it wins, the net profit is \$3. Table 2.31 follows this progression through a string of losses.

TABLE 2.31: Fibonacci progression used on a 2–1 wager

Wager	Past Losses	Amount	Win	Net Profit
1	0	1	2	2
2	1	2	4	3
3	3	3	6	3
4	6	5	10	4
5	11	8	16	5
6	19	13	26	7
7	32	21	42	10
8	53	34	68	15
9	87	55	110	23
10	142	89	178	36
n	$F_{n+2} - 2$	F_{n+1}	$2F_{n+1}$	$F_{n-1} + 2$

Once again, the net profit from n losses followed by a win has a compact expression in terms of the Fibonacci sequence: in this case $2F_{n+1} - F_{n+2} + 2 = F_{n-1} + 2$ units.

The advantage in using Fibonacci on a 2–1 wager is that any win returns a net profit. The disadvantage is that the probability of winning on any one spin has fallen from $\frac{18}{38}$ to $\frac{12}{38}$. It may take longer to notch that first win, and the likelihood of hitting the maximum bet limit increases. If the maximum bet is \$500, a Fibonacci progression starting at \$1 can absorb 13 losses. On 2–1 bets, this has probability

$$\left(\frac{26}{38}\right)^{13} \approx .0072,$$

just more than 1 chance in 139. The total loss in this event is \$975, which wipes out a lot of small wins.

Cancellation

The *cancellation* or *Labouchère* system could be sold to unsuspecting customers with the superficially reasonable claim that winning spins count more than losing ones, making it easier for a gambler to make a profit using it. A player using this system decides how much he wants to win and then chooses a list of numbers adding up to that amount. For example, a player out to win 15 units might write down $\{1, 2, 3, 4, 5\}$. The size of each bet is determined by adding the first and last numbers on the list, so this player's first bet would be for 6 units.

- If the bet wins, cross off the two numbers and make the next bet, which in this case would be $2 + 4 = 6$ units.

- If the bet loses, add the amount of the loss to the end of the list. The next bet would then be $1 + 6 = 7$ units.

Repeat this process until all of the numbers are crossed off. At this point, the net profit will be the sum of the original numbers, which was the target amount determined at the start. Since numbers are crossed off in pairs after a win while the list only ever gets longer by one number per losing spin, a roughly 50/50 split in wins and losses means that the list will eventually be empty.

"Roughly 50/50" is not guaranteed, of course. A short-term losing streak can result in a lot of large numbers at the end of the list, and thus a lot of large bets trying to win the initial target figure.

Example 2.42. In the example above beginning with suppose that the first 4 bets lose. The list now reads $\{1, 2, 3, 4, 5, 6, 7, 8, 9\}$, the gambler is currently down 30 units, and the next bet is 10 units. ∎

There are two problems here: First, the cancellation system frequently calls for risking far more than the target amount, as the bets get larger if the player encounters a string of losses. Second, a losing streak at the start can extend the session for many spins with no guarantee of success before a player's money runs out. The upper limit on bets may pose a third problem, but if the target is small compared to the upper limit, as would be the case if this sequence were played at a 500-unit limit table, this is less of a threat than it is with a martingale.

An alternative to the cancellation system that acts in part to keep the list from growing unwieldy in the face of a losing streak was proposed as the *Guaranteed Roulette System* by Wolfgang Bartschelly [157]. We will illustrate this system with an example. Suppose that the initial sequence is $\{1, 1, 1, 1, 1, 2, 2, 2, 2, 2\}$, pursuing a net win of $15 through a sequence of bets that start out smaller than in the above example. The first bet is for $3, and loses. Instead of appending a 3 to the end of this list, the 3 is distributed evenly over the terms of the sequence, beginning at the right end, so the new sequence is $\{1, 1, 1, 1, 1, 2, 2, 3, 3, 3\}$. The next bet is $4; if it loses, the new sequence is $\{1, 1, 1, 1, 1, 2, 3, 4, 4, 4\}$. The next bet is $5, as it would be in a standard cancellation system, but the list has 10 entries instead of expanding to 12 after the 2 losses.

The superficial attraction here is that the player crosses off two numbers after a win and *never* adds a number, so the entire sequence is guaranteed to be crossed off eventually. What the system obscures is that the bets will invariably rise, and a losing streak runs the same risk of depleting the gambler's bankroll as the cancellation system has.

Watching the Wheel

Other roulette betting systems rely on choosing bets that cover large sectors of the wheel. A Las Vegas croupier claims to have discovered a quirk of the wheel and used it to devise a system that encompasses 17 numbers, in two groups of 8 consecutive numbers on the wheel plus an extra number, in only 3 bets [129]. Figure 2.11 shows the wheel.

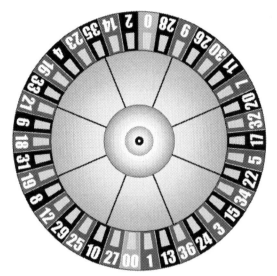

FIGURE 2.11: American roulette wheel [46].

The numbers involved in this system are the 8 numbers moving counter-clockwise from 2, the 8 numbers moving clockwise from 1, and 3:

$$2, 0, 28, 9, 26, 30, 11, 7; 1, 00, 27, 10, 25, 29, 12, 8; 3.$$

These numbers can all be bet with only 3 chips:

- A basket bet on 0, 00, 1, 2, and 3. Since an American roulette layout does not permit a corner bet on 0, 00, 1 and 2, the system must use this bet and include the 3—even as we know that the basket bet has a larger HA than any other bet on the board.

- A double street bet on 7, 8, 9, 10, 11, and 12.

- A double street bet on 25, 26, 27, 28, 29, and 30.

17 numbers—44.7% of the wheel—with only 3 bets and including 2 sections of 8 adjacent numbers on the wheel? What's not to like?

While this combination of bets may win 44.7% of the time, it loses more than half the time, and even when it wins, 2 of the 3 bets lose. That $2 must be deducted from the winnings. The PDF for the result of a spin is shown in Table 2.32.

TABLE 2.32: PDF for a 3-chip 17-number wager

Net winnings	Probability
3	.3158
4	.1316
−3	.5526

This bet combination has an expected return of

$$E = (3) \cdot \frac{12}{38} + (4) \cdot \frac{5}{38} + (-3) \cdot \frac{21}{38} = -\frac{7}{38}$$

and a HA of 6.14%—because it includes the basket bet among its wager, the player is giving away more than the standard 5.26%. If, by chance, the layout permitted the corner bet on 0, 00, 1, and 2, the system would cover 1 fewer number, but the HA would drop to 5.26%.

Roulette Dinner

Here's a system that might appeal to gamblers who feel that they are exceptionally unlucky: that every time they pick a number, it loses. Designed for use on European wheels offering the partage option, Roulette Dinner calls for the player to pick one number that they think will *not* appear and then to make a collection of 5 bets totaling 143 units that cover every other number.

If the number chosen is from the middle dozen, numbers 13–24, then the 5 bets are as follows [106]:

1. 72 units on High if the chosen number is 13–18 or on Low if it is 19–24.

2. 48 units on the dozen that does not overlap the first bet: First (1–12) if the number is 13–18 and Last (25–36) if it's 19–24.

3. 12 units on the adjacent street.

4. 8 units on the other two numbers in the street of the chosen number, either 4 units on each number or a split bet of 8 units.

5. 3 units on 0.

Example 2.43. If the gambler is convinced that the next number will not be 17, these bets are made:

1. 72 units on High, covering the numbers 18–36.

2. 48 units on the first dozen, covering 1–12.

3. 12 units on the street 13–15.

4. 4 units each on 16 and 18.

5. 3 units on 0.

■

The sizes of the various wagers are chosen to give uniform results:

- If the chosen number is spun, then the player loses 143 units.
- If *any other number* comes up, the player scores a net profit of 1 unit.

Example 2.44. If the spin turns up a 0, the player wins 105 units on bet #5 and pulls back 36 units on the High or Low bet due to the partage rule, for a total of 141 chips taken in. 140 chips are collected by the casino, and the net win is indeed 1 unit. ■

The probability of winning is $\frac{36}{37}$. The expected return for this collection of wagers is

$$E = (1) \cdot \frac{36}{37} + (-143) \cdot \frac{1}{37} = -\frac{107}{37}.$$

Dividing by the 143 units risked gives a HA of 2.02%—slightly less than the 2.70% HA common to European roulette due to the partage rule, which reduces the HA on the largest wager.

The reason that Roulette Dinner works for the middle dozen is that it's possible to cover 30 of the other 36 numbers with a High or Low bet and a dozen that does not overlap those numbers—but what if you're convinced that the next spin will not be 5? Is there a way to modify this cluster of wagers to omit a single number from 1–12 or from 25–36?

Most certainly. The trick lies in betting the opposite group of 18 numbers—High or Low—for 72 units and then looking at wagers 2 and 3 as ways to cover 15 numbers in 5 streets. This could be done with a single street bet and 2 double street bets or as 5 street bets. This works because the HA on every European roulette bet is the same.

Example 2.45. For the number 5, wager #1 would be on High. Wager #4 would be two bets of 4 units each, on 4 and 6. This leaves 5 streets. 1–2–3 would have to be bet as a separate street (wager #3), and the other 4 streets could be bet individually for 12 units each or as double street bets for 24 units apiece (wager #2). Wager #5, 3 units on 0, requires no modification. ■

While the mean outcome from a single spin with Roulette Dinner in play is –2.89 units on a 143-unit wager, this masks a high variance. The variance is

$$\sigma^2 = (1)^2 \cdot \frac{36}{37} + (-143)^2 \cdot \frac{1}{37} - (-2.89)^2 \approx 545.29 \text{ units}^2,$$

and the standard deviation is $\sigma = 23.35$ units, making this bet highly variable.

The casino can well afford to wait out a long string of 1-unit player wins, confident that its advantage will more than make up for them with the occasional 143-unit player loss. The probability that a player will win 143 times in a row on a European wheel is

$$\left(\frac{36}{37}\right)^{143} \approx .0199,$$

just under 2%.

2.8   Exercises

Solutions begin on page 325.

2.1. In 2018, traffic at the Venetian Casino seemed to favor Sands Roulette over American roulette, perhaps due to the lower table minimum at Sands Roulette. This player preference diminished any possible incentive to offer slightly better payoffs at Sands Roulette to compensate for the third green pocket, but suppose that casino management had offered a 36–1 payoff on a single number instead of 35–1. What would the new HA be? What would the HA be with a 37–1 payoff?

2.2. Consider a hypothetical roulette wheel with z zeroes, where $z = 1$ for a European wheel, $z = 2$ for an American wheel, $z = 3$ for Sands Roulette, and so forth.

a. If a wager on n numbers still pays off at $\dfrac{36 - n}{n}$ to 1, devise an expression for the expected value $E(z)$ of a \$1 bet on n numbers as a function of z.

b. *Keno* is a lottery-like game where players choose some numbers in the range 1–80, the casino draws 20 numbers in that interval, and payoffs are based on the number of the player's numbers that are "caught", or drawn among the casino's 20 numbers. Keno's popularity in American casinos has dwindled, in part because of its high house advantage: 25% is on the low end of keno HAs.

   How many zeros would have to be on a roulette wheel before the HA on a wager on n numbers paying $\dfrac{36 - n}{n}$ to 1 equals or exceeds 25%?

2.3. Find appropriate payoffs for digits bets (page 39) on the various numbers 0–9 using a standard American roulette wheel, and compute the corresponding house advantages.

New Wheels

2.4. Consider an alternate Alphabetic Roulette wheel with separate pockets for the letters Y and Z and the payoff schedule described in Table 2.4. Calculate the common house edge of the standard bets under this variation.

2.5. At the Luxor Casino in Las Vegas, the casino's name would correspond to a 5-letter bet in Alphabetic Roulette. Find the maximum integer or half-integer payoff for such a wager that would still give the casino an advantage, and calculate the corresponding house edge.

Royal Roulette is a relatively new variation on roulette, developed in Australia, that uses familiar playing cards as spots on the wheel [110]. The wheel contains 50 pockets: one for each of the 48 cards from 2 through king, a single ace, and a joker. The layout (Figure 2.12) is similar to a traditional roulette layout and offers the betting options listed in Table 2.33.

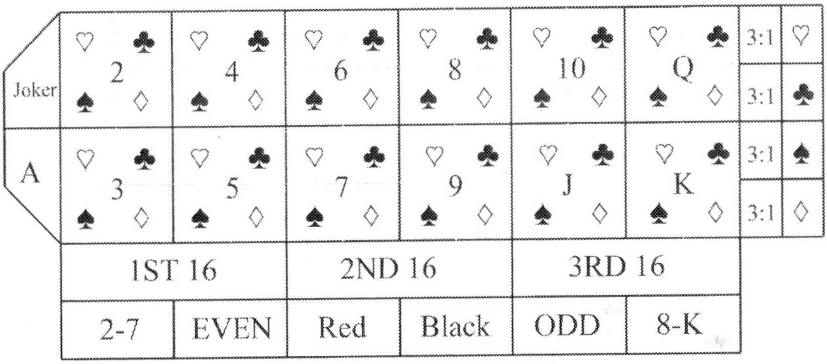

FIGURE 2.12: Royal Roulette layout.

TABLE 2.33: Royal Roulette betting options

Name	Payoff	Description
Straight Up	47 to 1	Any single card including Ace and Joker
Split	23 to 1	2 cards
Court	15 to 1	3 cards (Ace 3♠ 3♡ or Joker 2♠ 2♡)
Street/Corner	11 to 1	4 cards
Six Line	7 to 1	6 cards (Ace, Joker, 2♡, 2♠, 3♡, 3♠)
Eight Line	5 to 1	8 cards
Suit/Column	3 to 1	12 cards: One suit or one column
Sixteen Set	2 to 1	16 cards
Even Chances	1 to 1	24 cards: Red, Black, Odd, Even, 2-7, 8-K

2.6. How does Royal Roulette compare to the standard game? Find an expression for the payoff of a winning bet on n cards and compute the constant HA.

2.7. A second version of Royal Roulette removes the joker from the wheel and the layout—we might think of this, informally, as European Royal Roulette. The Six Line bet is eliminated, though the Court bet remains and must involve the Ace. Though the number of pockets on the wheel is now only 49, the payoffs are the same as in the jokered game. Find the common expectation of the wagers.

2.8. The inventors of Super 62 Roulette also devised roulette games with 61, 63, and 64 pockets on the wheel; the additional pockets correspond to 000 and 0000 spaces. Find the usual HA in Super 64 Roulette if bets on n numbers continue to pay off at $\dfrac{60 - n}{n}$ to 1.

2.9.

a. In Fast Action Roulette, find the probability that the 10 Bonus numbers include all 4 numbers in each of 2 rows.

b. Find the probability that exactly one row of the Fast Action Roulette layout contains 4 Bonus numbers.

2.10. *Roulette 73*, as the name suggests, offers a wheel with 73 pockets, numbered 1–72 and 0. The layout is a 6 by 12 grid topped with a rectangle for bets on 0 (Figure 2.13), so there are 6 possible column bets and street bets include 6 numbers.

0					
1	2	3	4	5	6
7	8	9	10	11	12
13	14	15	16	17	18
19	20	21	22	23	24
25	26	27	28	29	30
31	32	33	34	35	36
37	38	39	40	41	42
43	44	45	46	47	48
49	50	51	52	53	54
55	56	57	58	59	60
61	62	63	64	65	66
67	68	69	70	71	72

FIGURE 2.13: Roulette 73 layout.

Three-number bets are only available by combining 0 with two consecutive numbers from 1–6.

a. As in American and European roulette, a wager on n numbers has a payoff in inverse proportion to the area covered, in this case $\dfrac{72 - n}{n}$ to 1. Find the common HA of these wagers.

b. 24 of the numbers are colored gray on the layout, shown in Figure 2.13, though they are red or black on the wheel. These 24 numbers are the basis for a 14-chip combination bet. This bet includes 4 single-number bets: on the numbers 8, 13, 44, and 49, together with 10 chips split bets covering the other 20 numbers in pairs. Find the HA for this wager.

2.11. An early version of a roulette wheel contains only the numbers from 1–16 along with 0 and 00. The odd numbers and 0 are red; the even numbers and 00 are black. One such wheel is shown in Figure 2.14.

FIGURE 2.14: Early roulette wheel with 18 pockets.

Assume that the layout is a 4×4 grid with 0 and 00 in double-width spaces at the top as in Figure 2.15, with black numbers shaded.

0		00	
1	2	3	4
5	6	7	8
9	10	11	12
13	14	15	16

FIGURE 2.15: Antique roulette layout

If 0 and 00 are losing numbers when an even-money bet is made, devise a pay table for the various bets and calculate the house edge on each bet.

Electronic Roulette Games

2.12. The basket bet, however comparatively bad a wager it may be at American roulette, is an option in Double Bonus Spin Roulette. Find the HA of the basket bet in this game.

2.13. IGT manufactures and distributes an electronic roulette game, *Roulette Evolution*, that adds a host of group wagers to the standard collection of roulette bets. Several of these bets are based on the arrangement of numbers on the single-zero European wheel, and allowed gamblers automatically to select groups of numbers, which are shown in Figure 2.16 [44].

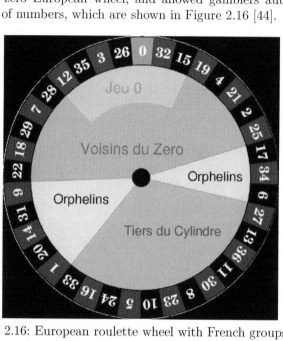

FIGURE 2.16: European roulette wheel with French groups labeled

It is noteworthy that, in casinos where Roulette Evolution is played with a physical wheel, *wheel clocking*, the practice of observing a croupier's spin patterns in an effort to predict where the ball will land, may provide a wagering edge on these bets. If the game is played with an electronic wheel, that advantage is less pronounced, if present at all. The following bets are available [107]. Find, for each, the probability of winning and the house advantage.

a. The *Tiers du Cylindre* ("third of a cylinder") bet calls for 6 split bets of 1 chip each, on 5/8, 10/11, 13/16, 23/24, 27/30, and 33/36.

b. The *Jeu 0* ("0-Series") bet places 4 chips on 26 and one each on the splits of 0/3, 12/15, and 32/35.

c. The *Orphelins* ("Orphans") bet puts a chip on the number 1 and one each on the splits 6/9, 14/17, 17/20, and 31/34. Note that 17 is covered by two split bets.

d. The *Grand Series* bet covers the 17 numbers shown in Figure 2.16 as "Voisins du Zero" (Neighbors of 0) as follows:

- 2 chips on the three-number combination 0/2/3.
- 2 chips on the four-number combination 25/26/28/29.
- Single chips on the five splits 4/7, 12/15, 18/21, 19/22, and 32/35.

e. A *Neighbors* bet is a 5-chip wager that places a single chip on a selected number and on the two numbers adjacent to it on either side of the wheel. For example, a Neighbors bet on the number 10 would cover the numbers 10, 5, 23, 24, and 8.

2.14. Find the other 100 To 1 Roulette bet besides single-number and Avenue bets that pays off at other than $\dfrac{100-n}{n}$ to 1 and compute its HA.

Roulette with Cards

2.15. Triple Flop Roulette is also available in a 55-card version that uses all 52 standard playing cards and 3 jokers. The layout is a 4×13 grid; each card rank appears in a separate row with aces at the top and deuces at the bottom. The three jokers are respectively colored yellow, green, and blue, and betting spaces for them are placed above the aces, where the zeros sit on a standard layout. Each column consists of the 13 cards of one suit, which allows for the suit bets that are not available in the 37- or 38-card versions of TFR.

a. Many standard roulette bets translate over to the 55-card game. Bets on 1, 2, 4, 13, or 26 cards follow the roulette pattern of paying off at $\frac{52-n}{n}$ to 1 odds, where n is the number of cards selected. Find the common house advantage of these bets.

b. A double street bet covers 8 cards, of 2 consecutive ranks, and pays 5–1 rather than the 5½–1 that the formula in part a. returns. Find the HA.

c. The equivalent of the basket bet in this game is a bet on 3 cards, which includes either 1 or 2 jokers together with 2 or 1 aces. Find the HA of this bet if the payoff odds are 16–1.

d. The relative orders of the Straight and Flush bets are reversed here: the Straight bet pays 11–1 while the Flush bet pays 10–1. Confirm that a 3-card straight is less common than a 3-card flush in a 55-card deck with three jokers. (Remember that both Straight and Flush bets pay off if the three cards form a straight flush or royal flush.)

e. Find the expected value of a 3 of a Kind bet, which pays 80–1. Note that an ace, king, or queen dealt with two jokers will be read as a royal flush, while a jack or lower with two jokers makes 3 of a kind.

2.16. Find the probability that a 3-card flop at 55-card Triple Flop Roulette is a high-card hand.

2.17. The Jackpot bet in 55-card Triple Flop Roulette pays off according to Table 2.34.

TABLE 2.34: Triple Flop Roulette: Jackpot bet pay table [142]

Hand	Payoff odds
3 jokers	1000–1
Natural royal flush	500–1
Jokered royal flush	250–1

Find the expected value of a $1 Jackpot bet.

Ca$h Card

Ca$h Card was developed by 18-year-old inventor Nicholas Kallabat and debuted in 2019. It is a card-based roulette game, suitable for the California market but also promoted as attractive to other states' casinos due to its higher payoffs, to attract players, and higher HA, to interest casino table game managers [18]. Ca$h Card is played with a standard 52-card deck plus 2 jokers and the "Ca$h Card," for a total of 55 cards. The layout is organized as a 13 × 4 grid; cards are arranged with each rank comprising a row and each column containing cards of one suit, to facilitate rank and suit bets. Spaces for joker and Ca$h Card bets are placed near the top of the grid, like the zeroes in American roulette, and outside bet analogs may be found on one long side of the grid.

Ca$h Card offers the following array of bets.

TABLE 2.35: Ca$h Card pay table [18]

Wager	Numbers	Payoff odds
Straight	1	50–1
Split or Any Joker	2	25–1
Street (Rank) or Corner	4	12–1
Suit (Column)	13	3–1
Range	13	3–1
Red/Black	26	1–1
High/Low	26	1–1

- Range bets are similar to dozens bets in standard roulette and cover all ranks from 2–4, 5–7, 8–10, or J–K. Every Range bet also wins if the Ca$h Card is drawn.

- In many card games, aces count both high and low, though not simultaneously. So it is with Ca$h Card. The High bet covers ranks from 8–K as well as either red ace, while the Low bet includes the black aces together with the 2s through 7s.

Each round of Ca$h Card involves 3 cards dealt from the top of a freshly shuffled 55-card deck. The second card is the winning card for the bets listed in Table 2.35; the first and third cards are used in the optional $5 side bet based on the 3-card poker hand that is formed.

2.18.

a. Find the common payoff formula for a Ca$h Card bet on n cards, where $n > 1$, and the common HA on these bets.

b. The single-card bets, including the Ca$h Card bet, have a higher house advantage than the other bets. Find it.

c. An earlier version of Ca$h Card used 3 jokers and the Ca$h Card instead of 2 jokers. Find a suitable payoff for a 3-card "Any Joker" bet that is in line with the HAs for other bets using this 56-card deck and the payoffs in Table 2.35.

2.19. Ca$h Card's lone side bet is the $5 bonus wager based on the 3-card poker hand formed by all 3 dealt cards. The jokers and Ca$h Card are not wild cards for the purposes of this bet; they only form a winning hand—the top-ranked hand—if they are all dealt together. This hand is unique among winning hands in that the payoff is an aggregate jackpot evenly divided among all winning bettors.

The pay table for this bet is shown in Table 2.36.

TABLE 2.36: Ca$h Card bonus bet pay table: $5 flat wager [18]

Hand	Payoff
Joker–Ca$h Card–Joker	$25,000 aggregate payoff
Royal flush	$5000
Straight flush	$500
3 of a kind	$350
Straight	$25
Flush	$20

a. Confirm that the hand rank in Table 2.36 is correct by showing that 3-card straights are less common than 3-card flushes.

b. Find the probability of a 3-card straight flush using this 55-card deck.

2.20. Find the house advantage on a $100 basket bet under Oklahoma roulette rules calling for a $1 ante on each spin.

Side Bets

2.21. Consider hypothetical Newar bets on the red odd numbers and the black even numbers, under the same payoffs given on page 65. Find the HA of these bets.

2.22. For the Prime Time roulette wager, what payoff on 7 straight primes would produce the 5.26% HA common to American roulette bets?

2.23. A second pay table for the Prime Time wager, Table 2.37, stopped at 5 straight prime spins, but paid off at 50–1 instead of 20–1 on 5 primes in an attempt to balance the pay table.

TABLE 2.37: Alternate Prime Time roulette side bet pay table [93]

Number of primes	Payoff
1	1–1
2	3–1
3	6–1
4	10–1
5	50–1

Compare the HA of this bet to the Prime Time bet described in Table 2.16 on page 66.

Roulette Betting Systems

2.24. Suppose that a past numbers sign at an American roulette wheel displays the results of the last 16 spins. Find the probability that *no* number appears twice in a displayed stretch of 16 numbers.

2.25. The first roulette system that stood up to mathematical scrutiny was posted on the World Wide Web in 1997 by gambling writer Mike Caro. The system was later published online at the *Gambling Times* Web site in 2001 [17], and was touted as applicable to American roulette wheels. This system requires familiarity with the standard American wheel layout, shown below.

> This system cuts the house advantage to literally nothing, if you believe in it enough to never get frustrated and switch tactics. What I'm going to say may seem strange, but here goes.
> First, never bet simply red or black. Also don't bet odd or even. These are equally poor, consistently losing wagers.
> Second, don't be suckered into betting zero or double zero, despite what some experts may suggest. This may seem like you're betting with the house, but for technical reasons you are actually betting against the house—and you are taking the worst of it.

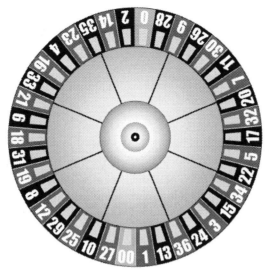

FIGURE 2.17: American roulette wheel [46].

So, in order to negate the house advantage, you must stick to straight non-green number bets. All odd red numbers turn out to be bad choices, based on over two trillion computer trials. Don't bet them.

All even black numbers fare poorly, and cannot be bet, for much the same reason, which I won't explain here.

Let's get straight to the money-saving advice. Any bet you decide to make must cover only even-red or odd-black numbers. There are no exceptions.

Finally, you need to be very disciplined in excluding the number 30 and the group of consecutive numbers that begins with 11 and continues clockwise through and including 14.

This system may seem mystical, but I take gambling quite seriously, and this works for me.

Explain why Caro's system works, and find the true house advantage of a bet using it.

2.26. For a gambler playing the Fibonacci progression until two straight wins appear, find the net win or loss after the 14-spin sequence WLWLLLWLLL-WLLW turns up. What is the amount of the next bet?

2.27. Prove the result stated on page 88: When using the Fibonacci progression on an even-money wager, the net profit after n losses followed by 2 straight wins is $F_{n+2} + 2$ units.

Chapter 3

Craps

Craps is the premier casino dice game. Dice have been found in the artifacts of many ancient civilizations. They have evolved from *astragali*: the heel bones of sheep or goats. An astragalus functions as an instrument of chance with 4 sides that are not equally likely. Figure 3.1 shows a collection of 8 Greek glass reproductions of astragali from the Metropolitan Museum of Art in New York City.

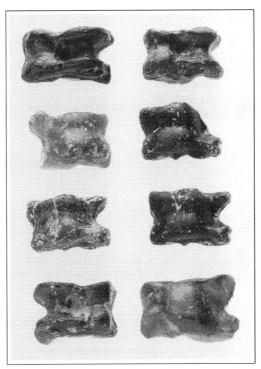

FIGURE 3.1: Glass reproductions of astragali, c. 3rd-2nd century BCE [58].

Astragali are irregularly shaped bones with 4 surfaces on which they can land when tossed on a flat surface. Florence Nightingale David conducted some experiments with astragali and estimated the probabilities of the 4 faces as .40, .40, .10, and .10. She cautioned, however, that since astragali are far from uniform, these probabilities might vary based on the species of animal and the physical condition of the bones [6].

Dice evolved over time before settling into the familiar cubical dice in common use today. Standard casino dice are perfect cubes measuring ¾inches (1.905 cm) on an edge, accurate to one 10,000th of an inch. This precision ensures that every face has an equal probability of landing upmost when the dice are rolled.

Most craps wagers are resolved based on the sum of the two dice; Table 3.1 shows the outcomes when 2d6 are rolled.

TABLE 3.1: Chart of sums when 2d6 are rolled

	1	2	3	4	5	6
1	2	3	4	5	6	7
2	3	4	5	6	7	8
3	4	5	6	7	8	9
4	5	6	7	8	9	10
5	6	7	8	9	10	11
6	7	8	9	10	11	12

Though each face of each die is equally likely, examination of Table 3.1 shows that the possible sums, from 2–12, are not equally likely. The most likely sum is 7, with probability $\frac{1}{6}$, and the least likely sums are 2 and 12. Each has probability $\frac{1}{36}$.

3.1 Craps Basics

There is a large collection of wagers available to a craps bettor. Figure 3.2 depicts all of the bets displayed on a standard craps layout.

A full-size craps table would include a second betting field for come, pass, C&E (Craps and Eleven), and field bets, placed symmetrically to the right of the illustration to duplicate the betting options at the other end of the table. The payoffs in Figure 3.2 are quoted in the form "*x* for 1" rather than "*x* to 1"; this is customary on many craps tables. There are other bets available which must be placed by dealers for players wishing to make them; we examine those beginning on page 114.

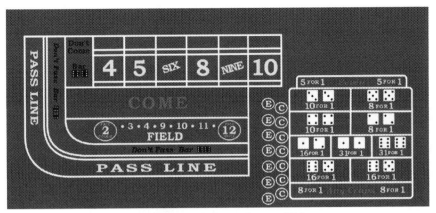

FIGURE 3.2: Craps layout [27].

Pass and Don't Pass

The most essential craps bets are the even-money *Pass* and *Don't Pass* wagers. The Pass bet is a bet with the player rolling the dice, who is known as the *shooter*. Don't Pass is a bet against the shooter. A round begins when the shooter rolls the two dice, which is called the *come-out* roll. If the come-out roll is 7 or 11, the Pass bet wins and the Don't Pass bet loses; if 2, 3, or 12, the Pass bet loses immediately. Don't Pass wins if the come-out roll is 2 or 3, but pushes instead of winning on an initial roll of 12; this rule allows the casino to maintain an advantage on both bets.

If the come-out roll is any other number, that number becomes the shooter's *point*. The shooter then continues rolling until he either rolls the point again or rolls a 7. All other rolls are disregarded for the purposes of resolving this main bet. The shooter and all players betting Pass win if he re-rolls the point before a 7, and loses, with Don't Pass bettors winning, if he rolls a 7 first.

Example 3.1. If the first roll is a 5, this becomes the point. From this roll forward, only 5s and 7s matter for Pass and Don't Pass bets. If the second roll is 12, the roll continues—Pass line bettors do not lose on a 12 once a point has been established. Nothing changes if the third roll is a 4. If the fourth roll is a 5, then Pass line bettors win and Don't Pass bettors lose. ■

Resolving a single craps decision could, in theory, last a very long time if the shooter manages not to roll either the point or a 7 over many throws. That said, $\frac{1}{3}$ of all Pass and Don't Pass decisions end on the first roll with a 2, 3, 7, 11, or 12. The average number of tosses to resolve a single decision is actually quite low. If we denote the probability that a single round uses k

throws as $P(k)$, then the average number of throws is

$$\sum_{k=1}^{\infty} k \cdot P(k).$$

We have $P(1) = \frac{1}{3}$; the probability for higher numbers of rolls depends on the point. We must establish a point, roll several times without resolving the bet by re-rolling the point or rolling a 7, and then settle all bets by making the point or rolling a 7. If $k \geqslant 2$, then

$$P(k) = p_1 \cdot p_2^{k-2} \cdot p_3,$$

where

$$p_1 = P(\text{Point is thrown}),$$
$$p_2 = P(\text{Neither the point nor a 7 is thrown}),$$
$$p_3 = P(\text{Either the point or a 7 is thrown}).$$

This product must be summed across all 6 possible point values.

Table 3.2 shows the values of these 3 probabilities for the 6 points.

TABLE 3.2: Craps: Probabilities p_1, p_2, p_3

Points	p_1	p_2	p_3
4 or 10	$\dfrac{3}{36}$	$\dfrac{27}{36}$	$\dfrac{9}{36}$
5 or 9	$\dfrac{4}{36}$	$\dfrac{26}{36}$	$\dfrac{10}{36}$
6 or 8	$\dfrac{5}{36}$	$\dfrac{25}{36}$	$\dfrac{11}{36}$

Inserting these values into the sum gives us

$$\text{Average} = \frac{1}{3} + \sum_{k=2}^{\infty} k \cdot 2 \cdot \left[\frac{3 \cdot 27^{k-2} \cdot 9}{36^k} + \frac{4 \cdot 26^{k-2} \cdot 10}{36^k} + \frac{5 \cdot 25^{k-2} \cdot 11}{36^k} \right]$$

$$= \frac{1}{3} + \sum_{k=2}^{\infty} \left(\frac{2k}{36^k} \right) \cdot \left[3 \cdot 9 \cdot 27^{k-2} + 4 \cdot 10 \cdot 26^{k-2} + 5 \cdot 11 \cdot 25^{k-2} \right]$$

$$= \frac{557}{165} \approx 3.38.$$

An average craps decision is settled in between 3 and 4 rolls.

Computing the HA on the Pass and Don't Pass bets is most easily done by summing across all point values as we have done here. For each of the 6 points, the probability of winning a Pass bet with that number as the point is the product of two factors: the probability of establishing that number as the point by rolling it on the come-out roll and the probability of then re-rolling that number before rolling a 7. If the point N can be rolled in k ways, we have

$$P(\text{Win with point } N) = \frac{k}{36} \cdot \frac{k}{k+6}.$$

The total probability of winning a Pass bet is then found by adding 6 terms of this form to the probability of winning with a 7 or 11 on the come-out roll, which is $\frac{8}{36}$. The 6 terms can be grouped into 3 pairs of identical values, since 4 and 10, 5 and 9, and 6 and 8 have the same probability of winning when rolled as the point. We have

$$P(\text{Win on Pass}) = \frac{8}{36} + 2 \cdot \left[\frac{3}{36} \cdot \frac{3}{9} + \frac{4}{36} \cdot \frac{4}{10} + \frac{5}{36} \cdot \frac{5}{11} \right] = \frac{244}{495} \approx .4929.$$

This probability is very close to ½, making the Pass bet with its 1.41% HA one of the best bets in any casino.

The probability of winning a Don't Pass bet, *without* the rule that a 12 on the come-out roll pushes, would be $1 - .4929$, a value greater than ½, and so the gambler would have the advantage. Barring the 12 on the come-out restores a casino advantage: the new win probability is

$$P(\text{win on Don't Pass}) = 1 - \frac{244}{495} - \frac{1}{36} = \frac{949}{1980} \approx .4793.$$

Though the probability of an outright win on Don't Pass is smaller than $P(\text{Win on Pass})$, Don't Pass has a slightly lower house advantage, 1.36%, due to the "push on 12" rule.

Some casinos push the Don't Pass bet when the come-out roll is a 2; this is mathematically identical to barring the 12. A casino which pushes the Don't Pass bet when a 3 is rolled on the come-out, as was the case in northern Nevada for a time, is raising its advantage from 1.36% to 4.14%.

Free Odds

Pass and Don't Pass bettors have, at many casinos, the opportunity to back up their bets with an additional *free odds* wager once a point has been established. Odds bets are made by placing an additional chip or chips immediately behind a player's bet. These bets are called "free" odds because they are paid at true odds, with no house advantage. The payoff odds depend on the point, as shown in Table 3.3. Since Don't Pass bettors have a mathematical advantage over the casino once a point has been established, Don't Pass bettors making odds bets must accept odds paying less than even money on the free odds portion of their wager.

As these bets decrease the overall HA on a Pass or Don't Pass bet, casinos place a limit on the amount of a free odds bet. Typically, this is scaled to

TABLE 3.3: Craps: Free odds pay table

Point	Pass	Don't Pass
4 or 10	2–1	1–2
5 or 9	3–2	2–3
6 or 8	6–5	5–6

the size of a player's original bet: "double odds" allow a player to bet twice his or her initial bet on an odds bet. A common casino practice is to allow "3X/4X/5X" odds: 3 times the bet if the point is 6 or 8, 4 times on a point of 5 or 9, and 5 times if the point is 4 or 10. In a competitive craps market or as a promotion, craps odds can soar: in 2012, the Riviera Casino in Las Vegas offered 1000X odds bets on any point in an effort to attract craps players to their casino.

Example 3.2. Suppose that the point is 4. With the possibility of winning or losing on the come-out roll eliminated, the probability of winning a Pass line bet is now $\frac{3}{9}$. On a \$1 bet, the expectation is now

$$E = (1) \cdot \frac{3}{9} + (-1) \cdot \frac{6}{9} = -\frac{1}{3}.$$

If this bet is backed up by a full odds bet of \$5 paying 2–1, the expectation is

$$E = (11) \cdot \frac{3}{9} + (-6) \cdot \frac{6}{9} = -\frac{1}{3}.$$

While the expected value is the same (since the free odds portion of the bet carries no house advantage), the HA has moved from 33% to 5.56%. ∎

Example 3.3. With 1000X odds backing up a Pass line bet on a point of 9, the combined bet has an expected value of

$$E = (1501) \cdot \frac{4}{10} + (-1001) \cdot \frac{6}{10} = -\$.2.$$

This minuscule edge results in a house advantage of only .02%.

The risk inherent in a full odds bet of this magnitude is in the variance of the total bet. Each decision carries a swing of thousands of dollars, and a short losing streak could bankrupt a player, while the casino is well-capitalized against a similar streak of losses. ∎

Come and Don't Come

The *Come* and *Don't Come* bets function just like Pass and Don't Pass, and may be made at any time other than when a come-out roll is imminent. A player making either of these bets is designating the next roll of the dice as a personal come-out roll. It wins and loses on the same initial rolls, and can lead to an individual point number that must be rerolled before a 7 if the Come bet is to win. Once a point has been established, craps dealers move the player's chips to the appropriate numbered spot above the Come bar. A player may take the same odds on Come and Don't Come that the casino permits on Pass and Don't Pass.

Field

The *Field* bet (Figure 3.3) just below the Come bar appears in different versions in different casinos. Every version bundles together a collection of less-

FIGURE 3.3: Field betting area [27].

likely outcomes and pays off if any one of the numbers is rolled. In Figure 3.3, the Field bet covers 2, 3, 4, 9, 10, 11, and 12. A casual observer might note that this includes 7 of the 11 possible sums, not noticing right away that these 7 numbers combine for a cumulative probability of only $\frac{16}{36}$, less than 50%. Some tables switch in the 5 for the 9, which does not change the probability of winning.

Field bets pay even money and customarily offer bonus payouts of 2–1 or 3–1 paid when a 2 or 12 is rolled. The bonus payout affects the house edge.

- A 2–1 payoff on either 2 or 12 drops the HA to 8.33%; offering this payoff on both 2 and 12 gives a 5.56% house advantage. This is half the HA at the Bahamas casino mentioned above.

- The same 5.56% HA occurs if one of 2 and 12 pays even money and the other pays 3–1.

- Figure 3.3 shows a 2–1 payoff on 2 and a 3–1 payoff on 12. This wager holds only 2.78%.

- If a casino offers 3–1 on both 2 and 12, the bet is fair, with no advantage on either side. This is unlikely in casino play unless someone has made an error in designing or printing the layout.

A simple 1–1 payoff on all Field numbers was in force at a casino in the Bahamas in the 1970s with no bonuses on either 2 or 12 [63]. This version of the Field bet wins even money on 16 rolls and loses on the other 20, raising the casino's advantage to 11.11%.

Center Bets

Other wagers shown in Figure 3.2 may be made before any roll of the dice. These are collectively called *center bets*, for their location in the middle of the table. It is customary for dealers to place these bets for the players, arranging the chips that the players give them on the layout in a way that makes it clear to whom the chips belong. Some are one-roll bets, resolved on the next roll, while others can continue over many rolls. On the layout, one-roll bets and multiroll bets are often printed in different colors as an aid to dealers and players alike.

An example of multiroll center bets is the set of "Hardway" bets. The Hardway 8 wager is shown on the layout as ⚃⚃ and pays 10 for 1. A player making this bet is wagering that an 8 will be rolled the "hard way", as doubles, before a 7 is rolled or an 8 is rolled any other way. This bet has a high HA: 9.09%. In London, England, casinos are required by law to offer slightly higher payoffs on center bets in craps than are shown in Figure 3.2 [73]. For the Hardway 8, the payoff must be 9½–1, or 10½ for 1. This brings the HA down to 4.5%.

The Four Queens Casino once offered a special promotion based on the hardway bets. When doubles were rolled on the come-out roll to set a point, all players making a Pass line bet of at least $1 were given a free $1 bet on the hardway just rolled [30]. Since the bet was free to the player, it had a positive player edge. If the shooter threw a hardway 4 or 10, which pay off at 7–1, the positive expected return to the player was

$$E = (7) \cdot \frac{1}{9} \approx \$.7778.$$

On a 6 or 8, this bet carries a player expectation of 81.8¢.

This free bet was a boon to low rollers, as its advantage negated the HA on the main Pass bet provided that the gambler's initial wager was $6 or less [30].

Example 3.4. The ⚃⚄ bet in the center is a one-roll bet that the next roll of the dice will sum to 11. At 16 for 1, its expected value is

$$E = (16) \cdot \frac{2}{36} - 1 = -\frac{1}{9} \approx -\$.1111,$$

and the casino holds an 11.11% edge.   ∎

Example 3.5. The worst center bet is "Any 7", also called "Big Red" for its power to send a player's bankroll well into the red. This one-roll bet pays 4–1 (5 for 1) if the next roll is a 7. True odds on this wager would be 5–1; by offering 4–1, the casino holds a 16.67% advantage.   ∎

Since the Any 7 bet remains on the standard craps layout, some gamblers evidently bet it, but in light of the HA, one cannot help but ask: Why? One possible answer is that Any 7 can be bet as a form of insurance on a Pass line wager. A small Any 7 bet could provide a cushion to the blow of losing the main bet when a 7 comes up after a point is established.

Even with that massive disadvantage, a gambling system built around Any 7 was promoted as "fun if you have extra money" [83]. This system is a quest for a string of 3 straight 7s, and is advised for use if trios of 7s appear to be coming up unusually frequently. The probability of rolling 3 straight 7s is $\frac{1}{216}$, so "unusually frequently" could still mean "very rarely". A 7 on the come-out roll is the trigger for a \$5 Any 7 bet. If that wins, the player is advised to withdraw the original \$5 bet and let the \$20 in winnings ride. If the third 7 comes up, the player takes down the entire \$100. If the second bet loses, the player breaks even across both bets.

It sounds simple and lucrative—but **there is no way to combine negative-expectation bets into a positive-expectation parlay**. This system ignores the fact that successive rolls of the dice are independent. A rolled 7, or a pair of them, does not change the probability of a 7 on the next roll up or down from $\frac{1}{6}$. One time in 36, this two-bet parlay wins \$100; the rest of the time, it breaks even or loses \$5 without reaching the second roll. The expected value of this system is

$$E = (100) \cdot \frac{1}{36} + (0) \cdot \frac{5}{36} + (-5) \cdot \frac{5}{6} = -\$\frac{50}{36}.$$

Since \$5 was risked, the HA is 27.78%.

The center of the layout contains a number of spaces labeled C and E, which stand for *Craps* and *Eleven*. These bets can be bet individually, where C pays 7–1 if any craps number—2, 3, and 12—is rolled and E pays 15–1 on a roll of 11. E is the same as the ⚅⚅ wager at the center of the table. Players making a C or E bet hand their chips to a dealer, who places them on the layout in the C or E circle that points to where they're standing around the table.

Taken together, C&E pays 3–1 if a craps number is rolled, and 7–1 if the dice come up 11. The house holds an 11.11% advantage on the combined C&E bet.

The ⚀⚀, ⚀⚁, ⚄⚅, and ⚅⚅ wagers are collectively known as the *Horn*. Players may make a horn bet for a multiple of \$4, which is equally divided among all 4 wagers, or may choose a *High Horn* bet for a multiple of \$5. In this case, the player specifies which wager should receive a second unit. Combining multiple bets with high HAs leads to a composite bet with a high HA; here we have

$$E = (27) \cdot \frac{2}{36} + (12) \cdot \frac{4}{36} + (-4) \cdot \frac{30}{36} = -\frac{18}{36}$$

for the \$4 horn bet itself (HA = 12.50%),

$$E = (57) \cdot \frac{1}{36} + (26) \cdot \frac{1}{36} + (11) \cdot \frac{4}{36} + (-5) \cdot \frac{30}{36} = -\frac{23}{36}$$

for a \$5 high horn with the 2 or 12 high (HA $=$ 12.78%), and

$$E = (26) \cdot \frac{2}{36} + (26) \cdot \frac{2}{36} + (11) \cdot \frac{2}{36} + (-5) \cdot \frac{30}{36} = -\frac{24}{36}$$

for the \$5 high horn with the 3 or 11 high, for a HA of 13.33%.

Example 3.6. The *World* bet combines the horn bet with Any 7, for a 1-unit bet on each of the 5 spaces. If the dice show a 7, the bet is a net push: \$4 won on the Any Seven and \$4 lost on the four losing horn bets. If another winning total is rolled, the net win is \$26 on a 2 or 12 and \$11 on 3 and 11. These two high-advantage bets come together for a composite HA of 13.33%.   ∎

Off The Layout

A number of additional craps bets are not marked on the layout in Figure 3.2 and must be made by handing chips directly to casino personnel, who set the chips in the correct place among the spaces labeled 4–10 near the top of the layout or label them with small tokens indicating the wager.

The simplest of these are *Place* bets, which allow a player to choose his or her point instead of adopting the point rolled on the come-out or when the player places a Come bet. Place bets win if the selected number, which can be 4, 5, 6, 8, 9, or 10 at the player's option, is rolled before a 7 is thrown. These pay off at slightly better than the even money odds on a Pass line or Come bet; this is in part because a place bettor sacrifices the $\frac{8}{36}$ probability of winning a Pass line bet on the come-out roll. A place bet on 4 or 10 pays 9–5, one on 5 or 9 pays 7–5, and the 6 or 8 pays 7–6. In each case, fractional winnings are rounded down to favor the casino, which should influence the gambler making place bets to choose multiples of 5 or 6, depending on the number selected [8].

Example 3.7. A \$6 place bet on the 8 pays \$7 if it wins and has a house advantage of 1.52%. If the gambler wagers only \$5, the proper payoff of \$5.83 is rounded down to \$5, making the effective HA 9.09%. If the craps table uses 50¢ or \$2.50 chips and so can pay the bet off at \$5.50, the HA is slightly smaller: 4.55%.   ∎

The opposite of a place bet, a bet that 7 will be rolled before the selected number, is called a *Lay* bet. As the odds on a lay bet favor the bettor, the casino ensures its profit in several ways.

- The payoff odds are the inverse of place bet odds, and all pay off at less than 1–1. Laying the 4 or 10 pays 1–2, the 5 or 9 pays 2–3, and the 6 or 8 pay 5–6. These are the true odds, so at this point, there is no advantage for either side.

- The casino charges a 5% commission (rounded down to the nearest dollar) on the intended win on every lay bet, which establishes a positive house advantage. This commission is collected when the wager is made, so a gambler wishing to lay the 10 for $40 in hopes of winning $20 must pay $41 up front. Some casinos charge the commission only on winning lay bets when the wager is on the 4 or 10.

- Finally, many casinos will not accept lay bets unless the expected win is at least $20. This makes it impossible to escape a commission through rounding down, since the minimum commission is then $1. This corresponds to minimum lay bets of $40 on the 4 or 10, $30 on 5 or 9, and $24 on 6 or 8. Requiring higher bets also increases the variance of the wager, so the average outcome per bet is a larger absolute win for the casino.

The minimum lay bet on 10 requires the gambler to risk $40 in an effort to win $20, and to pay a $1 commission up front. The expected value of this bet is

$$E = (20) \cdot \frac{6}{9} + (-41) \cdot \frac{3}{9} = -\frac{1}{3}.$$

The HA is .813%; a low HA arises from a large bet. If the commission is only collected when the bet wins, the expectation becomes

$$E = (19) \cdot \frac{6}{9} + (-40) \cdot \frac{3}{9} = -\frac{2}{3}.$$

The expectation doubles, and the HA is slightly more than double: 1.67%. What sounds like a lucrative opportunity for the gambler turns out to benefit the casino.

Hop bets are simple one-roll wagers on the exact numbers rolled rather than the sum. These extend the standard ⚅⚅, ⚅⚄, ⚀⚀, and ⚁⚀ bets in the center to every other way that the dice can fall. Some tables include spaces for hop bets at the top of the center wagering section. Following the pattern established in the four bets noted here, hop bets on doubles pay 30–1 on a 35–1 shot and hop bets on combinations consisting of 2 different numbers pay 15–1 on a 17–1 proposition.

Example 3.8. If a player with a fondness for the number 7 wants a better wager than the Big Red bet described in Example 3.5, she might place three equal hop bets on 1–6, 2–5, and 3–4. Since these bets pay off at 15–1, the net win when one bet wins and the other two lose is 13–1. For three $1 hop bets, the expected return is

$$E = (13) \cdot \frac{6}{36} + (-3) \cdot \frac{30}{36} = -\frac{12}{36},$$

and dividing by the $3 wagered shows that the HA is only 11.11%, less than the 16.67% of Big Red. ∎

House Advantage and Bet Frequency

In comparing craps bets, it is useful to consider not only the house advantages but also the frequency of decisions. The Field bet has action on every roll, while the hardway bet on ⚁⚁ is a multiroll bet that, on average, takes 4 rolls to resolve—meaning that the player is effectively making only ¼ the number of wagers. If the craps table can handle 60 rolls per hour, a Field bettor would make 60 bets, placing 60 units in action and incurring a net loss of $3\frac{1}{3}$ units every hour if the HA is 5.56%. The same player betting on ⚁⚁ averages 15 wagers at an 11.11% disadvantage and, in the long run, can expect to lose $1\frac{2}{3}$ units per hour—half the loss of the Field bettor even as the HA is doubled.

3.2 Hazard

Craps can be traced back to the game of *hazard*, which shares some game rules with casino craps and a format suggesting street craps (Section 3.3). Hazard pits the shooter against one or more gamblers, the *faders*, who propose the stakes by laying their wager in front of the shooter. The shooter accepts the challenge by matching the faders' bet [113].

The shooter is then tasked with determining two points: the first one for the faders and the second one for himself. He throws the dice until he rolls a number between 5 and 9, inclusive. Note that 7 can be a point in Hazard and that there are no automatic winning or losing rolls at this stage. Once the faders' point, or *main*, is determined, the shooter rolls to determine his own point, the *chance*, which must be a number from 4 to 10. In determining the chance, other rolls can trigger an immediate win or loss, as on the come-out roll in craps [113]. A winning roll for the shooter is called a *nick*, while shooter losses are called *outs*.

- If the shooter re-rolls the main before determining the chance, he wins at once.

- Additionally, the shooter wins if the main is 7 and he rolls an 11, or if the main is 6 or 8 and he rolls a 12.

- If the shooter throws a 2 or 3, known as a *crab*, he loses immediately regardless of the main—similar to craps.

- If the main is 5, 6, 8, or 9, the shooter loses when rolling 11.

- If the main is 5, 7, or 9, the shooter loses on a roll of 12.

If the shooter sets his own point, different from the main, he continues to roll the dice until one point or the other shows, and only rolls producing one of the two points matter. This determines the winner.

Example 3.9.

a. Suppose that the shooter's first roll is a 7. This becomes the main. If the next roll is also a 7, the shooter wins.

b. If the first roll is a 3, the shooter rolls again, since 3 cannot be the main or the chance. If the next roll is 7, the main is 7. If the third roll is a 2, the shooter loses.

c. Let the first roll, and the main, be 8. If the shooter then rolls a 7, the game continues until either an 8 or a 7 is rolled and the winner is determined.

d. If the first roll is a 10, the shooter must roll again, as 10 can only be the chance, not the main. A second roll of 6 sets the main. If the third roll is 11, the shooter immediately loses.

■

A principle of gambling mathematics is that the easier a bet is to understand, the higher the house advantage. If this holds true for hazard, we would expect that this complicated set of rules leads to an approximately even game.

Table 3.4 shows the probability distribution function for the main. Since only rolls of 5 through 9 set this point, the denominator in these fractions is 24, not 36.

TABLE 3.4: Hazard: PDF for faders' point (main)

Point	Probability
5	$\dfrac{4}{24}$
6	$\dfrac{5}{24}$
7	$\dfrac{6}{24}$
8	$\dfrac{5}{24}$
9	$\dfrac{4}{24}$

Example 3.10. Suppose that the main is 5. The shooter then has probability $\frac{6}{36}$ of losing immediately by rolling a 2, 3, 11, or 12. The probability of winning immediately by rolling another 5 is $\frac{4}{36}$. The remaining $\frac{26}{36}$ of the time, the shooter sets his chance: 4 or a number from 6 through 10.

The probability of a shooter win against a main of 5 is the product of 3 factors:

- $p_1 = \dfrac{4}{36}$, the probability of setting the point of 5.

- $p_2(k)$ = the probability of rolling the number k on the second roll.
- $p_3(k,5)$ = the probability of winning with 5 as the main and k as the chance.

The function $p_3(k,5)$ is shown in Table 3.5.

TABLE 3.5: Hazard: Probability $p_3(k,5)$ of winning with a main of 5 and a chance of k

k	$p_3(k,5)$
2	0
3	0
4	.4286
5	1.000
6	.5556
7	.6000
8	.5556
9	.5000
10	.4286
11	0
12	0

Summing the product $p_1 \cdot p_2(k) \cdot p_3(k,5)$ across all values of k shows that the probability that the main is 5 and the shooter wins is approximately .0821. Dividing out the initial factor of $\frac{4}{24}$ shows that the probability of a shooter win, given that the main is 5, is .4924—so this is very close to an even game. ∎

This analysis can be repeated for the other mains and for k running from 2–12. The probability of a shooter win with a main of m and a chance of k is denoted $p_3(k,m)$ and is shown in Table 3.6.

Summing the win probabilities $p_1 \cdot p_2 \cdot p_3$ over all point combinations shows that the shooter's probability of winning is .4908. The faders hold a slim 1.84% edge, which is in line with the notion that complicated rules lead to low-HA wagers. By making a chance of 12 an automatic winner against a main of 5 and 9, the game can be reduced to a more nearly even game where the shooter's edge is only .02%—merely 2¢ of every $100 wagered.

An alternate version of the rules of hazard gives the shooter the choice of the main, rather than leaving it to be decided by the dice. In this version, the shooter should always choose a main of 7, as this gives him the highest chance of winning: 49.29%. The 1.41% fader advantage when the main is 7 matches the HA of the Pass line bet in craps, and this is an indicator of how hazard may have evolved into craps as it is now played in casinos.

Once the main and the chance have been determined, players may make side bets among themselves on which of the two points will be thrown first.

TABLE 3.6: Hazard: Probability $p_3(k,m)$ of winning with a chance of k and a main of m

$p_3(k,m)$			m		
k	5	6	7	8	9
2	.0000	.0000	.0000	.0000	.0000
3	.0000	.0000	.0000	.0000	.0000
4	.4286	.3750	.3333	.3750	.4286
5	1.000	.4444	.4000	.4444	.5000
6	.5556	1.000	.4545	.5000	.5556
7	.6000	.5455	1.000	.5455	.6000
8	.5556	.5000	.4545	1.000	.5556
9	.5000	.4444	.4000	.4444	1.000
10	.4286	.3750	.3333	.3750	.4286
11	.0000	.0000	1.000	.0000	.0000
12	.0000	1.000	.0000	1.000	.0000

These bets are made at true odds, so that neither side has the advantage [113]. This betting option may be the source of free odds bets in craps.

Example 3.11. If the main is 6 and the chance is 9, side bets on the 9 stand to collect 5–4 odds, while corresponding wagers on the 6 pay off at 4–5. The follows from the fact that the probability that a 9 is thrown before a 6 is $\frac{4}{4+5} = \frac{4}{9}$. In practice, a gambler wishing to back the 9 would lay \$4 on the playing surface, seeking an opponent willing to cover his wager with \$5. The total of \$9 would belong to the winning bettor. ■

3.3 Street Craps

Craps as played in casinos shares some traits with the illegal game of *street craps*, which takes its name from its origins on street corners and in back alleys. The basic play of the game is the same in craps and in street craps, but the wagering options are different and the advantage often shifts between the two sides of a wager. Lacking a professional dealer, all bets are arranged between the various gamblers participating in a given game.

A round of street craps begins with a shooter offering a sum of money as a wager on the hypothetical Pass line. Players wishing to bet against him may, collectively, wager any sum of money up to the amount of the shooter's offer [114]. Such players are said to *fade* the shooter. This is, of course, the reverse of the order of betting in hazard. Players may also bet with or against the shooter privately among themselves; they may additionally make come, don't come, hardway, and various other proposition bets by mutual agreement. The odds taken or offered on these bets typically parallel those found in legitimate casinos where possible; for bets with no casino analog, the edge favoring one player can be significant.

As in casino craps, a roll of 7 or 11 on the first throw wins for the shooter, while a 2 or 3 is an instant loser. One important difference between street craps and casino craps is that street craps has no custom of barring the 12 on the come-out roll, so an initial 12 is also a losing roll for the shooter. It follows that the "wrong" bettor who wagers against the shooter by fading him has a 1.41% advantage. Any other number becomes the shooter's point, and he continues rolling until re-rolling the point or rolling a 7.

One-Roll Bets

Additional side bets available among street craps players are numerous, limited only by the ingenuity of the gambler offering a wager and the receptiveness of the bettor to a particular offer. These wagers are not limited to the array of bets available on a casino layout, although a craps player could offer a standard craps wager to any takers and switch the advantage in his favor, as when offering 4–1 that the next roll will be a 7. The other players might recognize this as a casino wager and think little of its player disadvantage; the hustler knows that that 16.67% edge now belongs to him. It's not necessary to stick to the standard craps odds, either—if other players will bet that the next roll will be ⚂⚃ with the lure of only a 25–1 payoff, the edge to their opponent is 27.78%, double the value when casino odds of 30–1 are paid.

Some of these wagers have immense advantages for one side; a street craps hustler who is well-versed in elementary probability and has some persuasion skills has the potential to take a lot of money from other players in high-advantage bets.

Example 3.12. A "hustler's hardway" bet offers even money that the shooter won't throw a 1 or 2 on either die during the next roll [113]. If only two of the six sides of the dice are losers, this appears at first glance to be a bet worth making.

Simple counting shows the error of this reasoning; the problem lies, perhaps, in the failure to consider that there are two dice. The chance of not rolling a 1 or a 2 (or any two selected numbers) on a single die is $\frac{2}{3}$, but if two independent dice are thrown, the chance of avoiding 1s and 2s is

$$\left(\frac{2}{3}\right) \cdot \left(\frac{2}{3}\right) = \frac{4}{9} \approx .4444.$$

The expected value of this 1–1 proposition is

$$E = (1) \cdot \frac{4}{9} + (-1) \cdot \frac{5}{9} = -\$\frac{1}{9},$$

and so the hustler's advantage is 11.11%. ■

As a general rule, if the probability of winning a $1 bet paying even money is p, the expected value of that bet is simply

$$E = (1) \cdot p + (-1) \cdot (1 - p) = 2p - 1$$

and the advantage accruing to the dice hustler or casino offering that bet is the negative of this quantity: $1 - 2p$, which is positive when $p < .5$.

Example 3.13. Consider a proposed wager offering even money if the shooter doesn't roll a 5, 6, 7, or 8. Shorn of the context of a craps table, it is less obvious that this is a Field bet, but without the bonuses on 2 or 12 that reduce the HA to 2.78% or 5.56%. Without those extra payoffs, the gambler who accepts this bet is facing off against an adversary with an advantage of

$$1 - 2 \cdot \frac{16}{36} = \frac{1}{9},$$

or 11.11%. ■

Two-Roll Bets

Another class of gimmick street craps wagers is *two-roll* bets: a proposition that certain numbers either will or won't be rolled in the next two rolls of the dice. A craps hustler offering these bets does so knowing that he holds an advantage over anyone accepting his terms.

Example 3.14. Suppose a gambler offers even money that the shooter will throw a 6 or 8 in the next two rolls. A casual observer might think that since 6 and 8 are the likeliest sums after 7, he is getting both numbers working for him, and the shooter has two rolls to produce either one, this is a good bet.

Of course, it isn't. The probability of rolling a 6 or 8 on any one roll is $\frac{10}{36}$, and so the probability of *not* rolling a 6 or 8 is $\frac{26}{36}$. Over two independent rolls, the probability of winning this bet when a 6 or 8 is rolled in 2 tosses is

$$\frac{10}{36} + \frac{26}{36} \cdot \frac{10}{36} = \frac{155}{324} \approx .4784,$$

giving the hustler a 4.32% advantage. ■

Example 3.15. Another two-roll bet is an even-money proposition that the shooter won't roll a 6 or 7 in two rolls [113]. A gambler who lost the bet in Example 3.14 might see this as attractive. Sure, it's easy to recognize that trading out the 8 for the 7 changes the probabilities, but swapping "will" for "won't" means that he's betting against the shooter, which he saw worked out for his opponent in the first example.

Nope. Once again, the hustler has an edge. There are 5 ways to roll a 6, 6 ways to roll a 7, and 25 ways not to roll either one. The chance of the shooter *not* rolling a 6 or 7 in two rolls is

$$\left(\frac{25}{36}\right)^2 = \frac{625}{1296} \approx .4823 < .5,$$

and so that edge is 3.55%. ∎

Three-Roll Bets

Escalating from two-roll to three-roll proposition bets requires no advanced mathematics, since the rolls are independent. An extra advantage to the hustler searching for bettors to accept one of these wagers is that the calculations become somewhat harder to evaluate mentally on the fly.

Example 3.16. A hustler offers an even-money proposition that the shooter will throw a 2, 3, 11, or 12 in three rolls. We can assess this position by looking at it from the hustler's point of view and computing his chance of winning—if that exceeds .5, he holds the edge.

The hustler is betting that none of the 4 specified sums are rolled in 3 rolls. This event has high probability on a single roll: $\frac{5}{6}$. Squaring this to produce the chance of avoiding these four numbers in 2 rolls gives $\frac{25}{36}$—still better than ½. The probability of three straight rolls without these numbers appearing is

$$\left(\frac{5}{6}\right)^3 = \frac{125}{216}.$$

This is the hustler's chance of winning this proposition, and it exceeds .5. The edge to the hustler is $\frac{34}{216} \approx 15.74\%$. ∎

If this bet were extended to 4 rolls, the probability of not rolling a 2, 3, 11, or 12 is

$$\left(\frac{5}{6}\right)^4 = \frac{625}{1296} \approx .4823.$$

Turning the wager around and offering an even-money bet that the shooter *won't* roll at least one 2, 3, 11, or 12 in the next 4 rolls would give the hustler an edge of 3.56%.

Open-Ended Bets

In casino craps, some wagers are open-ended and can continue for many rolls before being resolved. The Pass, Don't Pass, and Hardway bets are all examples of this class of wagers. On the street, the hustlers have devised their own such bets: some are fair and others are not.

Example 3.17. An example of a fair proposition bet is a wager that the shooter will (or won't) throw a 4 or 10 before rolling a 7 [113]. There are 6 ways to roll either a 4 or a 10 and 6 ways to roll a 7, and these are the only rolls that matter for resolution of this wager. Accordingly, if you choose the "will" bet, your chance of winning on any given throw of the dice is $\frac{6}{12} = 50\%$.

Changing this proposition to "The shooter will roll a 3 or a 10 before a 7" changes the bet from a fair wager to one carrying a 9.09% HA. The number of winning rolls drops from 6 to 5 while the number of losing rolls remains 6. ∎

In street craps as in casino craps, there are propositions that pay off at better than even money. The casino sets its payoffs to ensure its advantage but must not hold so great an edge that it drives patrons away; by contrast, the hustler has less need to consider long-range implications. Since many craps hustlers move freely among games, they do not depend on repeat business and can set odds to maximize their income.

Example 3.18. The hardway 6 wager at a legitimate casino pays 10 for 1, or 9–1 (see Figure 3.2), which generates a HA of 9.09%. A street craps hustler might offer odds as low as 5–1 on this event. If the odds offered are X to 1, the bet has an expectation of

$$E(X) = (X) \cdot \frac{1}{11} + (-1) \cdot \frac{10}{11} = \frac{X - 10}{11}.$$

This is tabulated in table 3.7.

TABLE 3.7: Street Craps Hardway 6 expectation and HA

X	$E(X)$	HA
5	−.4545	45.45%
6	−.3636	36.36%
7	−.2727	27.27%
8	−.1818	18.18%
9	−.0909	9.09%
10	0	0.00%

A hustler who can find takers when offering 5–1 odds can confidently expect to win about 45% of the money risked over the life of the proposition, which gives him or her a better shot at making a lot of money quickly, before the other gamblers catch on to how skewed against them this offer is. At the other end, a 10–1 payoff offers a fair game. ∎

First Roll Barred

Some craps hustlers seek to gain their advantage on open-ended bets by offering fair odds but barring the first roll: turning a winning outcome for the player into a push while a losing outcome for the player registers as a win for the hustler [113]. This can be done with true odds bets, as when a hustler offers 2–1 odds that a 10 will be thrown before a 7. This is a fair game until the first roll is barred; the hustler derives an edge whenever the first roll after the wager is established is a 10. In essence, this gives the hustler one free shot at the player before a zero-expectation game begins. A player considering this bet might think "There's only 3 ways out of 36 that a win turns into a push for me on the first roll, and it's fair odds otherwise." This line of thought masks a huge hustler advantage.

The probability that the hustler wins this bet is

$$\frac{6}{36} + \frac{30}{36} \cdot \frac{2}{3} = \frac{13}{18},$$

and the expectation to the hustler is

$$E = (1) \cdot \frac{13}{18} + (-2) \cdot \frac{5}{18} = \frac{1}{6},$$

which gives a 16.67% advantage.

3.4   Crooked Dice

Legitimate casinos have no need to use crooked dice on their craps tables. Casinos make their money from the rules of the games and pay tables that pay off at less than true odds; there's no reason for any casino to risk its ability to do business by cheating its customers. Any casino found to be using unfair dice of any sort would swiftly see its gaming license revoked, and since craps players are permitted to handle the dice during play, loaded dice would be quickly noticed.

Street craps games operate outside the law and hustlers seldom worry about long-term profitability, so neither of these factors deters the use of loaded dice by street craps hustlers.

Inside Work

To many people, the term "loaded dice" suggests what is known in the trade as *inside work*: insertion of weights inside the dice, invisible from the outside, that affect the probabilities when the dice are rolled. If the dice are opaque, then loading is a relatively simple matter, since with good craftsmanship, there is no way to detect the loads by mere visual inspection.

Most casinos, and many informal dice games, use transparent dice, which are harder to load. Casino dice have the pips made from material with the same density as the dice and set flush with the faces of the cube, unlike common dice which often have indented pips. These opaque spots provide a place where loads may be introduced without detection. A dice loader drills out some of the pips, inserts small dense metal disks—gold and platinum are good choices—and replaces the pips above the loads. A superior dice loader will drill all 21 pips on the die a little deeper and fill in the spots with paint to make it harder to detect the loaded pips.

Loading the 6

With that mechanism in play, the most effective way to load a die is to place loads under the 6, which gives a die that favors rolling 1s. A more extreme approach involves loading all of the pips adjacent to the edges of the 6: two on the 5 and 4, and one on the 3 and 2. Even this collection of 12 loaded pips cannot guarantee that the die always rolls a 1—nor would that be a desirable outcome, for players would surely notice such a surplus of 1s.

Suppose that a single die is loaded so that the probability of rolling a 1 increases by 50%, with the probability of a 6 decreasing by the same amount. This die has the PDF shown in Table 3.8.

TABLE 3.8: PDF for a single loaded d6

Roll	Probability
1	$\dfrac{1}{4}$
2	$\dfrac{1}{6}$
3	$\dfrac{1}{6}$
4	$\dfrac{1}{6}$
5	$\dfrac{1}{6}$
6	$\dfrac{1}{12}$

Switching two of these into a street craps game would increase the probability of low sums and decrease the probability of high sums, as shown in Table 3.9.

Example 3.19. With this pair of dice in play, the ⚀⚀ bet at casino odds of has a clear player advantage. The expectation is

$$E = (30) \cdot \frac{1}{16} + (-1) \cdot \frac{15}{16} = \frac{15}{16},$$

TABLE 3.9: PDF for the sum of 2 loaded d6

Roll	Probability	Roll	Probability
2	$\frac{1}{16}$	8	$\frac{1}{9}$
3	$\frac{1}{12}$	9	$\frac{1}{12}$
4	$\frac{1}{9}$	10	$\frac{1}{18}$
5	$\frac{5}{36}$	11	$\frac{1}{36}$
6	$\frac{1}{6}$	12	$\frac{1}{144}$
7	$\frac{11}{72}$		

giving the player a 93.75% edge.

This level of success with a known longshot bet would increase the risk of the loaded dice being discovered. ∎

Since a gambler introducing loaded dice into a street craps game would almost surely be offering proposition bets that play into his advantage, the word "lucrative" has a different meaning for the dice cheat. A craps hustler looking for players to accept losing propositions might look for opportunities that offer a smaller edge, but are more subtle and have a greater likelihood of extended viability. One possibility lies in Lay bets against the 6. Betting on a 6 appearing before a 7 on casino terms with a 7–6 payoff would yield a positive expectation of

$$E = (7) \cdot \frac{12}{23} + (-6) \cdot \frac{11}{23} = \frac{18}{23},$$

and the player's edge is the not-insignificant 13.04%. Even if this bet is paid at even money, the player still holds an edge: 4.35%—so a craps hustler would look for players wishing to bet *against* the 6. As the opposite of the casino wager, this might tempt players, but since 6s are more likely than 7s with these dice, the advantage is reversed.

At the same time, many bad bets get worse. The house advantage on Any 7 rises from 16.67% to 23.61%. A craps hustler armed with a pair of these dice can comfortably offer 120–1 odds on the ⚃ ⚃ and be assured of nearly a 16% edge. The mere offer of such high odds might arouse suspicion, so the truly clever hustler might settle for offering 40–1 odds and enjoying a 72.5% advantage.

Missouts

If loading the 6 face seems like it brings too much risk of discovery, there are other options. A more subtle form of inside loaded dice relies on loading near the corners rather than on the faces. There are 2 corners on a standard d6 where 3 spots come together: the 4–5–6 and 2–3–6 vertices, which are frontmost in Figure 3.4. Opaque dice are loaded through this vertex by drilling shallow intersecting holes though each pip, filling the holes with a dense material such as mercury, and carefully restoring the pips. Transparent dice loaders settle for inserting metal slugs under the 3 pips that meet at the corner.

FIGURE 3.4: 2d6 with the best corners for weighting frontmost.

This corner load is designed to favor the numbers away from the loaded vertex. In Figure 3.4, the left-hand die favors the 1, 2, and 3 while the right-hand die favors 1, 4, and 5. Taken as a pair, these dice are called *missouts*, as they tend to roll more 7s than fair dice, increasing the chance that the shooter will miss his or her point [36].

Suppose that this load changes the probabilities by one-third: every favored side has probability $\frac{1}{6} + \frac{1}{18} = \frac{2}{9}$ and every non-favored side has probability $\frac{1}{6} - \frac{1}{18} = \frac{1}{9}$. The probability of a 7 rises slightly, from $\frac{1}{6}$ to

$$\frac{2}{9} \cdot \frac{1}{9} + \frac{2}{9} \cdot \frac{2}{9} + \frac{2}{9} \cdot \frac{2}{9} + \frac{1}{9} \cdot \frac{1}{9} + \frac{1}{9} \cdot \frac{1}{9} + \frac{1}{9} \cdot \frac{2}{9} = \frac{14}{81}.$$

Example 3.20. Revisiting Example 3.19: The probability of rolling a 2 on this pair of dice is $\frac{4}{81}$, since both 1s are favored sides. The expectation on a \$1 $\boxed{\cdot}\,\boxed{\cdot}$ bet is

$$E = (30) \cdot \frac{4}{81} + (-1) \cdot \frac{77}{81} = \frac{43}{81},$$

giving the bettor a 53.09% advantage.

This is not quite as good as the 93.75% edge with the more heavily-loaded dice described above, but it's still a pretty substantial advantage. ∎

What do missouts do to a Pass Line bet? The probability of winning on the come-out roll with a 7 or 11 is

$$P(7) + P(11) = \frac{14}{81} + \frac{3}{81} = \frac{17}{81} \approx .2099.$$

The probabilities of setting and subsequently making the various points are shown in Table 3.10.

TABLE 3.10: Probability of setting and making points with a pair of missouts

Point	P(Set point)	P(Make point)
4	$\frac{8}{81}$	$\frac{8}{22}$
5	$\frac{10}{81}$	$\frac{10}{24}$
6	$\frac{13}{81}$	$\frac{13}{27}$
8	$\frac{10}{81}$	$\frac{10}{24}$
9	$\frac{7}{81}$	$\frac{7}{21}$
10	$\frac{5}{81}$	$\frac{5}{19}$

Multiplying and adding these probabilities to the probability of an immediate win gives P(Win Pass Line bet) $\approx .4710$, which gives the Don't Pass bettor an advantage of 5.80%. If you've managed to slip these dice into a street craps game, bet hard against the shooter.

Outside Work

There are alternate ways to produce crooked dice that are not as complicated as inside work requires; some of the most effective cheaters' dice are simply spotted irregularly. Dice that are altered in this manner are called *outside work*.

Tops

A class of crooked dice called *Tops* bears only three different numbers, each one appearing on a pair of opposite faces. Since no observer can see two opposite sides of a die at the same time, especially under low-light conditions where street craps is sometimes played, a craps hustler who can switch in these crooked dice or work with a confederate with sleight-of-hand skills can generate an advantage near or equal to 100%.

Figure 3.5 shows a pair of 1–3–5 Tops. Each die sits in front of a mirror so that 5 of the 6 faces may be seen; the number on top is repeated on the unseen bottom face. A pair of Tops bearing only 1, 3, and 5 can only roll even

FIGURE 3.5: 1–3–5 Tops.

numbers, so it's impossible to seven out when rolling them, or to win on the come-out roll with a 7 or 11. All the hustler has to do once the dice are in play is bet with the shooter; the only risk assumed is when the shooter loses with a 2 on the come-out roll. Due to the extra 1s on each die, this has an increased probability of $\frac{1}{9}$, four times greater than with fair dice. The hustler's edge with these dice is 77.8%.

Another version of Tops has the dice spotted 2, 4, and 6, with each number appearing twice, as in Figure 3.6.

FIGURE 3.6: 2–4–6 Tops.

These dice are also unable to roll an odd sum, and so the risk to a right bettor is a loss on the come-out roll with a 12. These carry the same 77.8% edge as 1–3–5 Tops.

Either of these pairs of crooked dice runs a high risk of discovery if the players notice a long run without any odd sums appearing. A more sophisticated but less lucrative approach uses one crooked die together with a standard die. If a 1–3–5 Top is switched in for a standard die, Table 3.11 shows the possible results.

TABLE 3.11: Chart of outcomes for a 1–3–5 Top and a standard die

	⚀	⚁	⚂	⚃	⚄	⚅
⚀	2	3	4	5	6	7
⚀	2	3	4	5	6	7
⚂	4	5	6	7	8	9
⚂	4	5	6	7	8	9
⚄	6	7	8	9	10	11
⚄	6	7	8	9	10	11

With these two dice in play, one lucrative hustler move would be to offer odds that the next roll will be a 12. Casinos offer this one-roll bet at 30–1; since a 12 is impossible, a hustler could offer better odds at zero risk ("33 to 1 he don't roll boxcars" or "3 gets you a hundred if she rolls a 12", for example) and enjoy a 100% advantage. Since sums of 12 are rare even on a pair of fair dice, it might be possible to avoid discovery for a considerable length of time.

Example 3.21. A more clever option would be to offer even money that an 8 will be rolled before a 6. With a pair of standard dice, this is a fair wager, since each sum has a $\frac{5}{36}$ chance of appearing. With these dice, Table 3.11 shows that the hustler's expected return is

$$(1) \cdot \frac{6}{36} + (-1) \cdot \frac{4}{36} = \frac{1}{18},$$

and his advantage is 5.56%. ∎

A slightly more sophisticated pair of Tops mis-spots the two dice differently. One die contains the numbers 1, 4, and 5 twice apiece and the other is a standard 2–4–6 Top. A pair of these dice is shown in Figure 3.7. This pair produces the rolls in Table 3.12.

The scam involving the impossible 12 described above would work with these dice, as would a similar wager on 2. Another more intricate hustle might be to lure gamblers to bet against a hardway 8, since there is no other way to roll a sum of 8 on these dice. Offering even money or 2–1 that an easy 8 will be thrown before a hard 8 (leaving the roll of a 7, which loses the corresponding

FIGURE 3.7: Mixed 2–4–6 and 1–4–5 Tops.

TABLE 3.12: Chart of outcomes for a 2–4–6 Top and a 1–4–5 Top, as shown in Figure 3.7

	⚀	⚀	⚂	⚂	⚄	⚄
⚀	3	3	6	6	7	7
⚀	3	3	6	6	7	7
⚃	5	5	8	8	9	9
⚃	5	5	8	8	9	9
⚅	7	7	10	10	11	11
⚅	7	7	10	10	11	11

casino bet, out of the wager) carries a 100% edge but could attract bettors for a short while, since an easy 8 is 4 times more likely than a hard 8 on a pair of fair dice.

Example 3.22. Figure 3.8 shows a pair of 1–4–5 Tops emblazoned with the name of the Riviera Casino, which was open from 1955–2015. These dice were not used by the casino, but stand as an example of the lengths to which a committed dice cheat might go in pursuit of cash. One successful operation hinging on casino logo-bearing Tops was described in John Soares' *Loaded Dice: The True Story of a Casino Cheat* [138]. A crew of crooked gamblers in the 1960s employed the following approach to cheating at craps tables throughout Nevada:

- Procure a die from the craps tables at a given casino. This can be easy to do, as many casinos sell used canceled dice in their gift shops.

FIGURE 3.8: 1–4–5 Tops bearing the Riviera Casino imprint.

- Have the casino's logo duplicated onto a pair of 4–5–6 Tops of the correct color and size. This typically cost about $30 per pair.

- Return to the casino and palm the crooked dice into play. One team member controlled the dice while the others placed Field bets. The Field bet at craps is a bet that the next roll of the dice will come up 2, 3, 4, 9, 10, 11, or 12. As seen in Figure 3.2, the bet pays 2–1 if a 2 or 12 is rolled; it pays even money otherwise.

  Given the numbers on the 4–5–6 Tops, the Field bet only lost if the roll was an 8, and the chance of a 2–1 payoff on a 12 rose from $\frac{1}{36}$ to $\frac{1}{9}$. The expectation on a $1 Field wager made against these dice was

  $$E = (1) \cdot \frac{7}{9} + (2) \cdot \frac{1}{9} + (-1) \cdot \frac{1}{9} = \frac{8}{9} \approx \$.8889,$$

  giving the team an 88.9% advantage. If the casino paid 3–1 on 12s, the team's advantage was 100%.

- After a few rolls, switch the legitimate dice back into play and leave the casino while attracting as little attention as possible.

The exit was an important part of this scheme, as the probability of winning a Field bet with fair dice is $\frac{4}{9}$, so a long string of Field wins is unlikely. It was also essential to the plan that the dealers not examine the dice too closely in between rolls, so scouting the tables for a dealing crew that was less than completely attentive in handling the dice was time well spent. ■

Example 3.23. A more subtle form of mis-spotted dice changes only one number on a die, as when the 2 on a standard d6 is changed to a second 5. When used in conjunction with a standard d6, such a die has the potential to avoid detection for longer than a die with 3 numbers each repeated twice. Table 3.13 shows the chart of outcomes when a "double-5" die is rolled with a standard d6.

TABLE 3.13: Outcomes for a double-5 die rolled with a standard d6

	⚀	⚃	⚂	⚄	⚄	⚅
⚀	2	6	4	5	6	7
⚁	3	7	5	6	7	8
⚂	4	8	6	7	8	9
⚃	5	9	7	8	9	10
⚄	6	10	8	9	10	11
⚅	7	11	9	10	11	12

We see that the probabilities of 3, 4, and 5 have gone down while those of 9, 10, and 11 have gone up. A \$5 Place bet on 10 now has expectation

$$E = (9) \cdot \frac{4}{10} + (-5) \cdot \frac{6}{10} = \$0.60,$$

a positive value giving the gambler a 12% advantage. ∎

Craps tables now have a short mirror at the bottom, running the length of the table on the side opposite the dealers, which allows casino personnel to see 5 of the 6 sides of each die when then dice are within range of the mirror. This functions as extra defense against mis-spotted dice entering a game. If you're shooting Street Craps on the literal street, though... you're on your own.

Shapes

Shapes are crooked dice where a small amount of the die, from 4–40 thousandths of an inch, is shaved off of two opposite faces [113]. The shaved die is then slightly more likely to land on one of the shaved faces and less likely to land on one of the other 4 faces, since the area of those 4 faces is slightly smaller.

Suppose that a die is shaved on the 1 and 6 faces—call this a *1–6 shape*. Let s be the thickness of the material removed, and define $p = P(1) = P(6)$ and $q = P(2) = P(3) = P(4) = P(5)$. Provided that s is small, we may assume that p and q are proportional to the respective areas A and B of their associated faces. If s is too large, the die may be so misshapen that it can only land on 1 or 6; this would, of course, be easily discovered. (A Scrabble tile would be an extreme example of a too-large s.)

We have the following equations:

$$A = \frac{9}{16} \text{ in}^2.$$

$$B = \frac{3}{4} \cdot \left(\frac{3}{4} - 2s\right) \text{ in}^2.$$

$$2p + 4q = 1.$$

$$\frac{p}{q} = \frac{A}{B} = \frac{3}{3 - 8s}.$$

Solving the system consisting of the last two equations for p and q gives

$$p = \frac{A}{2A + 4B}, \quad q = \frac{B}{2A + 4B}.$$

Using the values of A and B above yields

$$p = \frac{3}{18 - 32s}, \quad q = \frac{3 - 8s}{18 - 32s}.$$

Example 3.24. If $s = .004$ inches, the probability of rolling a 1 on a single die is

$$p = \frac{3}{18 - 32(.004)} = \frac{375}{2234} \approx .1679,$$

slightly higher than the $\frac{1}{6}$ probability of rolling a 1 on a fair die. Similarly, the probability of rolling a 5 has fallen from $\frac{1}{6}$ to

$$q = \frac{3 - 8(.004)}{18 - 32(.004)} = \frac{371}{2234} \approx .1661.$$

∎

At the other end of the limit stated above, where $s = .04$ inches, we have $p \approx .1794$ and $q \approx .1603$. Shaving a die much beyond .04 inches runs an increased risk that the modification will be noticed.

If a pair of these shapes is rolled, the probability of rolling a 2 or 12 is p^2, which is also the chance of rolling a 7 as 1–6 or 6–1. The probability of a sum composed of two numbers from 2–5 is q^2, and the probability of rolling a 1 or 6 together with a 2, 3, 4, or 5 is pq. Note that p and q vary with s. The PDF for the sum is shown in Table 3.14

The symmetry of the distortion—the probabilities of 1 and 6 go up while the probabilities of the other 4 numbers go down—means that the mean of a single throw of a 1–6 shape remains 3.5: the mean when a single d6 is rolled. It follows that the mean when two 1–6 shapes are rolled is 7, just as we see when rolling 2d6.

How does a pair of 1–6 shapes affect street craps? Increasing the probability of 1s and 6s on a single die means that sums of 2, 3, 7, 11, and 12 become

TABLE 3.14: PDF for the roll of two 1–6 shapes

Roll, X	$P(X)$
2, 12	p^2
3, 11	$2pq$
4, 10	$2pq + q^2$
5, 9	$2pq + 2q^2$
6, 8	$2pq + 3q^2$
7	$2p^2 + 4q^2$

more common, while all other sums are less common. We would, therefore, see more games resolved on the come-out roll and fewer points established. The extent of the changes would depend on the extent to which the dice are shaved.

Example 3.25. Consider a pair of 1–6 shapes with a thickness of $s = .04$ inches (the maximum stated above) removed from the 1 and 6 faces. The probability of winning a Pass line bet on the come-out roll is

$$P(7) + P(11) \approx .2247.$$

The corresponding probability with fair dice is approximately .2222, so the shapes move the probability only very slightly.

For Don't Pass, we must consider that street craps does not bar the 12 on the come-out roll, so the chance of winning changes from $P(2) + P(3) \approx .0833$ with fair dice to $P(2) + P(3) + P(12) \approx .1219$, a considerable increase, but one more attributable to the rule change than to the misshapen dice. A dice hustler who has introduced a pair of 1–6 shapes into a game of street craps always bets against the shooter and passes the dice to the next shooter rather than toss them himself and be forced to bet the Pass line.

Incorporating the probabilities of setting and then making or not making the various points shows that the probability of winning a Pass bet with two 1–6 shapes in play is .4896 and the chance of winning a Don't Pass bet with those dice is .5104. The expected value of a $1 bet on the Pass line is –2.07¢. A Don't Pass bettor holds that same 2.07% edge. ∎

With $s = .004$ inches, the edge when betting against the shooter falls to 1.48%—small, but nonetheless positive. A dice hustler using shapes rather than Tops is probably playing a long con: since shapes are harder to detect than Tops, the hustler is trading a smaller advantage for increased longevity at the game.

3.5   Controlled Shooting

Casino profits depend on the laws of probability working with real dice, cards, and wheels just as the theory predicts, in the long run. One such law is that the probability of rolling a 7 on 2d6 is $\frac{1}{6}$. If a craps shooter in a casino setting had the ability to control the dice in a way that either decreased or increased this probability, that might be an advantage that could be exploited by careful betting. This skill is sometimes called *rhythmic rolling*. If street craps is being played with the dice thrown on a blanket or other soft surface, controlled shooting might also be a weapon that a shooter could use—provided that he or she can control the dice well enough to gain a real edge.

Hardway Set

A common method of controlled shooting calls for setting the dice together with certain faces—say, the 1s and 6s—touching and to the outside, so that the two dice together form a rectangular prism and the faces coming together to form the longest side show the same number. The dice are then lightly tossed with a bit of backspin so they fly through the air as a single unit, rotating around the long axis. This is known as the *hardway set*; see Figure 3.9. The dice are thrown so that they bounce off the table very close to the

FIGURE 3.9: Standard casino dice showing the hardway set.

far wall, hit the wall, and fall to the felt with very little further movement. Given how the dice are set, 1s and 6s are said to be less likely to turn up.

In the most successful circumstance, the two dice rotate around the long axis so that the numbers 2, 3, 4, and 5 all have equal probability, ¼, and the numbers 1 and 6 are impossible. Table 3.15 shows the possible sums that would arise.

The probability distribution function for this roll, assuming that the dice rotate independently around the long axis, is shown in Table 3.16. (It should be noted that this is something of a worst-case scenario; the power of the

TABLE 3.15: Outcomes for an ideal pair of d6 tossed with the hardway set

	⚀	⚁	⚂	⚃
⚀	4	5	6	7
⚁	5	6	7	8
⚂	6	7	8	9
⚃	7	8	9	10

hardway set comes when the dice are not independent. We shall consider that shortly.) We note that the probability of rolling a 7 has risen, from $\frac{1}{6}$ to $\frac{1}{4}$, but that the probability of rolling a hardway combination has also risen the same way.

TABLE 3.16: PDF for an ideal pair of d6 tossed with the hardway set

Sum	Probability
4	.0625
5	.1250
6	.1875
7	.2500
8	.1875
9	.1250
10	.0625

Example 3.26. Consider the hardway bet on either the 4 or 10, which pays 8 for 1. With fair dice randomly thrown, the HA for this bet is 11.11%. If the dice can be controlled perfectly in accordance with Table 3.16, then there are *no* easy 4s or 10s, and the new expected value is

$$E = (7) \cdot \frac{1}{5} + (-1) \cdot \frac{4}{5} = \frac{3}{5},$$

giving a 60% edge to the shooter. ∎

This level of ideal behavior may be too much to hope for in the world of real dice and real tables. Suppose, though, that a skilled shooter could set and roll the dice using the hardway set so that the probability of a 7 *decreased* from $\frac{1}{6}$ to $\frac{1}{7}$. An obvious way to bet into this advantage would be to make Place bets on non-point numbers. A Place bet on 6 under these conditions, assuming for simplicity that the probability of a 6 remains $\frac{5}{36}$, has win probability of

$$P = \frac{5}{36} + \left(1 - \frac{1}{7} - \frac{5}{36}\right) \cdot \frac{5}{36} + \left(1 - \frac{1}{7} - \frac{5}{12}\right)^2 \cdot \frac{5}{36} + \cdots = \sum_{k=0}^{\infty} \frac{5}{36} \cdot q^k,$$

where $q = 1 - \frac{1}{7} - \frac{5}{36} = \frac{181}{252}$ is the probability of rolling a number other than 6 or 7. Summing the series gives

$$P = \frac{5}{36} \cdot \frac{1}{1-q} = \frac{5}{36} \cdot \frac{1}{1 - \frac{181}{252}} = \frac{35}{71}.$$

While this is less than ½, the bet pays off at 7–6, making the expected value

$$E = (7) \cdot \frac{35}{71} + (-6) \cdot \frac{36}{71} = \frac{29}{71} \approx .4085$$

—and giving the player a 6.81% edge. The same calculations apply to a Place bet on 8.

Example 3.27. One chance in 7 of rolling a 7 may seem like a high standard. To make a Place bet on 6 or 8 a fair wager, the probability of rolling a 7 needs only decrease from $\frac{1}{6}$ to p, where p is the solution to $E = 0$, or

$$(7) \cdot \frac{5}{36} \cdot \frac{1}{1 - \left(\frac{31}{36} - p\right)} + (-6) \cdot \left(1 - \frac{5}{36} \cdot \frac{1}{1 - \left(\frac{31}{36} - p\right)}\right) = 0.$$

The solution to this equation is $p = \frac{35}{216}$, a mere decrease of $\frac{1}{216}$ from the $\frac{1}{6}$ probability of rolling a 7 fairly and randomly. This corresponds to approximately one 7 in every 6.17 rolls. We conclude that a small change in the probability of rolling a 7 can change the advantage on certain bets from the casino to the players. ∎

Since the goal in using the hardway set is to have the dice rotate as a unit about the long axis, the technique is ineffective whenever the dice rotate independently too far. When the dice rotate together, they land in a hardway configuration. If one die gets a quarter-turn ahead of or behind the other, the upward faces sum to 5, 6, 8, or 9, which suggests that place bets on those numbers may be lucrative. However, if one die gets half a turn ahead, which is known as a *double pitch*, the dice sum to 7 no matter how they land [163]. Since betting into the hardway set relies on the shooter's ability not to seven out, double pitches are to be avoided. There will, of course, be some movement of the dice after they land; part of a successful throw from the hardway set is to minimize the amount of uncontrolled motion by tossing the dice softly, so that they hit the far wall as required and drop straight to the table.

3V Set

The dice arrangement shown in Figure 3.10 takes its name, the *3V set*, from the arrangement of the pips at the top of the dice: the two 3s are set in a V formation. In the configuration shown, the long faces show the sums 6 and 8, so this set, like the hardway set, is intended to reduce the percentage of 7s thrown. An advantage to the 3V set is that it can be set without a lot of searching for the key faces: just find the 3s and point them the right way.

FIGURE 3.10: Standard casino dice showing the 3V set.

All-Sevens Set

In the other direction, suppose you could control the dice to increase the probability p of rolling a 7. The Any 7 bet becomes a fair wager if

$$E = (4) \cdot p + (-1) \cdot (1 - p) = 5p - 1 = 0,$$

so you need to move the chance of a 7 to better than 1 in 5 to turn this bet in your favor. That's a lot further than in Example 3.27—this is the difference between making up a 1.51% HA and making up a 16.67% one.

An alternate way of setting dice to encourage 7s is the *all-sevens set*, which, as the name suggests, is designed to bring up more rolls of 7 than the 1 in 6 that arise from random rolling. Figure 3.11 shows a pair of dice in a version of all-sevens.

FIGURE 3.11: Standard casino dice showing the all-sevens set.

Setting dice in an all-sevens set is simple: just arrange the dice together, as with the hardway set, so the numbers on the long faces of the prism add up to 7 on each face [117]. This set is ideal for come-out rolls where 7 is an instant winner; if it turns up a point instead, a sharp dice controller will switch to the hardway set. This cuts down on the chance of sevening out. If, however, a shooter has multiple Come bets working on a come-out roll, the all-sevens set is a bad idea, as a roll of 7 will cause all Come bets to lose.

When perfectly executed, the all-sevens set drops the dice on the table showing either 5–2 or 4–3. This assumes that the 6 and 1 faces are set on the smallest faces of the prism, as they are placed with the hardway set. With this arrangement of the dice, the ideal probabilities when the dice are rolled are the same as in Table 3.16; the different outcomes arise from non-ideal rolls. In particular, double pitches in the all-sevens set turn up hardway combinations, not 7s. On a come-out roll, these are not losing rolls.

Is This Possible?

While we have clearly established that the ability to control a pair of dice, even a small amount, can affect the house advantage in theory, it is reasonable to ask if this can be done under casino conditions and while in compliance with casino rules for how dice are thrown. There are several self-described "controlled shooters" who profess that the ability to achieve non-random rolls can be easily learned. At the same time, some gaming experts have questioned the validity of these claims. Most observers, whether advocates or skeptics, will concede that mastering controlled rolling is considerably more challenging than becoming proficient at card counting in blackjack.

Non-random rolling is a question of physics: can the movement of the dice be sufficiently controlled to keep random movement, such as bouncing, to a minimum? Is that minimum low enough to affect how the dice land, especially in light of casinos' requirement that a pair of thrown dice hit the far wall of the table?

While some advocates of rhythmic rolling report this level of success consistently, skeptics in the craps community question the accuracy of their claims [16]. Unlike card counting in blackjack, the effectiveness of purported non-random rolling methods is not easily investigated by computer simulation. The variables involved in simulating the throw of two dice include initial orientation, launch angle, initial linear velocity, angular velocity, springiness of the table padding and felt, angle of impact with the table, and angle of rebound from the walls. Other than initial orientation, which is controllable by the shooter, these are not predictable with enough precision to construct sufficiently accurate simulations of real dice behavior. This forces us to rely on repeated trials of shooters rolling dice looking for statistically significant effects and hoping that success in a controlled trial translates into equal or greater success in a casino.

Example 3.28. Leaving the physics aside, what fraction of 7s under the expected $\frac{1}{6}$ would constitute evidence that controlled shooting to avoid rolling a 7 is effective?

In a sequence of n rolls of 2d6, the number of 7s is a binomial random variable with probability $p = \frac{1}{6}$. If the dice are rolling randomly, the expected number of 7s would be $np = \frac{n}{6}$. For convenience, assume a sample of 600 throws; the expected number of 7s would then be 100. A histogram of the results would be approximately bell-shaped and symmetric about the theoretical mean of 7. The standard deviation, which controls the precise shape of the histogram, would be

$$\sigma = \sqrt{np(1-p)} = \sqrt{600 \cdot \frac{1}{6} \cdot \frac{5}{6}} \approx 9.13.$$

The Empirical Rule states that approximately 95% of the rolls will be within two standard deviations of the mean: in the interval $100 \pm 2 \cdot 9.13$, or $[82, 118]$, where the interval is rounded inward to the nearest integer. Since the graph is symmetric, the outstanding 5% of the rolls is distributed equally between trials with fewer than 82 7s and trials with more than 118 7s. It follows that if the number of 7s in 600 rolls is fewer than 82 or greater than 118, we may reasonably conclude, with a high level of confidence, that the dice are not rolling randomly. ∎

At 81 or fewer 7s in 600 rolls, the proportion of 7s is less than or equal to 1 in 7.41; at 119 or greater, the proportion of 7s is at least 1 in 5.04. Based on our calculations above, we note that an abundance or deficiency of 7s need not be statistically significant, defined as 2 or more standard deviations removed from the mean, to represent a possible betting advantage. If we seek a ratio of 7s to rolls of 1 in 6.17, as suggested by Example 3.27, then the corresponding number of 7s in 600 tosses is 97 or fewer. This sits a mere .3043 standard deviations below the expected mean of 100, and the probability of a shooter achieving this ratio of 7s to rolls, or less, is about 38.05%.

This is not terribly unlikely, which suggests that advantage play at the craps table may be possible more frequently that some might think. Of course, this is after-the-fact analysis. The true challenge would be developing the ability to manipulate the dice to alter the probabilities and to bet appropriately on those rolls while it's happening.

Casinos bar blackjack card counters when they discover them; controlled craps shooters face considerably less scrutiny and are barred far less often. So long as they consistently hit the far wall when shooting, careful rhythmic rollers can evade casino attention. Mathematics and physics aside, this may be the strongest indicator of the relative strength of these two techniques and the concerns that they raise among casino management.

3.6   Variations

Crapless Craps

Crapless craps is a variation on craps that was developed by Bob Stupak, at the Vegas World Casino. The game continues at the Strat, Vegas World's successor; the Plaza in downtown Las Vegas also offers this variation. Crapless craps gets its name from its elimination of the losing numbers and one winning number on the come-out roll. If a 2, 3, 11, or 12 is rolled on the come-out roll, players neither lose nor win. These numbers become points like any other point, and the shooter must roll that number again before rolling a 7 to win. A 7 on the come-out roll remains an automatic winner.

To make up for this change in rules, the Don't Pass bet is not offered at crapless craps—you cannot bet against the shooter. On the face of it, this looks like a good deal for dice players: four automatic losing come-out rolls (1-1, 1-2, 2-1, and 6-6) are now points with a chance to win, while only two automatic wins (5-6 and 6-5) have been converted to possible losers. The mathematics tells a different story, though. For the various possible point numbers, Table 3.17 shows the probability of rolling those numbers on the come-out roll and then going on to win a Pass line bet with that point.

TABLE 3.17: Craps probabilities

Point	P(Point rolled)	P(Point wins)	P(Win on this point)
2 or 12	$\dfrac{1}{36}$	$\dfrac{1}{7}$	$\dfrac{1}{252}$
3 or 11	$\dfrac{2}{36}$	$\dfrac{2}{8}$	$\dfrac{1}{72}$
4 or 10	$\dfrac{3}{36}$	$\dfrac{3}{9}$	$\dfrac{1}{36}$
5 or 9	$\dfrac{4}{36}$	$\dfrac{4}{10}$	$\dfrac{2}{45}$
6 or 8	$\dfrac{5}{36}$	$\dfrac{5}{11}$	$\dfrac{25}{396}$

There are some important changes to the game's mathematics: the probability of winning after a come-out roll of 2 or 12 has risen from 0 to $\frac{1}{7}$, and the probability of winning after a 3 has risen from 0 to $\frac{1}{4}$. However, the probability of winning after an 11 has fallen from 1 to $\frac{1}{4}$. The player has given up far more than he has gained. Adding everything up reveals that the probability of winning a Pass line bet at crapless craps is

$$2 \cdot \left(\frac{1}{252} + \frac{1}{72} + \frac{1}{36} + \frac{2}{45} + \frac{25}{396} \right) + \frac{1}{6} = \frac{6557}{13,860} \approx .4730,$$

and thus that the house advantage has risen to approximately 5.39%—making this game about as advantageous for the casino as American roulette.

Example 3.29. The Strat Casino offers 10X odds on its crapless craps tables.

a. Remembering that odds bets are paid off at true odds with no house advantage, find the payoff odds for an odds bet when the point is 2 or 12.

   Only 7 of the 36 possible rolls matter: the one way to roll a 2 (or a 12) and the 6 ways to roll a 7, so the odds against this event are 6–1. An odds bet should pay 6–1 when the point is 2 or 12.

b. Find the payoff odds for an odds bet on 3 or 11.

   Eight rolls matter this time. The probability of rolling a 3 (or an 11) before a 7 is 2/8 or 1/4, so the odds against this event are 3–1. With a 3 or an 11 as the point, an odds bet should pay 3–1.

c. Find the house advantage of a $5 pass line bet backed up with maximum 10X odds.

   Using Table 3.17 and the prevailing odds for odds bets gives

$$E = (5) \cdot \frac{6}{36} + (305) \cdot \frac{2}{252} + (155) \cdot \frac{2}{72} + (105) \cdot \frac{2}{36}$$
$$+ (80) \cdot \frac{4}{45} + (65) \cdot \frac{50}{396} - (55) \cdot \frac{7303}{13,860}$$
$$\approx -.2691.$$

   Dividing by the $55 wager yields a HA of .489%. ∎

At the Hollywood Casino in Columbus, Ohio, this game is called *Craps Free Craps* and while still eliminating Don't Pass and Don't Come bets, allows a full range of other craps wagers, including several not printed on the layout and also available at a standard craps table. One such wager is a *Put* bet, which allows a craps player to designate any number except 7 as his or her point and treat the subsequent rolls as though they followed a come-out roll. Put bets are similar to Come bets without the need to establish a point by rolling dice; they also eliminate both the possibility of winning and the risk of losing (in standard craps) on the come-out roll.

Put bets pay even money, which distinguishes them from place bets. A Put bet on 6 or 8 has a win probability of $\frac{5}{11}$ and a HA of 9.09%. Making a Put bet on the 2 or 12, where the chance of winning is only $\frac{1}{7}$, has a HA of 71.4%—worse than buying most state lottery tickets.

Example 3.30. The power of a Put bet only shows when it can be immediately backed up with free odds, which is the case at the Hollywood. The odds portion of the combined wager pays off at 6–1 on the 2 or 12, but this does little to bring the high HA on the main wager down to a reasonable level unless a huge odds bet is made. With a $5 odds bet backing a $1 Put bet, the HA only falls to 11.9%. If a $1 Put bet on the 2 is backed up with $100\times$ odds of $100 (where permitted), the expectation is

$$E = (601) \cdot \frac{1}{7} + (-101) \cdot \frac{6}{7} = -\$\frac{5}{7},$$

which drops the HA to .707%. However, the variance on a sequence of these bets is huge, and it is possible that a player's bankroll will not withstand very many $101 losses on this low-probability event. ∎

As a general rule, it should be clear that a Put bet without substantial odds (perhaps $10\times$ at a minimum) is an inferior wager; moreover, Put bets should probably be avoided on the less-common numbers. A Put bet on 6 or 8, with good odds backing it up (Exercise 3.18), might be a sensible choice—beyond that, there are more reasonable bets available.

A better option than the Put bet for a rare event like the shooter rolling a 2 before a 7 is a *Buy* bet. A craps free craps player who buys the 2, betting that a 2 will be rolled before a 7, receives true odds of 6–1 on a win but must pay a 5% commission on the amount of his or her winnings. This is betting on the same event as the Put bet, but at a far smaller disadvantage. Buying the 2 for $1 has an expected value of

$$E = (6 \cdot .95) \cdot \frac{1}{7} + (-1) \cdot \frac{6}{7} = -\frac{.3}{7} \approx -.0429,$$

and so it carries a HA of 4.29%.

Example 3.31. Buying the 8 in this manner pays off at 6–5 with a 5% commission taken, for an effective payoff of 5.7–5, or 1.14–1. The expectation is

$$E = (1.14) \cdot \frac{5}{11} + (-1) \cdot \frac{6}{11} = -\frac{.3}{11} \approx -.0273.$$

∎

It should be noted that many bets at Crapless Craps are identical to their analogs on a standard craps table. Any bet that doesn't involve a come-out roll can be played under the same terms; this excludes only the Pass and Come bets. An example would be a Place bet on the 5, which is unaffected by the change in the come-out roll. For better or for worse, though, these unchanged bets tend to hold more player money by way of a higher HA than Pass/Don't Pass and Come/Don't Come wagers on the standard craps table.

Turn & Burn Craps

Turn & Burn Craps, was launched at the River Rock Casino in Geyserville, California, and is one of various workarounds have been devised to develop games resembling craps that can comply with California law by not using dice [151]. Turn & Burn is an electronic dice game with a pair of computer-generated dice and a selection of wagers that parallels many of the bets on a craps table, with the important change that every Turn & Burn bet is a one-roll bet. There are no Pass/Don't Pass or Come/Don't Come spaces available on the layout, and every roll of the simulated dice resolves all bets one way or the other. Effectively, this makes Turn & Burn Craps an electronic game, more akin to video poker or a slot machine under California law than to traditional craps.

Since Pass/Don't Pass and Come/Don't Come are among the best bets for the player in any casino, Turn & Burn does not simply repeat simple one-roll craps bets at the same odds, but changes the payoffs slightly to make the wagers on offer slightly more friendly to the gambler.

Example 3.32. On a standard craps table, the "Any Seven" bet that the next roll will be a 7 is one of the worst on the board. It pays off at 4–1 and carries a 16.67% house advantage. The corresponding Turn & Burn bet has the same winning condition, but the payoff is 4½–1, which cuts the house edge in half, to 8.33%—not great, but better for the gambler. Rounding is not an issue since all bets and payments are made electronically. ∎

Example 3.33. The Field bet at Turn & Burn works slightly differently than the craps Field bet. In standard craps, the field covers the numbers 2, 3, 4, 10, 11, and 12, typically paying 1–1 on all numbers except one of 2 or 12, which pays off at 2–1 or 3–1. Turn & Burn puts the same six sums in the field, but pays 2–1 on 2, 3–1 on 12, and even money on the other winning rolls. This drops the HA by a factor of 3: from 8.33% to 2.78%. ∎

Other Turn & Burn one-roll bets are compared to their craps counterparts in Table 3.18.

TABLE 3.18: Craps and Turn & Burn one-roll wagers compared

Wager	Craps Payoff	Craps HA	Turn and Burn Payoff	Turn and Burn HA
2	30–1	13.89%	33–1	8.33%
12	30–1	13.89%	33–1	8.33%
3	15–1	11.11%	16–1	5.56%
11	15–1	11.11%	16–1	5.56%
Any Craps	7–1	11.11%	7½–1	5.56%
Hop	30–1	13.89%	33–1	5.56%
Turn	15–1	11.11%	16–1	5.56%

Several new bets have been added to the Turn & Burn layout.

- *Hop bets* that allow the player to bet that the next roll will be a specified double have dedicated spaces on the table. They pay off at 33–1. On a standard craps table, hop bets may be available on player request.

- A collection of *Turn bets* are bets that the next roll will be a specified non-double—for example, 2–3 or 5–1. These pay 16–1. A conventional craps game also calls these hop bets, making no distinction between calling a double and calling a non-double.

- *Any Pair* collects all six doubles into one winning outcome. If the next roll is doubles, the bet pays 4½–1.

- *Big 6* and *Big 8* have been instituted as one-roll bets rather than a wager that 6 or 8 will be rolled before a 7, as in standard craps (see Exercise 3.4). These bets pay 6–1 if the next roll is a 6 or an 8, respectively.

In the final analysis, Turn & Burn bets are, in general, more favorable to the player—while still carrying sizable house advantages—than their analogs on a craps table. One notes, of course, that the most player-favorable craps bets—Pass, Don't Pass, Come, and Don't Come—are not present. It's unlikely that any casino would offer both games, though, so this difference is not likely to be of much practical use.

Pentagon Craps

In 1990, Bernard Bereuter received U.S. patent #US4900034A for 7-sided dice and the necessary other equipment for a game called *Pentagon Craps*. The game drew its name from the shape of the dice, which were pentagonal prisms. The pentagon-shaped ends of the dice fit within a 1-inch diameter circle and measured .588 inches (1.494 cm) on a side; the rectangular faces measured .588 by .753 inches (1.913 cm). These dimensions, it was said, would ensure that all 7 faces were equally likely when the dice were tossed on a craps table [96]. Two of these dice are shown in Figure 3.12.

FIGURE 3.12: Seven-sided dice used in Pentagon Craps.

The numbers 1–5 were inscribed on the edges of the prism rather than on faces so that one number would face upright when the die landed on a rectangle; 6 and 7 were assigned to the pentagonal faces.

Pentagon Craps is played, appropriately, on a pentagonal layout divided into 6 concentric five-sided rings. The equivalent of Pass Line bets are placed on the WIN-LINE located on the outermost ring. The come-out roll is resolved as follows:

- Rolls with either or both dice landing on a pentagonal face (one-die rolls of 6 or 7) are disregarded. This includes all rolls of 11 or higher. The dice are then thrown again.

- Rolls of 6 or 10 with both dice landing on a rectangular face are immediate winners. Rolls of 6/4 and 7/3 do not win or lose.

- Rolls of 2, 3, or 9 are immediate losers. Since they involve one die landing on a pentagon, rolls of 6/3 and 7/2 don't lose, and the come-out roll is repeated.

- Rolls of 4, 5, 7, or 8 with both dice landing on a rectangular face establish a point. The player must then reroll the point before throwing a 6 to win the bet.

Under these rules, 25 of the possible 49 ways for the two dice to land count on the come-out roll [91]. The numbers of ways that the various totals from 2–14 can be rolled and count are in Table 3.19.

TABLE 3.19: Pentagon Craps: Come-out roll possibilities

Roll	2	3	4	5	6	7	8	9	10	11	12	13	14
Count	1	2	3	4	5	4	3	2	1	0	0	0	0

The Don't Pass bet is one ring inside the WIN-LINE on what is appropriately called the LOSE-LINE. These bets lose when the WIN-LINE wins and win when it loses, except that the LOSE-LINE pushes on an initial roll of 3. This preserves an edge for the casino on the LOSE-LINE bet. Winning WIN-LINE and LOSE-LINE bets pay even money.

As in standard craps, one may designate the next roll as a come-out roll at any time by making a Come or Don't Come bet. These are resolved in accordance with the rules for the WIN-LINE and LOSE-LINE.

Example 3.34. What is the probability of winning a WIN-LINE bet?

We will focus only on the rolls that matter, so the denominator in many of our calculations will be 25 rather than 49. The bet can be won on the come-out roll with a 6 or 10. There are five ways to roll a 6 and one way to roll a 10 without using a pentagonal face—rolling a 5–5—so this probability is $\frac{6}{25}$.

To this we must add the probability of rolling a point number on the

TABLE 3.20: Pentagon Craps probabilities

Point	P(Point rolled)	P(Point wins)	P(Win on this point)
4	3/25	3/8	9/200
5	4/25	4/9	16/225
7	4/25	4/9	16/225
8	3/25	3/8	9/200

come-out roll and then successfully making that point. These probabilities are collected in Table 3.20.

The sum of the various winning probabilities is $\frac{17}{36} \approx .4722$. ∎

The probability of winning a LOSE-LINE bet is then $1 - \frac{17}{36} - P(3)$, since an initial roll of 3 is a push on a LOSE-LINE bet but is a loss on a WIN-LINE bet. The probability of rolling a 3 at Pentagon Craps is $\frac{2}{49}$, making the probability of winning a LOSE-LINE bet

$$1 - \frac{17}{36} - \frac{2}{49} = \frac{859}{1764} \approx .4870.$$

Comparing the HAs on the WIN-LINE and LOSE-LINE bets shows that the former has a HA of 5.56% and the latter's HA is 2.61%. As in standard craps, betting against the shooter has the lower HA, though the difference is more significant here.

The remaining rings on the layout are for placing one-roll bets. For these wagers, no rolls are ignored.

- Bets on a specified double from 1–1 through 7–7 appearing on the next roll pay off at 40–1 if they win.

- One may bet that the next roll will be a 6 or a 7 together with any other specified different number. This pair of bets each pays 20–1.

- A bet on "Any Pair" pays 5–1 when doubles of any number are rolled.

- The "6" bet pays off if the next roll sums to 6, paying 7–1.

- A double pentagon bet that both dice will land on a pentagonal side pays 10–1. This covers rolls of 12, 13, and 14.

  A $1 double pentagon bet has an expected return of

  $$E = (10) \cdot \left(\frac{2}{7}\right)^2 + (-1) \cdot \frac{45}{49} = -\frac{5}{49} \approx -\$0.1020.$$

- Betting that the next roll will include either a 6 or 7 together with any other number pays 3–1.

- The even-money Rectangle Bet pays when both dice land on a rectangular face, showing a number from 1–5, except when the dice show a 1–2, which pushes.

The probability of winning a Rectangle Bet is

$$\left(\frac{5}{7}\right)^2 - \frac{2}{49} = \frac{23}{49},$$

slightly less than ½. Its house advantage is 6.12%.

- The Pentagon Bet pays even money if one or both dice land on one of the two pentagonal faces.

This bet is best evaluated using the Complement Rule. The probability that both dice land on rectangular faces is

$$\left(\frac{5}{7}\right)^2 = \frac{25}{49},$$

so the probability of winning a Pentagon Bet is

$$1 - \frac{25}{49} = \frac{24}{49} \approx .4898.$$

Since the chance of winning this even-money bet is close to ½, the HA is relatively low, about 2.04%.

Pentagon Craps was marketed as a board game for home play, but never made many inroads into casinos.

Spider Craps

Craps using 8-sided dice was originally suggested by Richard Epstein in 1977 under the name "sparc" [40]. *Spider Craps* is a variation of craps that, like Pentagon Craps, used non-cubical dice: eight-sided regular polyhedra. Figure 3.13 displays two such dice, which are called *octahedra*. The game was developed by Jacob Engel in 2011.

FIGURE 3.13: Eight-sided dice used in Spider Craps.

Since the faces of 8-sided dice are all congruent equilateral triangles, the rules of Spider Craps do not exclude any rolls on the come-out or subsequent rolls, as is done in Pentagon Craps. The Pass line bet is resolved in direct analog with the standard craps wager:

- The bet wins if the come-out roll is 9 or 15—that is, if the most common or second-highest sum is rolled.

- The bet loses if the come-out roll is 2, 3, or 16: the lowest, second-lowest, and highest sums.

- Any other result of the come-out roll becomes the point, and the shooter rerolls until the point is rolled again or a 9—again, the most common sum—is rolled. Possible points are 4, 5, 6, 7, 8, 10, 11, 12, 13, and 14. The bet wins if the point is rerolled before a 9 occurs.

There are 8 ways to roll a 9 on 2d8, so if the point number can be rolled in k ways, the probability of winning a Pass bet once that point has been established is

$$\frac{k}{k+8} < \frac{1}{2}.$$

It's also possible to win this bet on the come-out roll; that has probability $\frac{10}{64}$.

Multiplying these probabilities by the chance of establishing each point and then summing across all ten point numbers, as was done for Pentagon Craps in Example 3.34, gives

$$P(\text{Win}) = \frac{10}{64} + 2\left[\frac{3}{64}\cdot\frac{3}{11} + \frac{4}{64}\cdot\frac{4}{12} + \frac{5}{64}\cdot\frac{5}{13} + \frac{6}{64}\cdot\frac{6}{14} + \frac{7}{64}\cdot\frac{7}{15}\right]$$

$$= \frac{37,319}{80,080} \approx .4660.$$

This gives a HA of 6.80%—which might drive players away.

In an effort to make this bet more attractive, a casino might consider paying off winning bets at 11–10 odds, which would cut the HA to 2.14%. This could be advertised as "10% Bonus On Winning Pass Line Bets!". There is some risk that this may be a little too complicated for dealers to administer, though. Paying 6–5 on winning bets gives the player a 2.52% advantage, and so is not a viable option.

We turn next to the Spider Craps Don't Pass wager. In light of the increased HA on the Pass line, reversing the Pass rules for Don't Pass would give the player a 6.80% edge, which rules that right out. The rules of the Don't Pass bet need to be changed in a way that gives the casino a small edge—large enough so the house makes a profit, but not so large that players are driven away.

The direct analog of the standard Don't Pass bet would be

- Win if the come-out roll is 2 or 3.

- Lose if the come-out roll is 9 or 15.

- Push if the come-out roll is 16, just as a craps Don't Pass bet pushes on an initial roll of 12.

- If a point is established on the come-out roll, reroll until the point or a 9 is rolled. If the point is rerolled before a 9, Don't Pass loses; if a 9 turns up first, the bet wins.

It turns out that excluding only 1 roll in 64 does not go far enough in balancing the game for the casino. Under these rules, we have

$$P(\text{Win at Don't Pass}) = \left(1 - \frac{37{,}319}{80{,}080}\right) - \frac{1}{64} \approx .5184.$$

The player would have the advantage, so once again, we reject this rule.

If both the 2 and the 16 are barred on the come-out roll, the probability of winning only falls to .5027, so this is also not enough. If instead the 2 and 3 are barred, the new Don't Pass wager wins with probability

$$\left(1 - \frac{37{,}319}{80{,}080}\right) - \frac{3}{64} = \frac{156{,}029}{320{,}320} \approx .4871.$$

This gives a HA of 2.58%—an edge that is reasonable without being overwhelming. The rules for the come-out roll on a Spider Craps Don't Pass bet are thus

- Win instantly on a 16.

- Lose immediately on a 9 or 15.

- Push if the come-out roll is 2 or 3.

Any other value becomes the point, and the Don't Pass bettor wins if the shooter rolls a 9 before re-rolling it.

This is an even money bet with a HA of 2.58%. For other Spider Craps bets, the payoff is part of the consideration. In designing a wager, several criteria are important.

- The bet should be attractive to players. A 55–1 payoff might trigger a "Wow!" response and inspire some table action.

- From a non-mathematical perspective, the new rules should be both easy for players to understand and easy for dealers to implement. A satisfactory HA for some side bet might be achievable by setting the payoff odds at 28.4 to 1, but that's too complicated for quick game play in a live casino. As with Turn & Burn Craps, though, implementing Spider Craps as an electronic game would permit such complicated payoff odds.

- As always, the HA should straddle the line between being lucrative for casinos but not so lucrative that players reject the bet.

152   *Craps*

Example 3.35. The one-roll Field bet at craps is effectively a wager that one of the less-common sums will be rolled. The exact rules can vary slightly among casinos; we shall consider the version that pays even money if the next roll is 3, 4, 9, 10, or 11 and 2–1 if the next roll is a 2 or 12. These rules produce a bet with a HA of 5.56%.

In adapting this wager to Spider Craps, we have two parameters that can be adjusted: which numbers are included as part of the field of winning rolls and which rolls might qualify for a bonus payoff exceeding 1–1. While the Field bet pays off on 7 of the 11 possible sums of 2d6, the probability of winning is nonetheless less than ½, and this should guide our choice of Spider Craps field numbers. Choosing to pay off on rolls from 2–6 and 12–16 provides a symmetric array of winning rolls (easy for dealers) and a .4688 probability of winning (attractive for players).

The HA depends on the payoffs. If the payoff is 1–1 on all winning numbers, the resulting HA is 6.25%. While this is close to the 5.56% HA on a craps Field bet, players who are accustomed to one or two bonus numbers that pay off at better than even money might be reluctant to move their action over to a Spider Craps table. By paying 2–1 on either the 2 or 16, the HA drops to 4.69%; a 2–1 payout on both 2 and 16 results in a 3.13% edge. This same 3.13% could also be achieved by paying 3–1 on either 2 or 16, but not both—a triple payout on both 2 and 12 makes this a fair wager. ∎

Example 3.36. "Big Red" (page 112) is a one-roll craps bet that the next roll will be a 7. The bet is known for a high HA: it pays off at 4–1 and carries a 16.67% casino edge. Spider Craps replaces the 7 by a 9—but what should the payoff be?

Suppose that the bet pays off at X to 1. The expected value of a $1 wager is then

$$E = (X) \cdot \frac{1}{8} + (-1) \cdot \frac{7}{8} = \frac{X-7}{8}.$$

From the casino's perspective, we'd like this to be negative, but not *too* negative. Table 3.21 shows the HA for several plausible values of X.

TABLE 3.21: Spider Craps Big Red payoff options

X	HA
4	37.5%
5	25%
5.5	18.75%
6	12.5%
6.5	6.25%
7	0%

The 6–1 payoff gives a bet that is better for the player than Big Red at standard craps—though it's still a pretty bad wager. ∎

The appeal of Spider Craps may lie in these other bets. Resolving a standard craps pass line bet takes an average of $\frac{557}{165} \approx 3.38$ rolls (page 108); for Spider Craps, this number rises, to 4.14 rolls. Gamblers want something to bet on—and casinos need something to take bets on—while they wait. Some other Spider Craps side bets, with comparable craps wagers, are listed in Table 3.22.

TABLE 3.22: Spider Craps side bets

Side bet	Payoff	Craps payoff	HA	Craps HA
2	55–1	30–1	12.50%	13.89%
3	28–1	15–1	9.38%	11.11%
15/Yo	28–1	15–1	9.38%	11.11%
16/12	55–1	30–1	12.50%	13.89%
Any Craps	14–1	7–1	6.25%	11.11%

Devil Dice

Devil Dice is a proposed variation on craps, incorporating a progressive jackpot, that is novel in its use of 3 dice. The standard version of the game only uses 3 dice on come-out rolls; a deluxe version rolls 3 dice on every roll and includes some new betting options based on 3-die combinations [70].

Standard Devil Dice

In the simpler version of Devil Dice, a shooter preparing for a come-out roll is handed 3 dice, with one of a different color than the other two. Suppose that the two ordinary dice are colored red and that the third is blue, for convenience. The shooter rolls all 3 dice. If the result is anything other than three 6s, the blue die is ignored and play continues as usual for craps. If the result is three 6s, then play temporarily halts while the dice are replaced by 3 gold dice, which the shooter then rolls. If the gold dice also show three 6s, then a progressive jackpot is equally divided among all participating players. The probability that the jackpot is awarded is

$$\left(\frac{1}{216}\right)^2 = \frac{1}{46,656}.$$

Following the resolution of the jackpot roll, the game resumes, using the 12 rolled on the two red dice as the first roll's outcome. The blue die is set aside and remains out of play until the next come-out roll.

In this configuration, all of the usual craps bets remain available at their customary payoff odds. No payouts are changed to accommodate the cost of the progressive jackpot; instead, players must make a $1 progressive wager together with a Pass or Don't Pass wager to be eligible for a share of the jackpot.

Suppose that there are n people at the table. How high must the jackpot J be for this wager to have a player advantage?

The expected value E of the \$1 progressive is

$$E(n, J) = \left(\frac{J}{n}\right) \cdot \frac{1}{46,656} + (-1) \cdot \frac{46,655}{46,656} = \frac{J - 46,655n}{46,656n}.$$

It is then easily seen that this is positive if $J > 46,655n$. A full craps table, part of which is shown in Figure 3.2 (page 107), has space for 14 players, so a gambler would have the edge at a full table only if the jackpot exceeds \$653,170.

Positive expectation, of course, is only part of the story of this jackpot. The variance of this bet, assuming a break-even jackpot ($E = 0$), is $\sigma^2 = 46{,}653.00$ dollars squared, and the standard deviation is $\sigma \approx \$215.99$. While the player may have a fair bet, or better, with a large jackpot, the very low probability of winning still makes this bet highly volatile and thus inadvisable.

Deluxe Devil Dice

Deluxe Devil Dice throws three dice on every roll. The progressive jackpot of the standard game remains, but it is joined by a collection of new wagers that take advantage of the extra d6. Some of these bets are taken from the Asian dice game *sic bo*, in which 3d6 are spun in a wire cage and players may bet on various combinations and sums that appear. A common feature of many of these 3-die bets is that they lose if the two red dice sum to 7, much like craps bets that call for a certain sum to be rolled before a 7 [65].

- *Triple Combination* bets can be made on any of the 6 triples: 1–1–1 through 6–6–6. These bets win if the selected triple is rolled before the red dice add up to 7. *Any Triple Combination* allows a bet that any one of the 6 triples will be rolled before a red sum of 7.

- *Straight* bets, like Triple Combination bets, are bets that one or all of the 4 straights 1–2–3, 2–3–4, 3–4–5, or 4–5–6 will be rolled before a red sum of 7. Note that if 3–4–5 is rolled with the 3 and 4 on the red dice, this bet loses. 3–4–5 rolled any other way before a red sum of 7 wins. These can be modified to specify that the blue die appear in the lowest, middle, or highest position of the straight.

  This class of bets may also include *skip straights*: 1–3–5 and 2–4–6, at the casino's option.

- *Red Sum = Blue* bets, which pay off if the red sum equals the single blue die. Since the blue die cannot show a 7, there is no possibility that this bet appears to win and lose on the same roll. These bets can be made on particular blue numbers from 2–6 or may cover all blue rolls in one bet.

As with the craps Pass and Don't Pass bets, we can assess these wagers by considering only the 3-die rolls that resolve them, ignoring all neutral rolls.

Example 3.37. For a bet on a specific triple, there is one winning roll and 36 losing rolls. The losing rolls are counted by noting that there are 6 ways to roll a sum of 7 on the red dice. Each one may be combined with 6 numbers on the blue die, and there are no triples with a red sum of 7. The probability of winning a bet on a specific triple is then $\frac{1}{37}$. Paying this bet at 35–1 gives the same 2.70% HA we see in European roulette. At 30–1, the HA rises to 16.22%.

For Any Triple Combination, there are 42 rolls of interest: 6 winners (one of which also triggers the progressive jackpot roll if it occurs on the come-out roll) and the same 36 losers. The probability of winning is now $\frac{1}{7}$. A 5–1 payoff gives the casino a 14.29% edge. ∎

Any of these wagers may also be made as a one-roll bet, with appropriately different payoff odds.

Example 3.38. If we recast Any Triple Combination as a one-roll bet, all that matters is whether or not a triple appears. This is the "Any Triple" wager from sic bo [9]. There are 6 winning rolls and 210 losing rolls. The bet pays 30–1 and has a HA of 13.89%. ∎

For the Red Sum = Blue bets, the probability of winning depends on the blue number. Consider a bet on 5. Of the 216 ways for 3d6 to land, only 36 even show a 5 on the blue die. Of those, there are 4 ways for the red dice to add up to 5, making the probability of winning $\frac{4}{216} = \frac{1}{54}$. A 50–1 payoff would be both easy to handle for dealers and attractive to players. The resulting expected value would be

$$E = (50) \cdot \frac{1}{54} + (-1) \cdot \frac{53}{54} = -\frac{3}{54} = -\frac{1}{18},$$

and we see the 5.56% HA which is familiar from the craps Field bet.

3.7 Card Craps

Origins

Another craps variation in states where dice may not be used to determine the final outcome of a wager is to use playing cards to simulate a roll of two dice. Paramount among game designers' goals in adapting games for California and Oklahoma is that the odds underlying the games must be the same whether they're played with traditional equipment or with cards. This can be tricky in some craps variations that use a multi-deck shoe of cards to simulate die rolls. However, the playing card approach has resulted in some bets added to the classic games that make use of the cards in new ways.

Card craps got its start in Jersey City, New Jersey in 1945 [112]. In the original version of this game, a deck of 48 cards, two aces through 6s in all 4 suits, was used. Two cards were dealt to the table and their values added to generate a craps roll. Subsequent rolls were found by returning the cards so dealt to the deck, reshuffling, and redealing—while this ensured that the probabilities were consistent from roll to roll, the frequent shuffles slowed the game down somewhat.

Example 3.39. One might reasonably ask, however, if dealing two cards at a time changed the probability of rolling doubles, which is $\frac{1}{6}$ with a pair of dice. In this version of Card Craps, the probability of doubles is

$$\frac{48}{48} \cdot \frac{7}{47} = \frac{7}{47} \approx .1489,$$

a value which is about 10.5% lower than $\frac{1}{6}$.

If a 72-card deck with 3 cards of each rank and suit combination is used, as is advised when a large group is playing Card Craps [112], the probability of doubles is

$$\frac{11}{71} \approx .1549,$$

which is 7.04% lower than $\frac{1}{6}$. Only in an infinite deck approximation, or a game where the first card is returned to the deck before the second card is dealt, can we reach $P(\text{Doubles}) = \frac{1}{6}$.

For any craps wager where the exact numbers on the dice did not matter, this has a minimal effect on gameplay. ∎

Players were permitted to bet both with and against the dealer once a point was established. These *right* and *wrong* bets, respectively, were made either taking or laying odds, as the game favors the player failing to make a point by drawing a 7 before repeating a point draw. Odds for the right and wrong bets are shown in Table 3.23.

TABLE 3.23: Card Craps odds

Point	Right bet odds	Wrong bet odds
4, 10	2–1	1–2
5, 9	3–2	2–3
6, 8	6–5	5–6

The cards inspired a new game concept, the *double hardway*. A player making a right bet on a dealer facing a even-numbered point of 4, 6, 8, or 10 received double the standard payoff if the dealer made the point by dealing two identical cards, as when a 6 is dealt as 3♣ 3♣, for example [112]. Payoffs on a 4 or 10 doubled to 4–1, while a 6 or 8 made the double hardway paid off at 12–5.

Example 3.40. Including the double hardway bonus, what is the house edge on a right or wrong bet with the dealer trying to make a 10?

Once the point of 10 has been established, the only rolls that matter are 10 and 7. The 10 may be dealt in $\binom{8}{2} = 28$ ways as a pair of 5s; 4 of those ways have two cards of the same suit. There are $8 \cdot 8 = 64$ ways to deal a 10 consisting of a 4 and a 6, and $(48 \cdot 8)/2 = 192$ ways to deal a 7. The relevant probabilities for a right bet on 10 are then

$$P(\text{Win with double hardway}) = \frac{4}{284} = \frac{1}{71}$$
$$P(\text{Win without double hardway}) = \frac{88}{284} = \frac{22}{71}$$
$$P(\text{Lose}) = \frac{192}{284} = \frac{48}{71}.$$

The expected value of a \$1 right bet on 10 is then

$$E = (4) \cdot \frac{1}{71} + (2) \cdot \frac{22}{71} + (-1) \cdot \frac{48}{71} = 0,$$

making this a fair wager.

For the wrong bet on 10, the payoff odds from Table 3.23 call for the wrong bettor to pay an extra betting unit if the dealer makes the 10 on a double hardway deal. Mathematically, the payoffs are recomputed, but the probabilities are the same. The expectation

$$E = (-2) \cdot \frac{1}{71} + (-1) \cdot \frac{22}{71} + (.50) \cdot \frac{48}{71}$$

remains 0.

The same calculations apply to a right or wrong bet on 4, which are also fair. ∎

In light of this fairness, a Card Craps operator, or *book*, levied a charge, the *vigorish* or "vig", on some or all bets. This was the source of the book's profit. For example, the book might charge 5% on right bets only, or charge 3% on all bets [112].

Modern Variations

Prior to 2018, some Oklahoma casinos offered craps variations using cards. Each game preserved standard craps probabilities by frequent reshuffling of the cards that determined the outcomes.

- At the Downstream Casino in Quapaw, casino officials sought to design a dice-free version of craps that still allowed active player participation.

Their game used two rows of 6 cards each, ace through 6 dealt face down, and asked gamblers simply to call out two numbers between 1 and 6. The numbers so called corresponded to cards which were turned over and constituted the roll. The game was installed at the Downstream in 2010 [59].

- Two rows of cards, of different colors, were also dealt in Durant at the Choctaw Casino. Players have no choice of cards: the first card of each color was used to build the roll [24].

- Card-Based Craps at the Winstar World Casino in Thackerville employed a 36-card deck consisting of the aces through 9s from a standard deck. Each card was barcoded and mapped to one of the 36 possible outcomes when two standard dice are rolled. For each simulated roll, three cards were dealt to the game table; the middle card was scanned and used as the roll [23]. The deck was reshuffled after each roll.

All three game variations were replaced by traditional craps using dice once Oklahoma law was changed in 2018.

In California, card rooms have also branched out into games that simulate dice with cards. In 2014, the 101 Casino in Petaluma dealt a game called *Party Craps*. This version of craps simply uses a deck consisting of as many as 432 cards: the aces through 6s from 1 up to 18 decks. The number of decks is not specified outside of that range. All standard craps wagers including Buy and Lay bets are available. Two cards dealt from the deck substitute for dice.

In moving from 2 decks to as many as 18, attention should be paid to reshuffling. A surplus of low cards coming out of the deck may shift the odds of a subsequent roll of 11 or 12 enough to give the player an edge—though since the HA on these bets is so high, the imbalance would have to be considerable. Consider the 30–1 bet that the next roll will be ⚅⚅. This bet has an expectation of

$$E = (30) \cdot \frac{1}{36} + (-1) \cdot \frac{35}{36} = -\frac{5}{36} \approx -.1389,$$

assuming that all results are equally likely.

In an 18-deck shoe, what excess of 6s over non-6s turns this expected value positive?

Let x denote the number of cards dealt from the shoe, and suppose that s of these cards are 6s. The probability of drawing two 6s is now

$$p = \frac{\binom{72-s}{2}}{\binom{432-x}{2}} = \frac{(72-s)(71-s)}{(432-x)(431-x)}$$

and the expected value is

$$E = (30) \cdot p + (-1) \cdot (1-p).$$

$E > 0$ whenever $p > \frac{1}{31}$.

If 34 cards are dealt without a 6 appearing, we have $p \approx .0324$ and $E = .0029$, so the player has a slim .29% edge. The probability of dealing 34 non-6s and no 6s is

$$\frac{\binom{360}{34}}{\binom{432}{34}} \approx .0023,$$

so this would occur approximately once in 437 shoes, on the average.

If we fix s and solve the equation $E = \frac{1}{31}$ for x, we find the minimum value of x for which a shoe with s 6s removed has positive expected value. The equation is quadratic; we choose the negative value of the square root in the quadratic formula to keep $x < 432$. The relevant solution is

$$x = \frac{863 - \sqrt{863^2 - 4 \cdot [186,192 - 31(72 - s)(71 - s)]}}{2}.$$

Table 3.24 shows the values of x corresponding to s from 0–10.

TABLE 3.24: Party Craps: Player edge for a 432-card shoe with s 6s and x cards removed

s	x	Player edge
0	34	0.29%
1	39	0.01%
2	45	0.23%
3	51	0.46%
4	56	0.17%
5	62	0.40%
6	67	0.10%
7	73	0.34%
8	78	0.02%
9	84	0.27%
10	90	0.53%

This relationship leads to a card-counting strategy. Cards from ace through 5 count +1 and 6s count –5, and this running count is maintained as each card is dealt. Convert the running count to a true count by dividing by the number of 24-card "decks" remaining—this part may require some careful retraining for a blackjack card counter accustomed to estimating full 52-card decks. If the true count reaches +3, the player has the edge provided that $s \leqslant 54$.

Casino defense against this counting technique is to deal the cards from a continuous shuffling machine. This was the strategy used in *Play Craps*, a version of Party Craps introduced in Nevada in 2009. The Viejas Casino outside San Diego hosted the game briefly. Play Craps uses 324 cards, each one emblazoned with one face of a d6, in a continuous shuffler which negates any card-counting strategies.

In addition to removing the threat posed by card counters, a CSM makes it viable to deal a card-based craps game with fewer decks. Of course, the more decks in use, the smaller the effect on the probabilities of removing the first card from the shoe. In Play Craps, the probabilities of the various bet outcomes are all within .004% of the probabilities occurring when dice are used [85]. Even this tiny effect can be eliminated by including a joker in the shoe. When dealt as the second card, the joker carries the value of the first card dealt. If the joker appears on the first card, it is disregarded and returned to the CSM.

Card Craps with Dice

The Card Craps games described so far eliminated the dice entirely. Another option is to retain the dice as an intermediary and use the die roll to identify cards which simulate a roll of two dice; this preserves the players' role in the game and the excitement of rolling dice. There are several ways in which casinos in California and Oklahoma map die rolls onto cards.

- At the Barona Casino in Lakeside, California, six cards from ace through 6 are shuffled and dealt face up to the table after every 7-out. The numbers rolled on the dice are then used to identify two cards; the card values are added to determine the result of the roll. Card values correspond to the same die rolls until the cards are redealt.

  Example 3.41. If the cards are dealt in the order $3\spadesuit, 2\spadesuit, 4\spadesuit, 5\spadesuit, A\spadesuit, 6\spadesuit$ and the dice land ⚁ ⚃, the craps roll uses the third and sixth cards and is then $4 + 6 = 10$. ∎

  This method amounts to a simple permutation of the set $\{1, 2, 3, 4, 5, 6\}$, and so does not change the game's probabilities.

- Harrah's Resort Southern California in Funner, California offers a game called *Rincon Craps*, named for the Rincon Band of Luiseño Indians who own the casino. Rincon Craps has craps shooters roll two dice, one red and one green. As at the Downstream, two rows of six cards each, Ace through 6 in red and green, are dealt face down before each roll. The numbers on the dice identify which cards are turned over and added. The cards are quickly shuffled and redealt after every roll.

  A similar game with red and blue dice is played at the Pechanga Casino in Temecula, California.

Other casinos use nonstandard dice to select the cards in play.

- The Firelake Grand Casino in Shawnee, Oklahoma had players roll a six-sided die with two 1s, two 2s, and two 3s. The roll of the die indicates how many cards are burned from the top of a deck. The next two cards determine the roll. Two decks are used on alternating rolls; they are shuffled between uses to preserve the probabilities.

- The Pala Casino uses a red die numbered $\{2, 2, 2, 5, 5, 5\}$ and a blue die numbered $\{3, 3, 3, 4, 4, 4\}$. Two cards from a special 36-card deck, which contains one card bearing each of the 36 ways in which two dice can land when rolled, are dealt: one each face down to a red space and a blue space. When the dice are rolled, the higher number determines which of the cards is flipped over.

A moment's reflection reveals that Pala's blue die is unnecessary. The card selection process can be streamlined by looking only at the red die:

> If the red die shows a 2, turn over the blue card.
> If the red die shows a 5, turn over the red card.

While this is certainly convenient for Pala's craps dealers, we might ask how many ways there are to renumber the red and blue dice so that the following criteria are met:

1. Only the numbers 1–6 are used.

2. No ties are possible.

3. Each die has a 50% chance of bearing the higher number when thrown.

4. Both dice need to be consulted on at least some rolls.

Definition 3.1. A pair of six-sided dice that satisfies these four criteria is called a *valid pair*.

In searching for valid pairs, changing the assignment of numbers to faces does not constitute a new die, so $\{2, 2, 2, 5, 5, 5\}$ and $\{2, 5, 2, 5, 2, 5\}$ are considered the same. For convenience, we list the numbers on a die in increasing order.

Example 3.42. One valid pair would have a red die numbered $\{1, 3, 3, 3, 5, 5\}$ and a blue die numbered $\{2, 2, 4, 4, 4, 4\}$. Table 3.25 shows the winner, **R**ed or **B**lue, for every combination of these two dice.

Each die wins in 18 of the 36 possible outcomes, and so each die wins with probability ½. On half of the rolls—whenever the red die shows a 3—it's necessary to read both dice to determine which card to turn over.

Two additional valid pairs of dice can be generated from this one by replacing one or both of the 5s on the red die with a 6. If both of the 5s become 6s, any of the 4s on the blue die can be replaced by 5s, giving 4 more valid pairs for a total of 7. ∎

As we approach the general question of counting the number of valid pairs of dice, we note that switching the colors of the dice in Example 3.42 does not produce a genuinely new dice configuration. Without loss of generality,

TABLE 3.25: Winning die in roll of two hypothetical Pala Casino dice

		Red					
		⚀	⚁	⚁	⚁	⚂	⚂
B	⚀	B	R	R	R	R	R
l	⚀	B	R	R	R	R	R
u	⚁	B	B	B	B	R	R
e	⚁	B	B	B	B	R	R
	⚁	B	B	B	B	R	R
	⚁	B	B	B	B	R	R

we will assume that the red die bears the lowest number appearing on the two dice.

Theorem 3.1. *A valid pair of dice requires that at least 5 numbers be used.*

Proof. A valid pair of dice using only 3 distinct numbers $A < B < C$ must have 6 copies of the number B on the blue die. The red die must then have 3 copies of A and 3 copies of C in order that both dice have a 50% chance of winning, but this contradicts the requirement that both dice need to be consulted sometimes, as the number on the red die decides the outcome.

If the valid pair bears the 4 distinct numbers $A < B < C < D$, then A must appear on the red die and so B must appear on the blue die, else the red die can never win. If C is then on the blue die, then only the red die matters, a contradiction. Hence, we must have A and C on the red die and B and D on the blue. (We note that the dice in use at the Pala Casino violate this condition.)

The red die wins only when it shows a C and the blue die shows a B. Let b denote the number of Bs on the blue die and c denote the number of Cs on the red die; it follows that

$$P(\text{Red wins}) = \frac{b}{6} \cdot \frac{c}{6} = \frac{bc}{36}.$$

Since b and c must be integers in the range 1–5, it is not possible to number the faces of the dice so that $bc = 18$, and so there are no valid pairs with 4 different numbers, completing the proof. $\qquad\square$

Example 3.43. To illustrate this result, suppose that the 6 sides on each die are split 3–3 between their 2 assigned numbers. We have the outcomes shown in Table 3.26. Since the red die wins only 25% of the time, this pair is not valid.

Fix the red die as $\{A, A, A, C, C, C\}$. Changing one D to a B on the blue die raises the probability of a red win to $33\frac{1}{3}\%$, and changing 2 Ds to Bs gives the red die a $41\frac{2}{3}\%$ chance of winning, so these cannot form valid pairs.

TABLE 3.26: Card craps winning die: $\{A, A, A, C, C, C\}$ vs. $\{B, B, B, D, D, D\}$

		Red					
		A	A	A	C	C	C
	B	B	B	B	R	R	R
B	B	B	B	B	R	R	R
l	B	B	B	B	R	R	R
u	D	B	B	B	B	B	B
e	D	B	B	B	B	B	B
	D	B	B	B	B	B	B

Turning all Ds to Bs gives a 50% chance of a red win, but this neither uses 4 numbers nor requires both dice: the blue die is $\{B, B, B, B, B, B\}$ and never affects the result.

If we fix the blue die as $\{B, B, B, D, D, D\}$ and increase the number of Cs on the red die, similar reasoning shows that it is still not possible to reach a 50% chance of the red die winning. This confirms that neither die's assigned 2 numbers can be split 3–3. ∎

In the five-number case, let the numbers on the dice be $A < B < C < D < E$, appearing respectively with frequencies a, b, c, d, and e all at least 1.

Lemma 3.1. *In a valid pair of dice using the five numbers $A < B < C < D < E$, the red die must contain A, C, and E.*

Proof. If the red die bears two consecutive numbers, it is possible to replace the smaller number of that pair by the larger on each side where it appears without changing the probability of winning from ½. This creates a pair of dice using only 4 numbers. By Theorem 3.1, this cannot be a valid pair. Similarly, the blue die cannot contain consecutive numbers.

Since A must appear on the red die, the only valid configurations have A, C, and E on the red die and B and D on the blue, and the proof is complete. □

Theorem 3.2. *There are 24 different valid pairs of dice that use 5 numbers.*

Proof. With the notation introduced prior to Lemma 3.1, we have the following probabilities:

$$P(\text{Red wins}) = \frac{e}{6} + \frac{b}{6} \cdot \frac{c}{6} = \frac{6e + bc}{36} = \frac{1}{2}$$

$$P(\text{Blue wins}) = \frac{a}{6} + \frac{c}{6} \cdot \frac{d}{6} = \frac{6a + cd}{36} = \frac{1}{2}$$

from which it follows that $6e + bc = 18$ and $6a + cd = 18$.

Lemma 3.1 means that the following equations connect the five values a, b, c, d, and e:

$$a + c + e = 6$$
$$b + d = 6.$$

These equations allow us to reduce the system above to the single equation

$$6a + 6c - bc = 18.$$

Since the blue die wins whenever the red die rolls an A, it is immediately seen that a must be 1 or 2. If $a = 3$, then the only red configurations yielding $P(\text{Red wins}) = .5$ must have 3 other faces that defeat everything on the blue die, and these render the blue die unnecessary. One such pair of dice would have red $\{A, A, A, E, E, E\}$ vs. blue $\{B, B, C, C, D, D\}$—which additionally violates Lemma 3.1. If $a > 3$, then the blue die wins at least as often as an A is rolled on the red die, and this probability is greater than ½. Hence, no value of a that exceeds 2 can lead to a valid pair.

When $a = 1$, we have

$$6c - bc = 12,$$

and when $a = 2$, we have

$$6c - bc = 6.$$

Inspecting all cases of these two equations with $1 \leqslant b, c \leqslant 5$ gives two solutions to each:

$$(a, b, c) = (1, 2, 3), (1, 3, 4), (2, 3, 2), (2, 4, 3).$$

Each solution leads to 6 valid pairs depending on which number does not appear on the dice, for a total of 24 valid pairs. These solutions are shown with the corresponding dice in Table 3.27, and the proof is complete. □

For each pair of dice in Table 3.27, the probability that both dice need to be consulted to determine the higher roll can be seen to be simply $\dfrac{c}{6}$.

If a valid pair of dice includes all 6 numbers, we let the 6 variables a–f denote the number of times the numbers 1–6 appear, respectively, where all 6 variables are positive integers and

$$a + b + c + d + e + f = 12.$$

This further constrains all 6 variables to the values 1, 2, 3, or 4, since every number from 1–6 appears at least once and the sums $a + c + e$ and $b + d + f$ both equal 6.

We begin by isolating the case where neither die bears consecutive numbers. Given the convention that the lowest number appears on the red die, this means that the red die bears the numbers 1, 3, and 5 while the blue die has 2s, 4s, and 6s.

TABLE 3.27: Valid 5-number pairs of dice for Pala Casino Card Craps

a	b	c	d	e	Red	Blue
1	2	3	4	2	{1,3,3,3,5,5}	{2,2,4,4,4,4}
					{1,3,3,3,6,6}	{2,2,4,4,4,4}
					{1,3,3,3,6,6}	{2,2,5,5,5,5}
					{1,4,4,4,6,6}	{2,2,5,5,5,5}
					{1,4,4,4,6,6}	{3,3,5,5,5,5}
					{2,4,4,4,6,6}	{3,3,5,5,5,5}
1	3	4	3	1	{1,3,3,3,3,5}	{2,2,2,4,4,4}
					{1,3,3,3,3,6}	{2,2,2,4,4,4}
					{1,3,3,3,3,6}	{2,2,2,5,5,5}
					{1,4,4,4,4,6}	{2,2,2,5,5,5}
					{1,4,4,4,4,6}	{3,3,3,5,5,5}
					{2,4,4,4,4,6}	{3,3,3,5,5,5}
2	3	2	3	2	{1,1,3,3,5,5}	{2,2,2,4,4,4}
					{1,1,3,3,6,6}	{2,2,2,4,4,4}
					{1,1,3,3,6,6}	{2,2,2,5,5,5}
					{1,1,4,4,6,6}	{2,2,2,5,5,5}
					{1,1,4,4,6,6}	{3,3,3,5,5,5}
					{2,2,4,4,6,6}	{3,3,3,5,5,5}
2	4	3	2	1	{1,1,3,3,3,5}	{2,2,2,2,4,4}
					{1,1,3,3,3,6}	{2,2,2,2,4,4}
					{1,1,3,3,3,6}	{2,2,2,2,5,5}
					{1,1,4,4,4,6}	{2,2,2,2,5,5}
					{1,1,4,4,4,6}	{3,3,3,3,5,5}
					{2,2,4,4,4,6}	{3,3,3,3,5,5}

Theorem 3.3. *There are four valid pairs of dice that use all 6 numbers where neither die bears consecutive numbers:*

- *Red {1,3,5,5,5,5} and Blue {2,2,4,4,6,6}.*
- *Red {1,3,3,5,5,5} and Blue {2,2,2,4,6,6}.*
- *Red {1,1,3,5,5,5} and Blue {2,2,2,4,4,6}.*
- *Red {1,1,3,3,5,5} and Blue {2,2,2,2,4,6}.*

Proof. With the frequencies a–f defined above, we have

$$a + c + e = 6$$
$$b + d + f = 6.$$

Furthermore,

$$P(\text{Red wins}) = \frac{c}{6} \cdot \frac{b}{6} + \frac{e}{6} \cdot \frac{b+d}{6} = \frac{bc + be + de}{36} = \frac{1}{2}$$

and

$$P(\text{Blue wins}) = \frac{b}{6}\cdot\frac{a}{6} + \frac{d}{6}\cdot\frac{a+c}{6} + \frac{f}{6} = \frac{ab+ad+cd+6f}{36} = \frac{1}{2}.$$

From these equations, we have

$$bc + be + de = 18$$
$$ab + ad + cd + 6f = 18.$$

We shall proceed by considering cases indexed by b.

If we take $b = 1$, the first equation becomes $c + e + de = 18$. Substituting $6 - a$ for $c + e$ gives

$$de = 12 + a.$$

Letting a run through its permissible values 1–4 results in a contradiction in each case.

a	de	**Reason for contradiction**
1	13	Neither d nor e can be 13.
2	14	$7 \nmid de$ for any permissible d and e.
3	15	d and e cannot be 5.
4	16	If $a = 4$, then $e = 1$, and $d \neq 16$.

If $b = 2$, then $d \leqslant 3$ since f must be positive. We have $2c + 2e + de = 18$, and similar reasoning as in the first case gives

$$de = 6 + 2a.$$

Substituting the values 1–4 for a gives a valid pair only if $a = 1$:

Red $\{1,3,5,5,5\}$ and Blue $\{2,2,4,4,6,6\}$.

This is the first valid pair listed above. If $a > 1$, we have contradictions.

a	de	**Outcome**
1	8	**Valid pair: Red $\{1,3,5,5,5\}$ and Blue $\{2,2,4,4,6,6\}$.**
2	10	Contradiction: $5 \nmid de$ for any permissible d and e.
3	12	Contradiction: $b = 2$ means $d < 4$, but $a = 3$ means $e < 4$.
4	14	Contradiction: Again, $7 \nmid de$.

In the case $b = 3$, we have $d \leqslant 2$, and $3c + 3e + de = 18$ becomes

$$de = 3a.$$

The second and third valid pairs listed above appear in two subcases; the other two result in contradictions.

a	de	Outcome
1	3	**Valid pair: Red {1,3,3,5,5,5} and Blue {2,2,2,4,6,6}.**
2	6	**Valid pair: Red {1,1,3,5,5,5} and Blue {2,2,2,4,4,6}.**
3	9	$a = 3$ implies $e < 3$. Contradiction.
4	12	$d, e < 3$; another contradiction.

Finally, if $b = 4$, the blue die must be $\{2,2,2,2,4,6\}$, so $d = f = 1$. Reducing the equation $bc + be + de = 18$ gives

$$e = 4a - 6.$$

a	e	Outcome
1	-2	Contradiction: e cannot be negative.
2	2	**Valid pair: Red {1,1,3,3,5,5} and Blue {2,2,2,2,4,6}.**
3	6	Contradiction: $1 \leqslant e \leqslant 4$.
4	10	Contradiction: Only 6 sides on the dice, so $e \neq 10$.

The only valid pair arising in this case is where $a = 2$. We have

Red $\{1,1,3,3,5,5\}$ and Blue $\{2,2,2,2,4,6\}$,

the final valid pair listed initially, and the proof is complete.

$\square$

A valid pair of dice using all 6 numbers with consecutive numbers on one or both dice may be derived from a valid pair using only 5 numbers.

Example 3.44. The valid pair Red: $\{1,3,3,3,5,5\}$ and Blue: $\{2,2,4,4,4,4\}$ can be modified into a 6-number valid pair by changing a 5 on the red die to a 6. This gives the valid pair Red: $\{1,3,3,3,5,6\}$ and Blue: $\{2,2,4,4,4,4\}$. This pair can also be modified by changing both 5s on the red die to 6s and then changing 1–3 of the 4s on the Blue die to 5s, generating three more 6-number valid pairs. ∎

Table 3.28 collects the 6-number results. The third column shows what percentage of the time it is necessary to read both dice to resolve the roll and identify a card.

3.8 Side Bets

It may reasonably be stated that every craps bet except Pass and Don't Pass is a side bet, because the primary game at a craps table is the shooter rolling the dice in pursuit of a point before a 7. Everything else on the layout, whether a single-roll wager like Any 7 or a multi-roll bet like a Place bet, is secondary to

TABLE 3.28: Valid 6-number pairs of dice for Pala Casino Card Craps

Red	Blue	%	Red	Blue	%
{1,3,5,5,5}	{2,2,4,4,6,6}	55.55%	{1,3,4,4,4,6}	{2,2,2,5,5,5}	66.67%
{1,3,3,5,5,5}	{2,2,2,4,6,6}	55.55%	{1,4,4,4,4,6}	{2,2,3,5,5,5}	66.67%
{1,1,3,5,5,5}	{2,2,2,4,4,6}	55.55%	{1,4,4,4,4,6}	{2,3,3,5,5,5}	66.67%
{1,1,3,3,5,5}	{2,2,2,2,4,6}	55.55%	{1,1,3,3,5,6}	{2,2,2,4,4,4}	50%
{1,3,3,3,5,6}	{2,2,4,4,4,4}	50%	{1,1,3,3,6,6}	{2,2,2,4,5,5}	33.33%
{1,3,3,3,6,6}	{2,2,4,4,4,5}	50%	{1,1,3,3,6,6}	{2,2,2,4,4,5}	33.33%
{1,3,3,3,6,6}	{2,2,4,4,5,5}	50%	{1,1,3,4,6,6}	{2,2,2,5,5,5}	33.33%
{1,3,3,3,6,6}	{2,2,4,5,5,5}	50%	{1,2,4,4,6,6}	{3,3,3,5,5,5}	50%
{1,3,3,4,6,6}	{2,2,5,5,5,5}	50%	{1,1,4,4,6,6}	{2,3,3,5,5,5}	33.33%
{1,3,4,4,6,6}	{2,2,5,5,5,5}	50%	{1,1,4,4,6,6}	{2,2,3,5,5,5}	33.33%
{1,4,4,4,6,6}	{2,3,5,5,5,5}	50%	{1,1,3,3,3,6}	{2,2,2,2,4,5}	50%
{1,3,3,3,3,6}	{2,2,2,4,4,5}	66.67%	{1,1,3,4,4,6}	{2,2,2,2,5,5}	50%
{1,3,3,3,3,6}	{2,2,2,4,5,5}	66.67%	{1,1,3,3,4,6}	{2,2,2,2,5,5}	50%
{1,3,3,3,4,6}	{2,2,2,5,5,5}	66.67%	{1,2,4,4,4,6}	{3,3,3,3,5,5}	66.67%
{1,3,3,4,4,6}	{2,2,2,5,5,5}	66.67%			

Pass and Don't Pass. Gamblers seeking extra action at a craps table already have myriad options, which may explain why *WIN Magazine* did not find it necessary to propose a new craps bet in the style of WINGO.

Nonetheless, since it is often easier to sell a new side bet for an established (and unprotected by patent) game than to invent a new game from scratch, new craps side bets have occasionally found places in casinos. A challenge to a casino table games director is that knowledgable craps players understand the game's odds and know where the best bets on the layout are, and will not be easily drawn to a new wager. The growth potential at a craps table lies more in drawing in new players, and this is perhaps best done with wagers that are easy to understand, even if they may have high HAs [26]. This being the case, we find that few craps bets besides Pass and Don't Pass carry house advantages anywhere near Pass and Don't Pass, whose HAs run under 1.5%— even less if free odds are taken.

Quick Side Bets

Some new craps side bets are designed to be settled rapidly, if not in one roll, then in a small fixed number of rolls. We shall refer to these as *quick* side bets. A number of quick side bets are keyed to the shooter trying to make a point; we consider those first.

7 Point 7

One quick side bet is *7 Point 7* which was introduced in 2008 and is resolved in no more than 2 rolls. The bet can be made just prior to a come-out roll, and takes its name from its two winning ways [86].

- *7*: If the come-out roll is a 7, the bet wins, paying 2–1. If the roll is a 2, 3, 11, or 12, the bet loses.

- *Point 7*: If a point is established on the come-out roll, the bet pays 3–1 if the next roll is a 7; otherwise it loses.

A player making the 7 Point 7 bet in conjunction with a Pass line wager starts out with both bets winning on a 7, but if a point is made, the two bets cannot win together, which might lead some players to think of 7 Point 7 as a form of insurance protecting the Pass bet. If 7 Point 7 is made to back up an equal Pass Line bet, the possible outcomes are detailed in Table 3.29.

TABLE 3.29: Outcomes for $1 7 Point 7 bet with $1 Pass line bet

Roll	Pass	7 Point 7	Net win
First roll			
2, 3, 12	Loses	Loses	−$2
7	Wins	Wins	$3
11	Wins	Loses	$0
4, 5, 6, 8, 9, 10	Point set, both bets continue		—
Second roll			
7	Loses	Wins	$2
Point	Wins	Loses	$0
Other	Continues	Loses	TBD: $0 or −$2

The probability of winning a 7 Point 7 bet is

$$\frac{1}{6} + \frac{24}{36} \cdot \frac{1}{6} = \frac{60}{216},$$

and the expected value of a $1 bet is

$$(2) \cdot \frac{1}{6} + (3) \cdot \frac{24}{216} + (-1) \cdot \frac{156}{216} = -\frac{1}{18},$$

giving a HA of 5.56%.

Hard Pass

The various hardway bets at the center of a craps table are multi-roll bets, and if a shooter holds the dice for many throws without rolling the even number in question or rolling a 7, they can remain unresolved for some time. A quicker wager relying on the shooter rolling doubles is *Hard Pass*, which is designed as a high-risk, high-reward gamble [75].

- The shooter must first establish a point—4, 6, 8, or 10—by rolling doubles. Any other roll on the come-out is a losing roll for Hard Pass.

- The shooter must then make that point by rolling doubles again. If the shooter sevens out or makes the point with an easy combination, Hard Pass loses.

The probability that Hard Pass is lost on the come-out roll is $\frac{32}{36}$—only 4 rolls move the wager on to the second stage. At that stage, the probability of winning depends on the point. This probability is $\frac{1}{9}$ if the point is 4 or 10, and $\frac{1}{11}$ if the point is 6 or 8.

The complete probability of winning a Hard Pass bet is

$$p = \frac{1}{36} \cdot \left[\frac{1}{9} + \frac{1}{11} + \frac{1}{11} + \frac{1}{9} \right] = \frac{10}{891} \approx .0112.$$

Given this tiny chance of winning, it is no surprise, perhaps, that Hard Pass has payoff odds of 80–1. The HA on Hard Pass is 9.09%.

Example 3.45. Some craps bettors might not like losing a variation on a Pass bet when a 7 or 11 is rolled on the come-out. If Hard Pass were revised so that it also won on an initial 7 or 11, what should it pay?

The new probability of winning would be

$$\frac{8}{36} + \frac{10}{891} = \frac{208}{891} \approx .2334,$$

just under ¼. Paying this bet at 3–1 gives a 6.62% house advantage, which might be reasonable. ∎

Point Right Back

Point Right Back is a proposed craps side bet that is resolved in no more than 2 rolls [75]. As with 7 Point 7 and Hard Pass, it must be made immediately before a come-out roll. The bet's name is drawn from how it wins: if the shooter establishes a point and then immediately makes it on the next roll.

This bet loses at once if the first roll is a 2, 3, 7, 11, or 12—this is only one-third of the time. The remaining $\frac{2}{3}$ of come-out rolls set a point. The probability of winning the bet then depends on the point. Including the come-out roll, we have

$$P(\text{Win}) = 2 \cdot \left[\left(\frac{3}{36} \right)^2 + \left(\frac{4}{36} \right)^2 + \left(\frac{5}{36} \right)^2 \right] = \frac{25}{324} \approx .0772.$$

This bet wins about 1 time in 13.

While a gambler does not select which point she chooses to back, payoffs for Point Right Back depend on the point. We consider this bet in complete generality, with payoffs of X to 1 if the point is 4 or 10, Y to 1 on 5 or 9, and Z to 1 if it's 6 or 8. It is understood that $X \geqslant Y \geqslant Z$. The expectation then becomes

$$E(X, Y, Z) = (X) \cdot \frac{18}{1296} + (Y) \cdot \frac{32}{1296} + (Z) \cdot \frac{50}{1296} + (-1) \cdot \frac{299}{324}.$$

Example 3.46. The original proposal for Point Right Back specified payoff odds of 10–1 if the point was 6 or 8 ($Z = 10$), and 12–1 odds on all other points ($X = Y = 12$). This combination of payoffs gave the casino a 7.41% advantage [75]. ∎

Example 3.47. A flat payoff of 10–1 regardless of the point yields a house advantage of 15.12%.

This HA gives the casino some room to run promotions which briefly increase one or more payoffs. Suppose, for example, that one of the 6 points is chosen each day to pay 15–1. This would result in another variable in the equation for E. The new HAs are shown in Table 3.30.

TABLE 3.30: Point Right Back: House edge with a bonus point paying 15–1

Bonus point	HA
4 or 10	11.65%
5 or 9	8.95%
6 or 8	5.48%

If the bonus points are restricted to 4, 5, 9, and 10, the bonus odds could be set at 20–1; a 20–1 payoff on 6 or 8 would result in a player edge of 4.17%.

Perhaps the Four Queens Casino would choose to establish 4 as a permanent bonus number paying 25–1 while retaining a 10–1 payoff on the other point numbers. Their edge under the new pay table would be 4.71%. ∎

Fielder's Choice

Fielder's Choice, a product of Aces Up Gaming, divides the numbers covered by the Field bet into Left, Center, and Right Field bets, allowing players to bet on a subset of the Field numbers and trade a lower probability of winning for a higher payoff. Just as a baseball diamond has some overlap among left, center, and right fields, the numbers 4 and 10 are each assigned to two of the three fields. The available wagers are shown in Table 3.31.

If rolls of 2 and 12 are both paid at 2–1 while other numbers pay 1–1, the Field bet carries a 5.56% HA. How do these new bets compare?

TABLE 3.31: Fielder's Choice target numbers and payoff odds

Wager	Numbers covered	Payoff odds
Left field	2, 3, 4	4–1
Center field	4, 9, 10	2–1
Right field	10, 11, 12	4–1

- The Left and Right Field bets have the same probability of winning, $\frac{1}{6}$, and the same payoff, so they have the same HA: 16.67%.

  At a 5–1 payoff, both of these are fair bets.

- Center Field has win probability $\frac{10}{36}$, slightly greater than probability of winning one of the other field bets. Its payoff of half the Left or Right Field payoffs and this probability combine to produce the same HA.

In summary, the HA on these new bets is triple the HA on the original Field bet.

Extended Side Bets

By contrast to quick side bets, *extended* craps side bets act, in part, to maintain interest in a craps game when the shooter holds the dice through many rolls without making a point or sevening out. These wagers can remain in action for dozens of rolls, which can keep players at the craps table where they might make additional bets.

Fire Bet

The *Fire Bet* is a craps side bet developed by Las Vegas casino supervisor Perry Stasi, whose name suggests its purpose: to cash in on a shooter having a "hot" hand and making many points before sevening out [10]. A bet on the Fire Bet pays off if the shooter makes a certain number of different points, at least three or four depending on the pay table, before tossing a 7. While the pay table for the Fire Bet varies from casino to casino, one version is Table 3.32:

TABLE 3.32: Fire Bet probabilities [121]

Points Made	Payoff
3	7 for 1
4	30 for 1
5	150 for 1
6	300 for 1

The points made must be different: if the shooter makes the points 4, 9, 4, and 6 before sevening out, this is only three different points for the purpose of the Fire Bet. A roll of 7 on a come-out roll does not interrupt a string of points. Mathematically, the Fire Bet is somewhat complicated to analyze, due to the different probabilities for establishing and making the six possible points. A good source for the computations required is [121]; there we find the probabilities shown in Table 3.33.

TABLE 3.33: Fire Bet probabilities [121]

Points Made	Probability
0	.5939
1	.2608
2	.1013
3	.0334
4	.0088
5	.0016
6	.000162

With Table 3.32 as the pay table and payoffs starting at 3 points made, the probability of winning anything with the Fire Bet is a mere .0440, or 4.40%. The corresponding expectation is

$$E = (7) \cdot .0334 + (30) \cdot .0088 + (150) \cdot .0016 + (300) \cdot .000162 - 1 = -.2136,$$

giving an HA of 21.36%.

In part, no doubt, because of the high house advantage, the Fire Bet was able to gain a place on the craps table in many casinos. Despite the high HA, some gamblers were drawn to it; they are perhaps attracted in part by the large payoff for a small bet. The maximum Fire Bet wager at most casinos offering this wager is $5.

Different Doubles

Different Doubles, like the Fire Bet, is a wager that a shooter will do something before rolling a 7. In this case, the challenge is to roll as many different doubles as possible, with a pay table, Table 3.34, that starts when 3 different doubles are thrown.

For the purposes of this bet, all we care about are 12 different rolls: the doubles and 7s. Every other possible roll has no effect on the wager or on the probability of rolling k doubles before a 7. The probability of rolling 0 doubles before a 7 is then $P(0) = \frac{6}{12} = \frac{1}{2}$. The probability of rolling exactly 1 double before a 7 is

$$P(1) = \frac{1}{2} \cdot \frac{6}{11} = \frac{3}{11},$$

since all we need consider is the probability of rolling a double followed by the

TABLE 3.34: Different Doubles pay table

Number of doubles	Payoff
3	4–1
4	8–1
5	15–1
6	100–1

probability of rolling a 7. The denominator of the second fraction is 11 rather than 12 because we remove the first double from the list of rolls that affect the wager.

Continuing in this manner, $P(2)$ is the probability of rolling two different doubles followed by a 7, or

$$P(2) = \frac{1}{2} \cdot \frac{5}{11} \cdot \frac{6}{10} = \frac{3}{22}.$$

As we proceed further, the denominators fall with each successive roll and the numerators corresponding to doubles fall along with them, The denominator on the final roll of 7 remains 6, although that term is absent from $P(6)$, since there is no need to close out a string of 6 doubles with a 7. Extending the pattern gives

$$P(3) = \frac{1}{2} \cdot \frac{5}{11} \cdot \frac{4}{10} \cdot \frac{6}{9} = \frac{2}{33}$$
$$P(4) = \frac{1}{2} \cdot \frac{5}{11} \cdot \frac{4}{10} \cdot \frac{3}{9} \cdot \frac{6}{8} = \frac{1}{44}$$
$$P(5) = \frac{1}{2} \cdot \frac{5}{11} \cdot \frac{4}{10} \cdot \frac{3}{9} \cdot \frac{2}{8} \cdot \frac{6}{7} = \frac{1}{154}$$
$$P(6) = \frac{1}{2} \cdot \frac{5}{11} \cdot \frac{4}{10} \cdot \frac{3}{9} \cdot \frac{2}{8} \cdot \frac{1}{7} = \frac{1}{924}.$$

The probability of winning a Different Doubles bet is

$$P(3) + P(4) + P(5) + P(6) = \frac{1}{11}.$$

Comparing the probabilities listed above to the corresponding payoffs in Table 3.34 suggests that the house advantage on this wager is quite high: paying 100–1 on a 1 in 924 shot is a way for the casino to retain a lot of the money. The HA on Different Doubles is 27.92%, so we have another bet like the Fire Bet whose high maximum payout masks a high HA. Changing the payoff for 6 doubles to 300–1 reduces the HA to 6.28%.

Example 3.48. *Double D is a variation on Different Doubles that eliminates the payoff on 3 doubles and increases the payoff on 4–6 different doubles to* make up for the deletion. If Double D pays 10–1 for 4 doubles, 50–1 for 5, and 250–1 for 6, find the new HA.

The expectation on Double D is

$$E = (10) \cdot \frac{1}{44} + (50) \cdot \frac{1}{154} + (250) \cdot \frac{1}{924} + (-1) \cdot \frac{32}{33} = -\frac{34}{231} \approx -\$.1472.$$

Notice that the probability of losing is now almost 97%. The HA is then 14.72%. ∎

All/Tall/Small

All/Tall/Small is a trio of bets that pay off if the shooter is able to throw various sums before rolling a 7. Unlike the Fire Bet, All, Tall, and Small do not count points made by a shooter, merely sums tossed. These bets can only be placed immediately prior to a come-out roll.

- *All* pays anywhere from 150–1 to 175–1 if the shooter rolls every sum from 2–12 before rolling a 7.
- *Tall* covers the top end of that range, paying 34–1 if the shooter successfully rolls 8, 9, 10, 11, and 12 before a 7.
- *Small* pays the same 34–1 as Tall, on a roll of 2, 3, 4, 5, and 6 before 7.

Some casinos pay as low as 30–1 on Tall and Small. Mathematically, these two outcomes are equally likely. Sums once rolled can be repeated with no effect on the wagers.

Attempting to compute the probability of winning any of these three bets using only simple arithmetic and algebra runs into a number of formidable obstacles.

- One small obstacle is that the number of rolls before the shooter sevens out is unbounded. We have dealt with this in several other circumstances by using infinite series.

- However, that approach is less fruitful here because as numbers are successfully rolled, the probability of a neutral roll that neither ends the bet nor advances the count increases.

  Example 3.49. If we are making the Small bet, the probability of a neutral roll (one of the Tall numbers) on the first roll is $\frac{15}{36}$. If the first roll is a 4, then another roll of a 4 becomes a neutral roll. The probability of a neutral roll is then $\frac{18}{36}$. This probability continues to increase as small numbers are rolled. ∎

This means that an infinite series approach to the probability would not be geometric, with a common ratio. This makes the series considerably harder to sum.

- The amount by which the neutral probability increases as the bet progresses depends on the order in which the numbers are rolled, which makes it tricky to devise a general elementary formula for the winning probability. This sets All/Tall/Small apart from Different Doubles: while the doubles in the latter bet may arrive in any order, each of them has the same probability, so the probabilities change predictably regardless of the order.

We could, of course, try to derive probabilities for every possible order and add them together, but this involves $5! = 120$ permutations for Tall and Small and $\frac{10!}{2^5} = 113,400$ distinguishable permutations for All. Before embarking on a mathematical challenge of that magnitude, we would do well to see if there's another route to the probabilities we seek.

Exactly computing the winning probabilities is most easily done by using either calculus or the theory of Markov chains, both of which are beyond the scope of this book [120]. However, a computer simulation can be used to approximate these probabilities efficiently and accurately. A Python program was written to simulate one million rounds of all three bets; a version of the program that simulates the Small bet is shown in Figure 3.14.

```
import random

def dieroll():
    return (random.randint(1,6))

wins = 0
for i in range (0,1000000):
        A = [0]*13
        crapsroll = 0
        outcome = 0
        while(crapsroll != 7 and outcome == 0):
            crapsroll = dieroll()+dieroll()
            A[crapsroll] = A[crapsroll] + 1
            outcome = A[2]*A[3]*A[4]*A[5]*A[6]
            if outcome >= 1:
                wins = wins + 1
                print('Winner!!')
        if (i % 100000 == 0):
            print ('trials = ', i)
print ('Total Wins = ', wins)
print ('Probability of winning = ', wins/(i+1))
print ('Probability of winning = ', wins/(i+1))
```

FIGURE 3.14: Python program for 1 million trials of the Small bet.

The line

```
outcome = A[2]*A[3]*A[4]*A[5]*A[6]
```

can easily be changed to simulate the All and Tall bets.

The results of one million simulations of each bet are shown in Table 3.35. Each of the three simulations, which rolled approximately 6.76 million

TABLE 3.35: All/Tall/Small: Experimental and theoretical probabilities

| | ——— Probability ——— | | |
Wager	**Experimental**	**Theoretical**	**% Error**
All	.005184	.005258	−1.40%
Tall	.026529	.026354	+0.66%
Small	.026382	.026354	+0.11%

dice per trial, took less than 5 minutes to execute. We note that all three simulations gave experimental probabilities that are close to the theoretical probabilities (which are computed using calculus in [120]) to an acceptable degree of accuracy. In the analysis that follows, we shall use the theoretical probability in all calculations.

Example 3.50. We know that the HAs on Small and Tall will be the same. How does the HA on All compare?

For Small and Tall with a 34–1 payoff, we have

$$E = (34) \cdot .026354 + (-1) \cdot (1 - .026354) \approx -\$.0776,$$

and the HA is 7.76%. For the All bet paying 175–1, the expectation is

$$E = (175) \cdot .005258 + (-1) \cdot (1 - .005258) \approx -\$.0746,$$

giving a 7.46% HA.

At the maximum payoffs for both bets, All has a slightly smaller HA, but varying the payoff odds could change the relative advantage of the two bets. If All pays 174–1 (possibly advertised as 175 for 1), its HA jumps to 7.99%, and Small or Tall are then better for gamblers. ∎

What are the win probabilities for All/Tall/Small when 8-sided dice are in use?

The Spider Craps version of Small requires rolling all of the numbers from 2–8 inclusive before rolling a 9; Tall requires rolling 10–17 before a 9. All, of course, requires rolling every other sum before a 9. These had small winning probabilities with 6-sided dice, so we expect smaller probabilities of winning—which give room for bigger payoffs—with d8s.

A small modification of the Python program on page 176 used for All/Tall/Small with d6s can be used to estimate these probabilities. The new program gave the experimental probabilities listed in Table 3.36.

As a quick check on the program, we note that Tall and Small have approximately equal experimental probabilities, as we expect.

TABLE 3.36: All/Tall/Small: Experimental probabilities in Spider Craps

Wager	Experimental Probability
All	.001155
Tall	.010664
Small	.010710

As a second check for accuracy, we follow the lead of Michael Shackleford, the "Wizard of Odds", who computed the theoretical probabilities for All/Tall/Small with standard dice using integral calculus [120]. Modifying Shackleford's work for Spider Craps gives theoretical probabilities that are compared to the above experimental probabilities in Table 3.37.

TABLE 3.37: All/Tall/Small: Experimental and theoretical probabilities in Spider Craps

| | ——— Probability ——— | | |
Wager	Experimental	Theoretical	% Error
All	.001155	.001554	−0.26%
Tall	.010664	.010693	−0.28%
Small	.010710	.010693	+0.16%

As we saw with 2d6, our experimental probabilities are all quite close to the theoretical values.

We turn now to determining appropriate payoffs for All and Tall/Small in Spider Craps. The theoretical probability for All is approximately $\dfrac{1}{643.5}$, which suggests that a 600–1 payoff might be convenient for dealers and give the casino a necessary advantage. With a 600–1 payoff, we have

$$E = (600) \cdot .001554 + (-1) \cdot .998446 \approx -\$.0660,$$

so the casino holds a very reasonable 6.60% edge at 600–1.

Tall and Small have a theoretical probability of .010693, or about 1 in 93.5. Reasonable rounding suggests setting the payoff odds at 90–1, which yields a 2.69% HA. Switching the payoff to 85–1 increases the house edge to 8.04%, somewhat closer to the proposed HA for All and a bit more likely to make money for the casino—too small a house advantage on bets like these that might not attract much action is not lucrative enough.

Repeater

Another craps bet designed to build excitement over many rolls is the *Repeater* bet, a wager developed and marketed by Aces Up Gaming, the company behind Fielder's Choice. To make the simplest Repeater bet, the gambler selects a number from 2–12, excluding 7, and is paid according to Table 3.38 if that number is thrown a specified target number of times, k, before a 7 is rolled. If n is the number chosen by the player, we have $k = n$ if n is from 2–6 and $k = 14 - n$ if n is between 8 and 12. An additional Repeater bet option covers the Any Craps numbers plus 11: 2, 3, 11, and 12, a set of numbers often abbreviated "C & E" for "Craps and 11". For C & E, we have $k = 7$. Craps dealers are supplied with markers to be placed on the modified layout to track how many times each number is rolled.

TABLE 3.38: Repeater bet target numbers and payoff odds

Roll	Target number k	Payoff odds
2, 12	2	40 for 1
3, 11	3	50 for 1
4, 10	4	65 for 1
5, 9	5	80 for 1
6, 8	6	90 for 1
C & E	7	100 for 1

The payoffs on Repeater bets are higher than the standard craps bets, whose payoffs top out at 30–1, offer. These high payoffs might draw new players to the craps table by offering the potential for big wins on a par with slot machines, where casual gamblers might be found and introduced to dice [26].

Repeater bets may be placed prior to a come-out roll, as with Pass and Don't Pass, or at any time during a string of rolls provided that the chosen sum has not yet appeared. On a roll of 7, all Repeater bets lose unless another come-out roll has been reached; all Repeater bets are "off", or temporarily inactive, during a come-out roll. Rolling a 9 on a come-out roll does not count toward the goal of five 9s, but neither does a come-out roll of 7 end the bet with a loss.

For a selected Repeater bet, let p be the probability of a successful roll, and let $q = \frac{5}{6} - p$ be the probability of a "neutral" roll that is neither a selected number nor 7. The probability of winning a Repeater bet is the probability that the chosen number is rolled the necessary number of times, k, before a 7 is rolled. Since the number of rolls is theoretically unlimited, the probability is given by an infinite series:

$$P(\text{Win}) = \sum_{i=k}^{\infty} P(\text{Win in } i \text{ rolls}).$$

To evaluate $P(\text{Win in } i \text{ rolls})$, we use a binomial model, adjusted for the fact that winning in exactly i rolls requires that the last roll be the chosen number, so we are seeking the number of ways to distribute $k - 1$ winning rolls among $i - 1$ rolls. This may be done in $\binom{i-1}{k-1}$ ways.

Once the k winning rolls—including the last—and the $i - k$ neutral rolls, which are independent, are identified, factors of p^k and q^{i-k} account for the probabilities of the individual rolls. Assembling everything gives

$$P(\text{Win in } i \text{ rolls}) = \binom{i-1}{k-1} \cdot p^k q^{i-k},$$

and summing gives

$$P(\text{Win}) = \sum_{i=k}^{\infty} \binom{i-1}{k-1} \cdot p^k q^{i-k}.$$

The factors p^k and q^{-k} do not depend on the summation variable i and so may be factored out of the sum, giving

$$P(\text{Win}) = \left(\frac{p}{q}\right)^k \sum_{i=k}^{\infty} \binom{i-1}{k-1} q^i.$$

This series is easily evaluated by a computer algebra system: For $|q| < 1$, we have

$$\sum_{i=k}^{\infty} \binom{i-1}{k-1} q^i = \left(\frac{q}{1-q}\right)^k,$$

and

$$P(\text{Win}) = \left(\frac{p}{1-q}\right)^k.$$

Example 3.51. For the Repeater bet on the number 10, we have $k = 4, p = \frac{1}{12}$, and $q = \frac{3}{4}$. The probability of winning is then $\frac{1}{81} \approx .0123$, giving the casino a 19.75% advantage. ∎

Table 3.39 shows the house advantages for Repeater bets on the various sums.

TABLE 3.39: Repeater bet house advantages

Roll	Target number k	HA
2, 12	2	18.37%
3, 11	3	21.88%
4, 10	4	19.75%
5, 9	5	18.08%
6, 8	6	20.62%
C & E	7	21.87%

An alternate pay table for Repeater calls for the number n to be rolled n

TABLE 3.40: Repeater bet alternate target numbers and payoff odds [26]

Roll	Target number	Payoff odds
8	8	400 for 1
9	9	2500 for 1
10	10	25,000 for 1
11	11	1,000,000 for 1
12	12	50,000,000 for 1

times before a 7 is rolled. The payoffs for the numbers 2–6 are unchanged; the new payoffs on 8–12 are shown in Table 3.40.

This Repeater bet on 12 is a wager that 12 will be thrown 12 times before a 7 is thrown once. At a 50 million for 1 payoff, this is among the highest-paying bets proposed for casino games. The probability of winning this bet is

$$\left(\frac{1}{7}\right)^{12} \approx 7.2248 \times 10^{-11},$$

approximately 1 in 13.8 billion, and the house advantage is 99.84%—also among the highest of any wager. If the payoff on this bet was raised to 10 billion for 1, the HA would drop to 27.72%—far more reasonable, but with a nonzero (admittedly tiny) chance of bankrupting the casino.

No Lose Free Roll

A craps wager found at Bob Stupak's Vegas World, the No Lose Free Roll was a variation on the Field bet and appeared in the 1980s. "No Lose" was, of course, a misnomer, as the bet had a $\frac{26}{36}$ probability of losing. For the lucky gambler who managed to overcome these odds—over 2 to 1 odds—against winning, the potential for great riches loomed.

Or so the promotional materials, which called this bet "the greatest proposition in the history of gambling", indicated [90].

A standard Field bet covers the numbers 2, 3, 4, 9, 10, 11, and 12, with bonus payments of 2–1 on 2 or 12, and occasionally a 3–1 payoff on 12. The No Lose Free Roll took some combinations away from the bettor and added in some others: the bet paid off at even money if the following roll was a 2, 3, 11, or 12, as well as a hard-way 4, 6, 8, or 10.

Example 3.52. The house advantage on a $5 bet on the standard Field bet is 5.56%. What is the house edge *on the first roll* of a $5 No Lose Free Roll wager?

The first roll either wins at even money, with probability $p = \frac{10}{36}$, or loses, with probability $q = 1 - p = \frac{26}{36}$. The expectation is then

$$E = (5) \cdot \frac{10}{36} + (-5) \cdot \frac{26}{36} = -\frac{80}{36} \approx -\$2.22,$$

which, when divided by the $5 initial bet, gives a very high house advantage of 44.44%, eight times higher than the HA on the original Field bet. ∎

Ten of the 36 possible rolls were winners, while 26 were losers. How, then, could this be billed as a "no lose" bet? The attraction was the ability, once the No Lose Free Roll bet had won once, to let one's winnings ride for subsequent rolls, without risk of loss. Winnings could accumulate until the first losing roll, at which time all of a player's collected chips were returned. There was a ceiling on accumulated winnings: an initial bet of $5 had a maximum payout of $20,000, for example. This means that the $5 win in Example 3.52 was a minimum, as the player had the potential to win far more if the designated numbers kept coming up on the dice.

To reach this maximum, a $5 wager would have to hit 12 times in a row, for a total win of $20,480—only $20,000 of which would be paid out. The probability of this event is

$$\left(\frac{10}{36}\right)^{12} = \frac{1,000,000,000,000}{4,738,381,338,321,616,896} \approx 2.1104 \times 10^{-7},$$

approximately once in every 4,738,381 games.

This is, as one might expect, unlikely, but what if you could string together a shorter run of lucky rolls? How does the No Lose Free Roll, in its entirety, compare to other craps side bets?

The amounts to be won on this roll are $5, $15, $35, ... $10,235, $20,000, or $10 \cdot 2^{n-1} - 5$ as n runs from 1 to 11, followed by a final payoff of $20,000. To win $10 \cdot 2^{n-1} - 5$, it is necessary to hit a winning number on n consecutive rolls and follow that up by rolling a losing number, which gives

$$P(10 \cdot 2^{n-1} - 5) = p^n \cdot q$$

With $p = \dfrac{10}{36}$ and $q = \dfrac{26}{36}$, the expectation is then

$$E = \sum_{n=1}^{11} \left[(10 \cdot 2^{n-1} - 5) \cdot \left(\frac{10}{36}\right)^n \cdot \left(\frac{26}{36}\right) \right] + (20,000) \cdot \left(\frac{10}{36}\right)^{12} + (-5) \cdot \left(\frac{26}{36}\right).$$

This sums to approximately −$.4889. Dividing by the $5 wager gives a house advantage of about 9.78%.

Example 3.53. Suppose that the No Lose Free Roll had no cap on winnings— so long as the shooter kept rolling winning numbers, a running NLFR wager was matched. Would that tip the edge to the player?

This question asks us to balance a payoff that's headed toward infinity with a probability of winning that's approaching 0. The new expectation is

$$E = \sum_{n=1}^{\infty} \left[(10 \cdot 2^{n-1} - 5) \cdot \left(\frac{10}{36}\right)^n \cdot \left(\frac{26}{36}\right) \right] + (-5) \cdot \left(\frac{26}{36}\right).$$

The infinite series evaluates to $\dfrac{25}{8}$, giving an expectation of

$$E = \frac{25}{8} - \frac{130}{36} = -\frac{35}{72} \approx -\$0.486.$$

The corresponding HA is 9.72%, and so the edge would still rest with the casino. ∎

One might reasonably ask, given this last result, why Vegas World capped player winnings on a \$5 initial bet at \$20,000. A possible explanation is that the casino wanted to insure itself against a long run of winning tosses, which, while very unlikely, could bankrupt the casino. Twenty-five straight winning NLFR rolls, with a \$5 initial bet, would result in a prize of \$335,544,315. While this has a very low probability (about 1 chance in 81 trillion), the consequences if it happened would be catastrophic—for the casino.

Hard Hardway

In 1997, Donald Catlin and Leonard Frome proposed a new craps wager called the *Hard Hardway* bet. This is a simple wager that each of the 6 even sums, 2 through 12, will be rolled the hard way twice before a 7 is rolled. As one might expect, this is an extremely unlikely event. Mathematical analysis going well beyond elementary probability leads to a chance of winning of

$$p = \frac{9,198,254,528,424}{1,458,015,678,282,240,000},$$

which is approximately 1 chance in 158,510 [20].

Catlin and Frome noted that an easier route to a good estimate of the probability of winning a Hard Hardway bet was a computer simulation of the wager; they indicated that 1 billion trials would produce a sufficiently accurate approximation. The Python program in Figure 3.15 executes that simulation.

When this program was executed, the first success occurred on trial number 233,593—slightly later than expected—and the last win was on trial 999,921,022. In 1 billion trials, there were 6405 wins, for a win probability of

$$6.405 \times 10^{-6} \approx \frac{1}{156,128},$$

a figure within 1.53% of Catlin and Frome's theoretical value.

On one hand, this seems so unlikely that no one would ever make this bet. On the other hand, the probability of winning a Powerball lottery jackpot is considerably less—1 in 292,201,338—and people still buy lottery tickets, so it's possible that, as with the Fire Bet, a large potential payoff might attract bettors in spite of the unfavorable odds. Suppose that a casino offers the Hard Hardway bet with a payoff of 150,000 to 1 for convenience—theirs. The expected value for a \$1 Hard Hardway bet with this payoff is

$$E = (150,000) \cdot p + (-1) \cdot (1 - p) \approx -\$.054,$$

```
import random

def dieroll():
    return (random.randint(1,6))

wins = 0
for i in range (0,1000000000): # 1 billion trials
        A = [0]*6
        crapsroll = 0
        amin = 0
        while(crapsroll != 7 and amin < 2):
            first=dieroll()
            second=dieroll()
            crapsroll = first + second
            if first == second:
                A[first-1] += 1
            amin = min(A)
            if (amin == 2):
                wins = wins + 1
                print (A[0],A[1],A[2],A[3],A[4],A[5],i)
        if (i % 100000000 == 0):
            print ('trials = ', i)
print ('Total Wins = ', wins)
print ('Probability of winning = ', wins/(i+1))
```

FIGURE 3.15: Python program for 1 billion trials of Hard Hardway.

where p is the theoretical value above, and so the house advantage is about 5.4%. While this is more unfavorable to a player than a Pass or Don't Pass bet, the house advantage with this payoff is safely on the low end of the HAs of other craps side bets.

3.9 Exercises

Solutions begin on page 327.

3.1. One Roman game of chance played with astragali used 4 bones, which were tossed at once. The best toss was "Venus", which occurred when one astragalus fell showing each of the 4 faces. Find the probability of a Venus.

3.2. When tossing 4 astragali, find the probability that all 4 show the same face.

3.3. Show that the formula on page 108 for the probability of a round of craps lasting k rolls, $P(k) = p_1 \cdot p_2^{k-2} \cdot p_3$, can be written in terms of the single probability p_1 and k.

3.4. Place bets on 6 or 8 are a better gamble than the *Big 6* and *Big 8* bets, which were available on older craps layouts. New Jersey state law, however, forbids the Big 6 and Big 8 bets. In Figure 3.2 on page 107, spaces for these bets might be found at the lower left corner, near the Pass line and Field space as shown in Figure 3.16. Like Place bets on 6 or 8, these bets win when the

FIGURE 3.16: Big 6 and big 8 betting spaces on a craps layout.

chosen number is rolled before a 7 is rolled. Unlike Place bets, Big 6 and Big 8 pay even money. Compare the HA on Big 6 and Big 8 to the HA of Place bets on 6 or 8.

3.5. As shown in Exercise 3.4, the Big 6 and Big 8 bets have higher HAs than a simple Place bet on 6 or 8, even though the bets involve the exact same events. The decline of player interest in Big 6 and Big 8 has opened up some scarce room on craps tables, and so some casinos have taken the space at the corner of their craps layout formerly devoted to those bets and installed spaces for other bets rather than continuing the Pass line around the corner.

a. The game designers behind the Hard Pass side bet also devised a use for the corners of the craps layout with the *Even* and *Odd* bets [51]. These are one-roll bets that the next roll with be either even or odd. Since 18 of the 36 sums when 2d6 are tossed are even and 18 odd, a 1–1 payoff would provide a fair bet. Paying less than even money on every win would be likely to discourage action, so the bets pay less than even money when one of the craps numbers—2 and 12 for Even, and 3 for Odd—are rolled. What payoff odds on these numbers give a 1.11% house edge?

b. Harrah's and the Rio in Las Vegas have installed two bets in this space: *Under and Over 7*.

Under and Over 7 has its roots in a simplified two-die game that is sometimes seen at carnivals and charity casino night fundraisers where craps might be considered too complicated or too time-consuming. It can also be played as a street game, with the simple three-square layout, Figure 3.17, drawn with chalk on concrete [113].

Under	7	Over
2,3,4,5,6		8,9,10,11,12
Even money	5 for 1	Even money

FIGURE 3.17: Layout for Under and Over 7.

Three wagers are available: Under 7, 7, and Over 7. Under 7 and Over 7 pay off even money if the sum of two rolled dice is, respectively, less than or greater than 7. The 7 bet pays off 5 for 1 if the next roll is a 7. Find the HA for each wager.

c. At the Cannery Casino in North Las Vegas, a modified version of the Under and Over 7 bet occupied the outside corner in 2018. Two one-roll bets were defined: Low Dice and High Dice. The Low Dice bet pays even money if the next roll is a 3–6, and 5 for 1 if the next roll is a 2. High Dice works similarly: paying 1–1 on an 8–11 and 5 for 1 on a 12.

The bonus payouts on 2 or 12 make these bets better for the gambler than Under and Over 7. Find the new HA for these bets.

d. In 2020, the Gold Coast Casino in Las Vegas gave this spot over to *Muggsy's Corner*, a bet that was advertised as bringing a Roaring 20s feel to the craps table [155]. Muggsy's Corner pays off 2–1 when the come-out roll is a 7 and 3–1 if the shooter sets a point on the come-out roll and then rolls a 7 on the next throw. Find the probability of winning a bet at Muggsy's Corner.

3.6. Calculate the cumulative house advantage on a $10 Don't Pass bet backed up with 3× odds and a point of 8.

3.7. A different twist on craps odds bets came with a proposal for a "Hardway Odds" bet, available immediately after an even number was set as the point [70]. All hardway odds bets paid off no matter how the point was made, but paid off more if the point was made as doubles. On a point of 4 or 10, hardway odds paid 4–1 on a pair and even money otherwise. For hardway odds on 6 or 8, the hardway payoff was 2–1 and an easy point paid even money.
Confirm that these are indeed zero-edge wagers.

3.8. Suppose that the Hardway Odds bets described in Exercise 3.7 were reconfigured so that the odds bet only paid off if the point was made the hard way. Find appropriate payoff odds on both bets so that these new bets would retain a 0% HA.

3.9. Suppose that the main m and the chance k have been set in a round of hazard. If the main can be thrown in M ways and the chance can be thrown in K ways, derive a formula in terms of M and K for the probability function $p_3(k, m)$ where the chance is neither a nick nor a crab.

Street Craps and Crooked Dice

3.10. Suppose a street craps hustler offers even money that the shooter will roll a 7 before rolling either a 9 or 11. Assuming standard dice, is this a fair wager?

3.11. What is the street craps hustler's edge when offering an even-money bet that the shooter will roll a 6 before rolling a 7, with the first roll barred?

3.12. A craps hustler has inserted a single 1–2–3 Top die into a street craps game alongside a standard die. Devise a plausible bet where the hustler has a 100% advantage.

3.13. Find the player advantage for a \$5 Place bet on 9 using the dice from Example 3.23 on page 133.

3.14. Confirm that the sum of the 11 probabilities in Table 3.14 for a pair of 1–6 shapes is 1.

3.15. Find the player advantage on a hardway 6 or 8 bet, paying 10 for 1, if the shooter is using a hardway set with perfect control and the dice fall according to Table 3.15.

Variations

3.16. Show why Crapless Craps does not offer a Don't Pass wager by finding the player advantage if that bet was available. Note that this hypothetical bet cannot win on the come-out roll.

3.17. Suppose that Crapless Craps was modified so that the 11 and 12 were barred on the come-out roll [89]. As is the case in standard craps where the 12 is barred on Don't Pass bets, if either of these numbers was rolled first, Pass bets would be "no decision". and a new come-out roll would follow. This would mean than 11 and 12 would not be possible points for the shooter.

Find the HA on a Pass line bet for this modification.

3.18. Find the house advantage on a \$1 Put bet on 8 with 10× odds.

3.19. Find the minimum integer odds multiplier that takes the HA of a Put bet on 4 under 2%.

3.20. The minimum bet at Turn & Burn Craps is \$5. Use this value to compute the house advantage of the Any Pair wager.

3.21. Figure 3.18 shows a pair of 5-sided dice.

FIGURE 3.18: Five-sided dice.

These dice are triangular prisms with 3 edges flattened out and rounded so that the dice cannot land on them. The numbers 1 and 5 are on the flat ends of the prism, while 2, 3, and 4 are on the curved edges. There are 3 blank faces on which the dice can land; when this happens, one of the curved edges is uppermost.

Find reasonable rules for Pass and Don't Pass bets for a craps game using 2d5.

3.22. Find the probability of winning a Red Sum = Blue Deluxe Devil Dice bet covering all blue numbers from 2–6, and suggest a payoff that gives a viable house advantage.

3.23. Suppose that the casino gives up a large fraction of its advantage on the Red Sum = Blue bet by declaring a push whenever the blue die shows a ⊡. Find an integer-to-1 payoff that gives a HA between 1 and 7%.

3.24. Find the probability of winning the following Straight bets at Deluxe Devil Dice:

a. A bet on the single straight 1–2–3.

b. A bet on the straight 3–4–5.

c. A single bet covering all of the straights, with skip straights excluded.

d. A single bet covering all possible skip straights.

e. A wager covering all possible straights and skip straights, with the blue die in the middle.

f. A bet covering all skip straights with the blue die in the lowest position.

Card Craps

3.25. Confirm that a right bet on 6 in the original New Jersey Card Craps game is fair.

3.26. The Pala Casino offers an additional Card Craps bonus bet that pays off if the two cards dealt to the table are the 2 and the 12. Only one of these cards determines the roll; the second card is turned over if necessary to resolve this bet. Find the probability of winning this bet.

Side Bets

3.27. In an average hand of craps, how many different doubles are rolled before a 7 turns up?

3.28. Suppose that you make the Tall and Small bets at the same time. Using the theoretical probabilities, find the probability of winning exactly one of the two bets.

3.29. Show that the probability of winning a Repeater bet can be rewritten without reference to q, as

$$P(\text{Win}) = \left(\frac{6p}{6p+1} \right)^k .$$

3.30. In 1978, the first legal mainland American casino outside of Nevada opened in Atlantic City, New Jersey. The Resorts International Casino briefly offered a 3-Way Seven bet on its craps tables. This one-roll wager allowed players to bet on any way to roll a 7 on two dice: 6–1, 5–2, or 4–3, and paid off at 15–1 [118]. While this wager was less player-unfriendly than the notorious Any Seven bet, its house advantage was still very high. Compute the HA on this bet.

3.31. The most probable roll for a pair of standard dice is 7. It follows that the most likely sum when rolling 2d6 twice is 14. In light of this, the *14* side bet at craps is a simple wager that two successive rolls of the dice will add up to 14. This bet pays 7–1 (or 8 for 1).

a. Find the probability of winning the 14 bet.

b. Use the probability found in part a. to compute the HA.

c. The HA found in part b. may be too large to make the 14 bet popular. Suppose that the payoff is doubled, to 14–1, if the 14 is formed by a 2 and a 12, in either order. Find the new HA.

3.32. The *Jackpot!* side bet pays off if the shooter rolls a string of consecutive rolls of 7 or 11 before establishing a point and before rolling a 2, 3, or 12 [74]. The proposed pay table for a $1 bet is based on the length of the string and is shown in Table 3.41. Seven straight rolls totaling 7 or 11 trigger a progressive jackpot split among all winning players.

TABLE 3.41: Jackpot! payoffs

Number of 7s or 11s	Payoff
4	$50
5	$200
6	$1000
7	Jackpot

a. Find the probability of 4, 5, 6, and 7 consecutive rolls of 7 or 11.

b. What jackpot amount rounded to the next dollar, gives a fair wager when won by a single gambler?

3.33. The *Midway* bet was a one-roll craps side bet in place at some casinos in Atlantic City. A Midway bet paid off if the next roll was a 6, 7, or 8—the "middle" of the range of outcomes. The payoff was even money unless the shooter rolled a hard 6 or hard 8, in which case the bet paid 2–1. This bet might be seen as attractive because the gambler was getting the three most commonly rolled numbers working in his or her favor. Find the HA on the Midway bet and comment on how good a bet this was or wasn't.

3.34. As noted on page 66, a *prime* number is a integer $p > 1$ that is only divisible by itself and 1. The first few prime numbers are 2, 3, 5, 7, and 11. Consider a hypothetical one-roll craps side bet that the next roll will show a prime number. If the bet pays 2 for 1, what is its house advantage?

3.35. If the prime number bet in Exercise 3.34 was implemented in Spider Craps, what would the new HA be?

Chapter 4

Baccarat

One of the most glamorous casino table games is *baccarat*, a favored game of James Bond in six movies. Baccarat is a card game that is often played in the high-limit area of casinos and is the game of choice for many high rollers, among whom it is not unusual to risk tens of thousands of dollars on a single hand. Outside high-limit rooms, it is sometimes seen in a scaled-down form known as *mini-baccarat* or "mini-bac" for short.

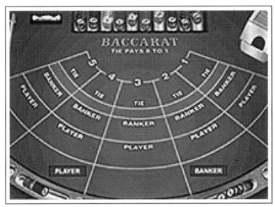

FIGURE 4.1: Electronic mini-baccarat table [21].

4.1 Baccarat Basics

Whether the full-size game or mini-baccarat, the rules of baccarat are simple: gamblers bet on which of two hands of cards—called the *Player* and the *Banker*—will be closer to 9. All players bet on one hand or the other; while a full-size baccarat table may have room for 14 gamblers, players do not receive individual hands. Players may also choose the Tie bet, which wins if the two hands have the same value.

Two cards, from a shoe containing six to eight decks, are initially dealt to each hand. The Riviera Casino once used a 16-deck shoe in an effort to encourage players to gamble longer—some players leave almost reflexively at the conclusion of a shoe while the cards are being shuffled. The Union Plaza Casino in downtown Las Vegas (now the Plaza) briefly used a 144-deck novelty shoe.

Baccarat hand values are computed by counting each card at its face value, with aces counting 1 and face cards 0. If the sum of the cards exceeds 9, then the tens digit is dropped. Unlike in blackjack, a baccarat hand does not "bust" if it exceeds the highest possible hand value of 9.

Example 4.1. A baccarat hand dealt a 9 and a 5, totaling 14, would be valued at 4. A pair of 9s dealt together make a hand of 8.   ■

If either hand has an initial 2-card value of 8 or 9, the hand is called a *natural*, and no further cards are drawn. If both hands are naturals, a 9 beats an 8. If there are no naturals, the game rules specify when either hand receives a third card; there is no player or dealer choice involved. The Player hand, which always goes first, has simpler rules: it is dealt a third card if its first two cards total 5 or less. If the Player hand stands with 6 or 7, the Banker hand draws a third card if its value is 5 or less. If the Player draws a third card, the Banker hand's action is determined by its value and the value of the Player's third card. Table 4.1 shows the standard rules.

TABLE 4.1: Baccarat rules for the Banker hand

Banker's hand	Banker draws if Player's 3rd card is
0–2	Anything
3	Not an 8
4	2–7
5	4–7
6	6 or 7
7	None
Banker draws on 0–5 if Player stands on 2 cards.	

Since the rules for drawing a third card are fixed, baccarat is a game of pure chance; there is no skill involved in the game.

Example 4.2. If the Player hand is 7♣ K♣, its value is 7. If the Banker hand draws 4♣ 3♣, it also scores 7. The rules of baccarat in Table 4.1 mean that no additional cards are drawn, and the hand ends in a 7–7 tie.

On the next hand, suppose the Player draws Q♡ 3♠ and the Banker hand is K♣ 2♣. Player leads 3–2, but each hand will now draw a third card. If the Player's third card is the 9♡, the total of 12 drops the 1, and the hand is valued at 2. If the Banker's third card is the 7♡, Banker wins 9–2.   ■

These rules for drawing and standing by the Banker hand include several combinations where the Banker hand is required to draw another card even though it's beating the Player hand already. For example, suppose that the Player hand is 2-2, totaling 4. The Player takes a third card and draws a 7, bringing that hand's total to 1. If the Banker's hand is Q-5, for a total of 5, the Banker leads but must nonetheless take another card since the Player's third card was a 7. It is also possible for the Banker hand to be denied a third card even if it's tied with the Player hand, as in the case where the Player has A-4 and draws an 8 for a total of 3, and the Banker holds 3-K. Because the Banker has 3 and the Player's third card was an 8, no further card will be drawn to the Banker hand, and the round ends in a tie.

Table 4.2 shows the PDF for the final value of the Player and Banker hands when dealt from an 8-deck shoe. Examination of this table shows that the complicated rules for drawing lead to considerable uniformity among the probabilities for both the Player and Banker hands.

TABLE 4.2: Probability distribution for baccarat hand values [119]

Value	Player	Banker
0	.0940	.0888
1	.0745	.0693
2	.0743	.0691
3	.0745	.0728
4	.0743	.0934
5	.0745	.1007
6	.1332	.1211
7	.1337	.1284
8	.1332	.1280
9	.1337	.1284

How did game designers settle on these rules?

Assume that the infinite deck approximation is in force. Consider the rule that the Banker hand always draws when holding 0–2, regardless of the Player total. The probability of a Banker 0 is

$$P(0) = \left(\frac{4}{13}\right)^2 + 9 \cdot \left(\frac{1}{13}\right)^2 = \frac{25}{169}.$$

Here, the first term computes the probability of drawing two 0s and the second is the probability of drawing two cards that add up to 0 without drawing any 0s. There are 9 ways to pick the ranks of the first and second cards that produce a sum of 0: A-9, 2-8, ..., 9-A.

The probability of any other 2-card total of $n, 1 \leqslant n \leqslant 9$, is then

$$P(n) = 2\left(\frac{1}{13} \cdot \frac{4}{13}\right) + 8\left(\frac{1}{13}\right)^2 = \frac{16}{169}.$$

In this expression, the first term computes the probability of a total of n

arising from drawing one n and one 0, in either order, and the second term accounts for the 8 ordered ways to draw 2 nonzero cards that add up to n, either by themselves or after subtracting 10. The probability that the Banker hand is between 0 and 2 and thus will always draw a third card is then

$$P(0) + P(1) + P(2) = \frac{57}{169}.$$

From this 2-card hand, we can construct a PDF for the final hand. The probability of drawing a 0 is $\frac{4}{13}$ while the probability of drawing any other rank is $\frac{1}{13}$. Table 4.3 shows the distribution.

TABLE 4.3: Probability distribution for final Banker baccarat hand with a 2-card total of 0–2

Final hand	Initial 2-card hand		
	0	1	2
0	.3077	.0769	.0769
1	.0769	.3077	.0769
2	.0769	.0769	.3077
3	.0769	.0769	.0769
4	.0769	.0769	.0769
5	.0769	.0769	.0769
6	.0769	.0769	.0769
7	.0769	.0769	.0769
8	.0769	.0769	.0769
9	.0769	.0769	.0769

The probability that the Banker hand improves when starting with a 0, 1, or 2 is $\frac{9}{13}, \frac{8}{13}$, or $\frac{7}{13}$, respectively. While the chance of improvement is significant, it is not 100%. Both hands have a chance to win or tie no matter their value; nothing is certain until the final card is dealt. The notion that "every card counts" is an attractive one.

Example 4.3. Consider the rule where the Banker hand takes a third card when holding 0–5 if the Player hand stands on 2 cards. If the Player hand stands on its initial cards, its value must be either 6 or 7, and so the Banker hand cannot win without drawing.

The probability that Player draws an initial 6 or 7 is $\frac{32}{169}$.

Under infinite deck, the probability of a Banker 6 or 7 in 2 cards is also $\frac{32}{169}$. These sums are equally likely, which means that in a faceoff between 6s or 7s, each hand has an equal chance of winning. If both hands stop after 2 cards with totals of 6 or 7, the probability of a tie is ½. ∎

Example 4.4. If the Banker's hand is 3 and the Player's third card is an 8, the Banker hand does not draw a third card. When Player draws an 8 as its third card, it follows that the hand's value is 8, 9, 0, 1, 2, or 3, according as the initial two-card hand was 0, 1, 2, 3, 4, or 5. The probability of a Player 0–5 that draws a third card (because the Banker does not have a natural 8 or 9) is

$$\frac{105}{169} \cdot \frac{137}{169} \approx .5037.$$

The 3-card Player hand has the probability distribution shown in Table 4.4.

TABLE 4.4: Player hand PDF when drawing an 8 as the third card

Value	Probability
0	.1524
1	.1524
2	.1524
3	.1524
8	.2381
9	.1524

Forcing Banker to stand on 3 gives the following outcomes:

$$P(\text{Banker wins}) = \frac{48}{105} \approx .4571.$$

$$P(\text{Player wins}) = \frac{41}{105} \approx .3905.$$

$$P(\text{Tie}) = \frac{16}{105} \approx .1524.$$

If the Banker were allowed to draw when holding 3 and giving 8, the probabilities shift slightly:

$$P(\text{Banker wins}) \approx .4755.$$
$$P(\text{Player wins}) \approx .4125.$$
$$P(\text{Tie}) \approx .1121.$$

Considering the game as a whole makes the rule not to draw to 3 when giving 8 the right decision to balance the game. ∎

In an eight-deck game, the probability of winning a resolved bet (that is, when the hand is not a tie) on Player is .4932 and the probability of winning a resolved bet on Banker is .5068 [8]. Winning Player bets are paid at 1 to 1, but since the probability of a Banker win is greater than .5, an even-money payoff would give gamblers a 1.36% advantage over the casino. As with the craps Don't Pass bet, some adjustment must be made to preserve a casino advantage. To do this, standard baccarat rules charge a 5% commission on winning Banker bets. The effective payoff on winning Banker bets is then .95–1, or 19–20. The Tie bet pays off at 8 to 1 if the two hands tie, and this occurs with probability .0955, slightly less than once every 10 hands [8]. If the hand is a tie, bets on Banker and on Player are also ties, and no money is won or lost.

As a convenience to everyone involved, players and dealers alike, some casinos use $20 chips at their baccarat tables; this is a denomination not ordinarily used elsewhere in the casino. Typical casino chips are produced in $1, $5, $25, $100, and $500 denominations, with higher denominations available if needed to streamline play at high-limit tables. $20 chips make the calculation of commissions easy: $1 in commission for every $20 chip in play on a winning Banker wager. In practice, to avoid complicated change-making (and time-wasting) transactions, all winning Banker bets are paid at even money, and commissions owed the casino are simply counted during the play of a shoe. Dealers often conduct the accounting using *lammers*: small tokens, slightly smaller than a casino chip, imprinted with a numerical value. Some lammers are shown in Figure 4.2. When a player wishes to leave or the shoe

FIGURE 4.2: Casino lammers, similar to those used in tracking baccarat commissions.

is exhausted and the cards are being shuffled or replaced, the casino collects the accumulated commissions owed.

The expected values of the three baccarat bets are computed here. In the case of Player and Banker bets, hands ending in a tie are disregarded in finding the probability of a bet winning or losing.

$$E(\text{Player}) = (1) \cdot .4932 + (-1) \cdot .5068 = -.0136$$
$$E(\text{Banker}) = (.95) \cdot .5068 + (-1) \cdot .4932 = -.0117.$$
$$E(\text{Tie}) = (8) \cdot .0955 + (-1) \cdot .9045 = -.1405.$$

Some casinos pay 9–1 on a winning Tie bet; this raises the expected value to –.0450, and the HA drops to 4.50%.

Example 4.5. A 5% commission on winning Banker bets is standard, but not mandatory. As a promotion, suppose that a casino reduces the commission to 4%. How much does this reduce the HA?

The probabilities of winning and losing a Banker bet are not changed by the change in commission, of course. The effective winning payoff is .96–1, and the expected value becomes

$$E = (.96) \cdot .5068 + (-1) \cdot .4932 = -.0067,$$

cutting the HA to .67%. ∎

Cutting the standard commission in half to 2.5% results in an essentially even game, with the player holding a tiny edge of .093%.

Example 4.6. In December 1989, the Sahara Casino in Las Vegas ran a baccarat promotion that eliminated the commission entirely [161]. The expectation of a $1 Banker bet was then

$$E = (1) \cdot (.4584) + (0) \cdot (.0955) + (-1) \cdot (.4461) = \$.0123,$$

giving gamblers a 1.23% edge on Banker bets. ∎

4.2 Card Counting in Baccarat

Baccarat, like blackjack, is a game in which cards are not replaced after each hand, and so successive hands are not independent: prior hands affect future hands. This naturally leads to the question of whether card counting as employed in blackjack (Section 5.3) can be used in baccarat to give then player an advantage over the casino.

There are two questions to consider:

1. Is it possible to devise a card-counting system for baccarat?

2. Does card-counting a baccarat shoe give the gambler an advantage?

The answer to the first question is "Of course"—several card counting systems have been developed. That said, devising a card-counting system is a comparatively simple task. Verifying that such a system gives gamblers a meaningful advantage under casino conditions is a much more complicated problem.

Example 4.7. Given a particular block of six cards ready to be dealt, it is possible for the sharp player to gain an advantage. For example, if it is known that the next six cards to be dealt are all 7s, which leads to both hands totaling 4, drawing a third card, and tying at 1, a large Tie bet is guaranteed to win—but it is extremely unlikely that any baccarat player would encounter such a setting.

The probability of 6 consecutive 7s occurring anywhere in an 8-deck shoe can be computed by thinking of the 7s as glued together and moving through the shuffle as a single card, making the effective deck size 411 cards. There are

$$\binom{32}{6} = 906,192$$

different ways to choose the 7s. Since the order that the cards fall in is important here, we count using permutations rather than combinations. Once the six 7s have been chosen, there are $6! = 720$ ways to arrange them in order. The deck of 411 cards, with six 7s together, can be arranged in 411! ways, and the complete shoe of 416 cards can be arranged in 416! ways. The probability of 6 consecutive 7s is then

$$(906,912 \cdot 6!) \cdot \frac{411!}{416!} \approx \frac{1}{18,625}.$$

This qualifies as a rare event—and it's at least as unlikely that a card-counting system would detect this concentration of 7s. ∎

The most significant counting systems each focus on one of the three available bets. A gambler looking for shoe configurations that favor the Player bet should use the count shown in Table 4.5.

TABLE 4.5: Baccarat card-counting values for the Player bet [67]

Card	A	2	3	4	5	6	7	8	9	T
Value	0	−1	−1	−2	−2	1	2	2	1	0

This counting system follows the model of the High-Low count in blackjack by keeping a running count as cards are exposed. This is easy since a single hand of baccarat consists of no more than 6 cards, and all are eventually turned face up. The running count is then converted to a true count by dividing by the number of decks remaining in the shoe, which gives a per-deck accounting of the balance between positive and negative cards. In each case, the true count measures a change in the HA: positive true counts move the edge toward the Player hand while negative TCs increase the house advantage on a Player bet.

Why were these values chosen? An initial look at Table 4.5 shows that prospects for the Player hand improve when big cards are dealt. When the shoe contains a surplus of small cards, the Player hand has a good chance of a two-card total of 5 or less, which means that it will draw a third card. That surplus of small cards then makes it more likely that the third card improves the hand.

For the Banker bet, a different counting system, Table 4.6, is used.

Where low cards in the shoe favor the Player bet, the reverse is true for the

TABLE 4.6: Baccarat card-counting values for the Banker bet [67]

Card	A	2	3	4	5	6	7	8	9	T
Value	0	1	1	1	2	−1	−2	−1	−1	0

Banker hand: high cards dealt to the table decrease the count. High cards in the shoe favor the Banker hand, since the higher its initial total, the less likely it will have to draw a third card and risk running the sum over 9, resulting in a low final total.

What's good for the player is not exactly what's bad for the Banker. We note that the values of the various ranks in the Player and Banker counts are not merely negatives of each other: the 4 and 8 break this pattern. Fours are worse for the player than they are good for the Banker, and 8s are better for the Player than they are bad for the Banker.

An alternate counting system combines the two methods above. The system in Table 4.7 produces a true count without division: simply add or subtract the indicated values as cards are dealt. There is no need to consider the number of decks that remain undealt.

TABLE 4.7: Combined baccarat card counting system [98]

Card	A	2	3	4	5	6	7	8	9	T
Value	1	1	1	2	−1	−2	−1	−1	0	0

Using Table 4.7, a gambler should bet on Banker if the true count is +15 or less, and bet on Player if it's +16 or greater. This imbalance from "bet Banker if the count is negative and Player if it's positive" reflects the slightly higher probability of a Banker hand winning. However, this only indicates which of the two bets, Player and Banker, is favored relative to the other and says nothing about whether or not the bettor holds an edge over the house.

A count for the Tie bet focuses on odd-numbered cards. If they are all dealt out, then only even sums are possible: 0, 2, 4, 6, and 8. The chance of a Player-Banker tie increases in this admittedly rare circumstance. The Tie count is simple: count +1 when an odd-numbered card is dealt. There's no need to divide: when the count reaches +160 (indicating that all of the odd cards in an 8-deck shoe have been dealt), make the Tie bet. The player advantage here is 62% [80]. This is, as one might surmise, a highly unlikely event.

Example 4.8. If 350 cards have been dealt from an 8-deck shoe, the probability that all 160 odd cards have been dealt is simply

$$\frac{\binom{256}{190}}{\binom{416}{350}},$$

the number of ways to choose the remaining 190 cards from the 256 even cards divided by the number of ways to deal 350 cards from a 416-card shoe. This number is approximately 2.847×10^{-16}, too small to be worthy of serious consideration. If a full 8-deck shoe was shuffled and dealt every day since the Big Bang (13.8 billion years ago), the probability that no shoe deals all 160 odd cards in the first 350 cards is about

$$\left(1 - \frac{\binom{256}{190}}{\binom{416}{350}}\right)^{5.0403 \times 10^{12}} \approx .998881.$$

The Tie count is thus destined to be more a mathematical curiosity than an effective tool for gamblers. ∎

Example 4.9. Consider the following sequence of baccarat hands dealt from the top of a shoe:

1. Player: 8♡ K♣.

   Banker: 10♢ 8♠.

   The hands tie at 8, and no further cards are drawn. The four counts are as follows:

   - Player count: +4. Since there are essentially 8 decks left, the true count is ½.
   - Banker count: −2. The true count is −¼.
   - Combined count: −2. This is also the true count. The combined count favors a bet on Banker for the next hand.
   - Tie count: 0—no odd cards were dealt.

2. Player: 10♠ 7♢.

   Banker: A♣ 10♣.

   Player leads, 7–1. The player hand stands and the Banker draws a third card, the 6♡. This brings its total to 7 for a second straight tie. The counts are updated:

   - Player count: +7. The true count is $\frac{7}{8}$, since we estimate decks remaining to the nearest half-deck.

- Banker count: –5. The true count is $-\frac{5}{8}$.

- Combined count: –4. A bet on Banker is still favored. Betting on Banker has lost nothing so far.

- Tie count: 2.

3. Player: $9\heartsuit$ $5\clubsuit$.

   Banker: $J\heartsuit$ $8\spadesuit$.

   Banker wins with a natural 8. Here are the counts:

   - Player count: +8. The true count is +1.

   - Banker count: –5. This has not changed from the last hand, since the four cards have values –1, 0, 2, and –1 respectively.

   - Combined count: –6. While we're nowhere near –200, Banker is still a better bet than Player.

   - Tie count: 4.

4. Player: $10\heartsuit$ $A\spadesuit$.

   Banker: $6\heartsuit$ $7\heartsuit$.

   Banker leads, 3–1, but Player will draw another card. It's the $8\spadesuit$, bringing the total to 9. Since the Banker has 3 and the third Player card was an 8, the Banker does not draw. Player wins, 9–3. The new counts are then

   - Player count: +12. If we round the number of decks to $7\frac{1}{2}$, the true Player count is +1.6, which we would round to +1.

   - Banker count: –9. The true count is approximately –1. A negative count here discourages Banker betting.

   - Combined count: –7. We would still bet on Banker, although sitting out a few hands, which is not unusual at a baccarat table, might also be a good option.

   - Tie count: 6. The waiting continues—as it must, since only 4 hands have been dealt. The minimum number of dealt hands to use up all of the odd-numbered cards is 27.

No matter which count system is used, the values move very slowly; the hands with a player edge will almost invariably be toward the end of the shoe. ∎

We see that devising a card counting system for baccarat is easily done. To the second question above, however, the answer must be "No: not to any useful extent." There are a number of reasons for this:

- Baccarat has simple rules which are rigidly prescribed, with no place for player decisions. Card counting in blackjack often calls for different player action under different deck compositions, including such actions as doubling down or splitting pairs that are not available in baccarat.

- An important part of the strategy behind card counting in blackjack is standing on a 12–16 against certain dealer upcards when the deck is rich in high cards in the hopes that the dealer will bust—the "let the dealer bust first" approach. Since baccarat hands do not bust if they go over 9, there is nothing to be gained by looking for bust opportunities.

- There is no bonus for a natural 8 or 9 in baccarat, whereas a natural 21 in blackjack pays better than even money (3 to 2 or 6 to 5, depending on the casino). Knowing when natural 8s or 9s are more likely to occur might have some value as far as increasing a wager, but even a natural pays only 1 to 1. Since sums of 8 and 9 may involve any card in the deck, there is no short list of ranks that could be easily tracked in search of two-card naturals.

- Bettors may wager on either the Banker or the Player hand. Since the rules are approximately symmetric for the play of the two hands, there are no cards that can be said to be much more favorable to one side than the other. By contrast, with ten-count cards in blackjack, an excess favors the players over the dealer and a deficit enhances the dealer's chance of winning.

Not long after the introduction of card counting methods to blackjack, mathematicians naturally examined how information gained from counting cards might be used to gain an edge at baccarat. Edward O. Thorp soon concluded that "no practical card counting systems are possible" for baccarat [56]. The advantage that a counter might gain is, at best, microscopic. A later computer simulation of 100,000,000 8-deck shoes using the Player count system described above showed that the Player bet generated a player edge of .33% when the true count was +30 or greater. This translates into a 15¢ advantage per shoe for a player betting $1000 only when the shoe turned favorable, a situation which could be expected to occur once in 1786 hands. A similar experiment with the Banker count showed that, even at a true count exceeding +30 (once in 5542 hands), the casino still held an edge of .19% [67]. While the combined count tells player which bet, Player or Banker, is favored relative to the other, betting with an edge over the casino requires waiting until the true count is +235 when betting Player and –200 for Banker [98].

Practically speaking, this suggests that a high enough Player or Banker count to turn the edge the gamblers' way would come very near the end of a shoe, with less than 1 deck remaining. Dividing by a fraction less than 1 gives a true count larger than the running count, making it easier—though not easy—to reach +30. While the combined count requires no division, its index values also show that a lot of cards need to be dealt before the gambler has even a tiny edge over the house.

For the Tie count, it is clear that a +160 count is also likely only near the end of a shoe. Frequently, baccarat dealers will place a cut card 14 cards from the end of the shoe and deal one more hand when the cut card appears. What is the probability that the last 14 cards are all even?

Mathematically, the position of the cards in the shoe does not matter; we are simply looking for the number of ways to pick 14 even cards out of a shoe containing 416 cards. Since there are 256 even-numbered cards in an 8-deck shoe, the probability is

$$\frac{\binom{256}{14}}{\binom{416}{14}} \approx 9.704 \times 10^{-4} \approx \frac{1}{1031},$$

so we would expect to see this a little less often than once every thousand shoes.

These tiny edges are hardly worth the effort, or the risk of exposure that is associated with card counting in blackjack—but that risk of exposure may not be all that great at a baccarat table. Most casinos are happy to provide scorecards and pens—often with two colors of ink, black (or blue) and red—to baccarat players who wish to track the winning hand from deal to deal in an effort to find and exploit patterns. Figure 4.3 shows one such double-sided pen.

FIGURE 4.3: Baccarat pen from the FireKeepers Casino in Battle Creek, Michigan. The left end of this pen writes in blue ink and the right end in red.

The patterns certainly exist, but this is only because in an extended gaming session, there will always be apparent patterns in the sequence of winning hands. These patterns, though, have no predictive value. For all practical purposes, the identity of the winning hand in a round of baccarat is independent of the previous rounds' winners. Counting cards is not a path to riches betting Player, Banker or Tie in baccarat. Side bets may be a different matter; some of these are discussed in Section 4.4.

4.3   Variations

Like blackjack, baccarat has been used as a jumping-off point for different versions of the same basic game. Some of these games, and some of their additional wagers, have been shown to be susceptible to card counting in ways that traditional baccarat is not.

Commission-Free Baccarat

The standard 5% commission charged on winning baccarat Banker bets is problematic on multiple levels. Without the commission, the Banker bet carries a player advantage, which casinos seek to avoid. Calculating and collecting the amount due can take considerable time, which is time that the dealers are not spending dealing hands and making money for the casino.

With that in mind, several baccarat variations have attempted to change the rules slightly to preserve a casino advantage on the Banker bet while eliminating the need for a commission.

* *EZ Baccarat* and *Dai Bacc* simplify the game by providing that if the Banker wins with a 3-card total of 7, Banker bets push instead of winning. Since this turns a gambler's win into a push, players may resist this change.

   The probability that the Banker has a 7 in 3 cards is approximately .0335 [119]. For this to be a winning hand, the Player hand must not total 7, 8, or 9 in 3 cards or 7 in 2 cards. (A 2-card Player 8 or 9 would stop the game before a third card is drawn.) The combined probability of these 4 hands is .2112, which means that the probability of a winning 3-card Banker 7 is

$$(.0335) \cdot (1 - .2112) = .0070.$$

   This is enough of a shift to balance out the lost commissions.

* *Nepal Baccarat*, a version offered at the Marina Bay Sands Casino in Singapore, cuts the payoff on a winning Banker hand of 6 from 19–20 to 1–2. Turning a win (or, more accurately, 95% of a full win) into half a win rather than a push may make this variation more palatable to gamblers.

   The probability of a Banker final hand of 6 winning against a Player hand of 0–5 is .0539 (see Table 4.2 on page 193). This is the probability that this new rule will be invoked; since it decreases the payoff from .95 to .50, the net effect is to increase the HA on Banker by 2.43%, for a net HA of 1.07%.

- *2-1 Baccarat* is a slightly stronger modification than EZ Baccarat or Nepal Baccarat. In this variation, winning hands of 8 or 9 pay 2–1 instead of even money, but both Banker and Player bets lose if the hand ends in a tie [67]. A tie in standard baccarat means that Banker and Player bets push; turning them into losing hands represents a bigger change than EZ Baccarat imposes. Paying 2–1 on a small proportion of winning hands may not be seen by players as adequate balancing—of course, the two changes are not intended to offset each other exactly, because the casino needs to make up for not collecting commissions.

Baccarat with 3-Card Hands

The rules for drawing a third Banker card at baccarat, Table 4.1, are complicated. One avenue for game designers working on baccarat variations is to simplify the game by dealing 3 cards to each hand from the outset. Several variations on this idea have been introduced in casinos.

The Resorts World Genting Casino in Malaysia is home to *Three Pictures*, a single-deck variation on baccarat that is played with 3 cards per hand. There is no opportunity for either hand to draw additional cards. Unlike standard baccarat where only 2 hands are dealt regardless of the number of players, each player has their own three-card hand and plays against the dealer's hand. Standard baccarat scoring is used, with face cards counting 0, all other cards counting their face value, and the tens digit on any total over 10 discarded. A hand made up of 3 picture cards is called "Three Knights"; this is the highest-ranking hand despite scoring 0 [97]. All other hands are ranked by their score, with hands closer to 9 winning.

Example 4.10. If a player is dealt $Q\clubsuit\ 3\diamondsuit\ J\diamondsuit$, her total is 3. If the dealer holds $5\diamondsuit\ 2\spadesuit\ 5\spadesuit$, her total is 2, and the player wins. ∎

The casino takes a 5% commission on all winning bets, as is the case in traditional baccarat when a gambler wins a Banker bet. At the Crown Casino in Melbourne, Australia, this game is simply called Three-Card Baccarat. Both games have been marketed in a commission-free version where a winning player total of 6 pays off at 1–2 instead of 1–1, as in Nepal Baccarat.

Three Pictures allows players to bet on their hand in a showdown with the dealer's, or to make a Tie bet that, like in baccarat, pays off at 8–1. An additional option is the *Three Pictures* bet. In Australia, the bet pays 20–1 on 3 picture cards and 5–1 if the player holds 2 picture cards.

The probability of drawing three picture cards in a 1-deck game is

$$\frac{\binom{12}{3}}{\binom{52}{3}} = \frac{220}{22,100} \approx .0010,$$

and the probability of drawing exactly 2 pictures is

$$\frac{\binom{12}{2} \cdot 40}{\binom{52}{3}} = \frac{2640}{22,100} \approx .1195.$$

The expectation of the Australian version of the Three Pictures bet is

$$E = (20) \cdot \frac{220}{22,100} + (5) \cdot \frac{2640}{22,100} + (-1) \cdot \frac{19,240}{22,100} = -\frac{1640}{22,100} \approx -\$.0742.$$

In Macau, Three Pictures is known simply as *Three-Card Baccarat*. This game includes a side bet, Three Faces, which pays 16–1 if the Banker hand contains 3 face cards. The absence of a payoff for 2 face cards makes this a weaker version of the Australian bet. As we saw above, the probability of drawing 3 face cards is $\frac{220}{22,100}$ from the top of a single deck. This yields an expected value of

$$E = (16) \cdot \frac{220}{22,100} + (-1) \cdot \frac{21,880}{22,100} = \frac{18,360}{22,100} \approx -.8308$$

and thus Three Faces carries a HA of 83.08%.

If Three Faces attracts players, then there's no real incentive for casinos to change the payoffs on this bet, which is one of the worst wagers to reach a casino floor. If, however, this bet isn't receiving enough action, it can certainly be restructured with any HA the casino wishes. If Three Faces pays off at X to 1, its expected value as a function of X is

$$E(X) = (X) \cdot \frac{220}{22,100} + (-1) \cdot \frac{21,880}{22,100} = \frac{220X - 21,880}{22,100}.$$

Setting this equation equal to the desired expectation and solving for X produces the correct payoff.

Example 4.11. For a 30% HA, we are looking at

$$E(X) = \frac{220X - 21,880}{22,100} = -.3000,$$

whose solution is $X = 69.32$. A 70–1 payoff yields an exact HA of 28.96%—close enough to 30% for all practical purposes, and much easier to pay off. ∎

Example 4.12. In order to match the 7.42% HA found on the Australian bet, it would be necessary to raise the Three Faces payoff to X to 1, where X is the solution to

$$\frac{220X - 21,880}{22,100} = -\frac{1640}{22,100},$$

or

$$220X - 21,880 = -1640,$$

from which we find that $X = 92$. ∎

7 Up Baccarat

7 Up Baccarat, a twist on baccarat that debuted at the Marina Bay Sands Casino in Singapore, gives the Player hand a 7 as its first card. This card is printed on the layout, so the initial deal only involves 3 cards [79]. With this rule in play, the Player hand has an enhanced probability of .3077 of drawing a 7 in 2 cards. This increased chance of a Player 7 is incorporated into gameplay in several ways.

- A winning Banker hand of 7 pays 9–5 instead of even money. A winning Player hand of 7, however, pays only 1–2.

- The Tie bet is reconfigured from a fixed 8–1 payoff. If the two hands tie at 7, the payoff is 9–1; Tie pays 7–1 for ties on any other number.

Though starting with a 7 seems to favor the Player bet, these changes give a house advantage of 2.59% for Player and 2.58% for Banker, more than double the edges from standard baccarat [67].

With the higher house edges, 7 Up Baccarat may only have appeal if counting cards can reduce the casino's advantage. A counting system focuses on aces and 2s, since these cards lead to Player naturals. When these cards are removed from the shoe, the Banker bet is preferred; when other cards are removed and the fraction of aces and 2s increases, a bet on Player is favored. There are separate counting systems for each wager, which are given in Tables 4.8 and 4.9.

TABLE 4.8: 7 Up Baccarat: Card-counting values for the Banker bet [67]

Card	A	2	3	4	5	6	7	8	9	T
Value	2	2	0	0	–1	–1	–2	–2	–2	1

TABLE 4.9: 7 Up Baccarat: Card-counting values for the Player bet [67]

Card	A	2	3	4	5	6	7	8	9	T
Value	–3	–3	0	0	1	1	1	2	1	0

The Banker count signifies an advantage at a true count of +3, while the Player count requires that the TC reach +4 before the bet has a positive player expectation.

Scarney Baccarat®

John Scarne (1903–1985), in addition to having a distinguished career as a magician, gambling consultant to the United States armed forces, and advisor to casinos and governments around the world, also developed a number of games for play in and out of casinos. *Scarney Baccarat®* was an intricate combination of elements of blackjack, baccarat, and craps that debuted in Curaçao before spreading to other casinos in the 1970s, though it is not currently offered in any casino [113, p. 447].

As the name suggests, Scarney Baccarat is first a variation on baccarat, and the game is won by the hand—player or dealer—coming closer to a total of 9, using baccarat counting rules. The game borrows from blackjack in dealing one two-card hand to each player rather than a single Player hand that faces the dealer's hand, and by allowing players to draw a third card, split pairs, double down, and insure against a dealer natural 9. From craps, the game offers a Field bet and additional side bets on the dealer's hand.

Scarney Baccarat is dealt from a four-deck shoe; two cards are dealt to each player and to the dealer. The players' cards are dealt face up, as is one of the dealer's cards. Based on their total and the dealer's upcard, players may elect to take a third card. As in baccarat, hands going over 9 are simply valued at their units digit. The dealer's action is fixed: he must stand on 5–9 and draw a third card if his hand totals 0–4. However, if the dealer's first two cards are 10s or face cards, giving a total of 0, and his first draw is another 10, he continues drawing cards until his total exceeds 0. This three-card hand of 0 consisting of three 10s is called a Scarney Baccarat; the subsequent draw is the only time when a Scarney Baccarat hand consists of more than three cards.

Example 4.13. The hand $K\diamondsuit$ $10\clubsuit$ $J\spadesuit$ is a Scarney Baccarat; a hand of $6\heartsuit$ $4\diamondsuit$ $Q\diamondsuit$ is a zero, not a Scarney Baccarat, and does not draw a fourth card. ∎

Since this rule changes a certain losing dealer hand into one which is guaranteed to beat some possible player hands, it's the source of the casino's advantage.

The probability of a three-card Scarney Baccarat, computed from the top of a fresh shoe, is

$$\frac{\binom{64}{3}}{\binom{208}{3}} = \frac{64}{208} \cdot \frac{63}{207} \cdot \frac{62}{206} = \frac{868}{30,797} \approx .0282 \approx \frac{1}{35.5}.$$

Scarney Baccarat also includes a pair of optional bets with analogs in craps. Players can make a one-hand wager that the value of the dealer's first two cards will be either 5, 6, 7, 8, or 9. Each of these bets pays off at 10 for 1. The Field bet at Scarney Baccarat merges these one-number proposition bets into a single wager, paying even money, that the dealer's first two cards sum to any value from 5 through 9.

These bets can be analyzed by looking at the various two-card combinations. There are $\binom{208}{2} = 21,528$ possibilities for the dealer's initial hand when drawn from a 4-deck shoe. These are sorted by their total in Table 4.10.

TABLE 4.10: Distribution of 2-card hands in Scarney Baccarat

Value	0	1	2	3	4
Count	3160	2048	2032	2048	2032
Value	5	6	7	8	9
Count	2048	2032	2048	2032	2048

Hands totaling zero are more numerous than any one nonzero hand because of the relative abundance of 10s in a standard deck; other even-numbered totals lag behind the odd numbers because hands made up of two identical cards (which must have an even sum) are slightly less common than pairs of two different cards. For example, two-card hands consisting of the $5\heartsuit$ and $J\spadesuit$ can be formed in $(4 \cdot 4)/2 = 8$ ways—we divide by 2 since the order does not matter—while a hand consisting of two $8\spadesuit$s can be made in only $\binom{4}{2} = 6$ ways.

For a side bet on 5, 7, or 9, the expectation on a \$1 bet is

$$E = (9) \cdot \frac{2048}{21,528} + (-1) \cdot \frac{19,480}{21,528} = -\frac{1048}{21,528} \approx -\$0.0487,$$

for a house edge of 4.87%. Side bets on 6 or 8 have a house advantage of 5.61%.

The probability of winning the Field bet is

$$\frac{2048 + 2032 + 2048 + 2032 + 2048}{21,528} \approx .4742;$$

the corresponding house edge is 5.17%.

In blackjack, an *insurance* bet (page 241) is a separate bet offered to gamblers when the dealer's first card is an ace. The bet pays off at 2–1 if the dealer subsequently draws a 10 or face card to complete a natural 21. Insurance in Scarney Baccarat comes in two forms, both of which have a maximum bet of half the player's main bet and pay off at 2–1, as in blackjack:

- *Scarney insurance*: Like blackjack insurance, this is offered when the dealer's upcard is a 9, and is a side bet for half the player's main bet that the dealer will turn over a 10 and thus have a natural 9.

- *Scarney baccarat insurance*: This is offered when the dealer's hand consists of two 10s and he is drawing a third card. Once again, the bet is capped at half the player's wager, and it pays off if the third card is also a 10. If this happens, the dealer will draw a fourth card, and a new round of insurance bets may be made. The opportunity to make Scarney baccarat insurance bets persists as long as the dealer keeps drawing 10s.

As in blackjack, insurance is only a bet worth making if the player is counting cards and has determined that the fraction of 10s to non-10s remaining in the undealt cards exceeds $\frac{1}{3}$.

With the blackjack options of splitting pairs, doubling down, and insurance available in Scarney Baccarat, it is possible to develop a version of basic strategy for the game; this is shown in Table 4.11. Using this strategy correctly allows a player to reduce the house advantage from the 2.44% obtained from the Scarney Baccarat rules to under 2% [113].

TABLE 4.11: Basic strategy for Scarney Baccarat [113, p. 447-50]

	Player's Hand	\multicolumn{10}{c}{Dealer's upcard}
	Player's Hand	0	1	2	3	4	5	6	7	8	9
No pair	6+	S	S	S	S	S	S	S	S	S	S
	5	H	H	S	S	S	H	H	S	H	H
	4	D	D	D	D	D	H	H	H	H	H
	3	D	D	D	D	D	D	H	H	H	H
	2	D	D	D	H	H	H	H	H	H	H
	0–1	H	H	H	H	H	H	H	H	H	H
Pair	A/A	D	D	D	H	H	H	H	H	H	H
	T/T	H	H	H	H	H	H	H	H	H	H
	9/9	S	S	S	S	S	S	S	S	S	S
	8/8	S	S	S	S	S	S	S	SP	S	S
	7/7	SP	SP	SP	SP	SP	SP	SP	SP	H	H
	6/6	SP	SP	SP	SP	SP	SP	SP	H	H	H
	5/5	H	SP	SP	SP	SP	SP	H	H	H	H
	4/4	S	S	S	S	S	S	S	S	S	S
	3/3	S	S	S	S	S	S	S	SP	SP	SP
	2/2	D	D	D	D	D	H	H	H	H	H

Never take insurance.
Key: D: Double down. H: Hit. S: Stand. SP: Split.

Aside from the insurance bets, can a player do better by counting cards? On one hand, we have seen that card counting has essentially no value in standard baccarat. On the other hand, this is in part because of the symmetry of the standard game; since a player may bet on either Player or Banker, there are no cards that preferentially benefit one side or the other. This is not a feature of Scarney Baccarat. However, there are no rules dictating when a player may draw a third card in Scarney Baccarat, and so the decision to stand or draw to an initial hand might be well-informed by knowledge of the composition of the remaining cards. At the same time, card counting's effects are necessarily limited because there is no opportunity to use any count to draw a fourth or fifth card.

The other feature of baccarat that limits the usefulness of card counting remains in Scarney Baccarat, though, and that is the fact that hands do not bust by going over 9. There's no incentive to stand on a comparatively low total in hopes that the dealer will bust, and this consideration means that counting cards at Scarney Baccarat was destined to be futile, except perhaps for the insurance bet.

Great 8 Baccarat

In 2015, the M Resort in Henderson, Nevada launched a new table game based on baccarat: *Great 8 Baccarat*. The game is played with six standard decks, each one augmented by two "Great 8" cards, for a total of 324 cards in play. The Great 8 is the highest-ranking card in the new deck. The goal of Great 8 baccarat is to have the hand closer to 8, either above or below. Unlike in conventional baccarat, hands totaling more than 9 are possible. Cards from ace through 10 count their face value; face cards each count 10. Two hands, Banker and Player, are dealt one card each, and as in baccarat, a set of fixed rules governs when either hand takes a second and final card [55].

Draw if first card is:	1, 2, 3, 4, 5
Stand if first card is:	6, 7, 9, 10
Game ends (no further cards):	8, Great 8 (naturals)

If the second card dealt to either hand is a Great 8, the first card is disregarded and the hand is valued as a Great 8 Natural.

Example 4.14. Suppose that the following cards are dealt:

> Player: 5♣
> Banker: J♡

The Banker stands with 10; the Player's 5 draws a second card, the 2♠. The player wins, 7–10.

In the next hand, we have these cards:

> Player: 3♠
> Banker: A♣

The Player leads 3–1, but both hands draw a second card. The Player draws the K♡; the Banker the 10♡. The Banker wins, since her hand of 11 is closer to 8 than the player's 13. ∎

A bet on the winning hand pays off at even money, with two exceptions [55]:

- If the winning hand is a Great 8 Natural, the payoff is 6–5.

- If the winning hand is a 6, bets on that hand push. This is the *Tiger 6* hand, and is the source of the casino's advantage. This rule eliminates the need to charge a commission on any winning hands.

Example 4.15. Under the infinite deck approximation, find the probability that one hand will end with a total of 6. This is only a Tiger 6 hand if it beats the other hand—against a competing 7, 8, Great 8, or 9, a 6 loses, and it pushes against a 10 or another 6.

A 6 on the first card ends that hand; the probability of an initial 6 is

$$P(6) = \frac{24}{324}.$$

An initial 7–10 draws no second card, and so cannot lead to a total of 6. If the first card dealt to a hand is an ace through 5, a second card is drawn which may bring the total to 6. Regardless of the first card, this probability is

$$p = \left(\frac{24}{324}\right)^2 = \frac{576}{324^2}.$$

The probability of a hand of 6 is then

$$P(6) + 5p = \left(1 + \frac{120}{324}\right) \cdot \frac{24}{324} = \frac{10,656}{104,976} \approx .1015.$$

∎

A hand of 6 wins as a Tiger 6 hand against opposing hands of 2–5 and 11–15. Using the infinite deck approximation, the probability distribution for the final value of a Great 8 baccarat hand is given in Table 4.12.

TABLE 4.12: Great 8 Baccarat: Final hand probability distribution

Value	Probability
2	.0055
3	.0110
4	.0165
5	.0219
6	.1015
7	.1015
8	.1015
Great 8	.0508
9	.1015
10	.3237
11	.0439
12	.0384
13	.0329
14	.0274
15	.0219

Using this table, the probability that a hand of 6 becomes a Tiger 6 hand by winning can be found by summing the probabilities of hands from 2–5 and 11–15; this sum is approximately .2195. Multiplying this value by the probability of a 6 gives the probability of a winning Tiger 6 hand as approximately .0223.

A 50-Card Deck

In 2013, Empire Global Gaming, Inc. began marketing a 50-card deck designed to provide new opportunities in old card games [39]. This new deck consisted of aces through tens in five suits or colors.

How would this deck affect baccarat? The fifth suit, of course, has no effect, since suits play no role in baccarat. Changing the ranks, and eliminating so many 10-count cards, has a meaningful effect on the casino edge in only one case, however. Since the gambler may bet on either the Player or the Banker hand and the shortage of 10s affects both hands in the same way, the probabilities of either hand winning are nearly unchanged. It should be noted that changing to the 50-card deck requires no adjustment of the rules for drawing a third card to either hand or to the 5% commission paid on winning Bank bets [15]. Table 4.13 compares the house edge on the three bets in the two games.

TABLE 4.13: 50-card deck: Baccarat house advantages [15]

Deck size	52 cards			50 cards		
Decks	Player	Banker	Tie	Player	Banker	Tie
4	1.24%	1.05%	14.59%	1.27%	1.01%	8.61%
6	1.24%	1.06%	14.44%	1.27%	1.01%	8.60%
8	1.24%	1.06%	14.36%	1.27%	1.01%	8.59%

The changes in the HA for the Player and Banker bets are less than .1%, and so are nearly insignificant. The major difference that the deck size introduces to baccarat lies in the house edge for the Tie bet, which drops from about 14.5% to 8.6%. While still too high to recommend this bet, the HA is a bit less unreasonable.

21st Century Baccarat

Native American casinos in the state of California are governed by the Indian Gaming Regulatory Act, a federal law passed in 1988 which served as the catalyst for Native casinos across the USA. These casinos may offer traditional baccarat and blackjack without state government interference, thanks to California Proposition 5, passed by voters in 2000. This proposition gave the Native casinos a chance to compete more easily with Nevada casinos, which is why the Nevada casino industry poured considerable resources into advertising opposing the measure [136]. In northern California, the San Francisco

Bay area is only about a 3½-hour drive from the casinos in Reno, while to the south, greater Los Angeles sits less than 4 hours from Las Vegas. On the I-15 freeway, the primary route between Los Angeles and Las Vegas, the border town of Primm, Nevada is home to several casinos just over the border in Nevada to capture California gamblers on their way to or from Las Vegas. This accounts for the interest of Nevada casinos in California gambling.

In the other direction, Nevada has no state lottery, but The Lotto Store at Primm, 50 feet from the state line on the California side, has the highest ticket sales of any lottery sales agent in the world.

Commercial non-tribal card rooms in California, however, are subject to State Penal Code 330, passed in 1873, which outlaws house-banked games throughout the state [156]. House-banked games are games where gamblers compete against the casino, such as craps, roulette, baccarat, and blackjack. Penal Code 330 restricts commercial casinos to card games where players compete directly against each other; this encompasses a variety of poker games as well as some card games developed specifically for the California market. Private poker games where the host takes no fee, or "rake", from the bets for the service of hosting the game, are not illegal under Penal Code 330.

Since the card room is merely hosting the game, it has no stake in the action and so cannot count on a house advantage to make a profit. Card rooms make money by charging a collection fee on every hand, . This fee is set by the card room; commonly, the collection fee is 1% of a player's wager with a minimum of $1. This mandatory fee makes many low-HA wagers, including Banker and Player bets, much worse for gamblers, especially low-stakes gamblers.

Consider the three basic baccarat bets: Player, Banker, and Tie. A $25 wager on each of these options carries a house edge of 1.36%, 1.17%, and 14.05%, respectively (page 196). If the card room charges a $1 collection fee per hand, this amounts to a 4% charge. A Player bet now effectively pays 25 for 26, and its new expectation (assuming the same probability of winning) is

$$E = (25) \cdot .4932 + (-26) \cdot .5068 = -.8468.$$

The new HA is 3.39%, nearly triple what is was without the collection fee. Similarly, the HA on Banker rises to 3.15%.

If the card room uses a fee structure of 1% of the wager with minimum fee of $1, betting less than $100 means that the player is giving the card room a higher edge. Over an hour-long session, paying an additional $1 per hand takes an inevitable toll on the gambler's bankroll.

Some California card rooms offer modified versions of baccarat that conform to the law while still retaining much of the game's essence. Because players must compete against each other and not the house, there are provisions for a player/banker, who functions as the dealer, faces off against the other players, and must pay all winning hands even as he or she collects from all losing players. A casino employee deals all of the cards; the player/banker designation is in name only.

At the Ocean's Eleven Casino in Oceanside, California, *21st Century Baccarat* fills the void. 21st Century Baccarat is a hybrid game with elements draw from blackjack that retains the flavor of baccarat and simplifies the draw/stand rules considerably. Two hands are dealt; one is assigned to the players and both of its cards are dealt face up. One of the player/dealer hand's cards is dealt face down, as in blackjack. Players must draw if holding 0–4, must stand on 7–9, and individually have the option of drawing a third card or standing with an initial 5 or 6. If a player opts to hit a 5 or 6, he or she so indicates by moving their wager from the Stand Line to the Hit Line on the layout. The player/dealer must hit on 5 or less and stand on 6 or more. Only one additional card is dealt to the player/dealer hand, regardless of the number of players requesting a third card.

Recalling that the draw rules for baccarat have been crafted to make the game nearly even, one might wonder how 21st Century Baccarat balances game play when player choice is added and the rules have been simplified. Part of the answer lies in how ties are handled. In standard baccarat, both Banker and Player bets push on ties. Here, ties at 0 or 1 are wins for the player/dealer. All other ties remain pushes.

Example 4.16. Robin, Alex, Chris, and Terry are playing a hand. The player hand is 10$\diamond$ 5$\spadesuit$ for a 5, while the player/dealer's upcard is the 7$\diamond$. Suppose that Robin, Alex, and Terry opt to hit the Player hand while Chris chooses to stand. A third card is then dealt to the Player hand: the 4$\diamond$. The three players who took a third card enter the showdown with a hand of 9, while Chris remains at 5.

The player/dealer turns over her hole card: the 7$\clubsuit$. Since her hand totals 4, she draws a third card. It is the 5$\diamond$, bringing her total to 9. Robin, Alex, and Terry tie the player/dealer, while Chris loses. ∎

Players, however, have two options for tie bets, which must be covered by the designated player/dealer. An Early Tie bet is made before any cards are dealt and works just as the Tie bet does in standard baccarat, paying 8–1. The new Late Tie betting option is available after the player hand and dealer's upcard are visible and comes in two forms:

- Players who have made an Early Tie bet may make a Late Tie bet in addition. This bet may not exceed the initial tie bet and also pays 8–1.

- A Late Tie bet may be made by any player who hs not made the Early Tie bet. This late bet is limited to $25 and only pays 6–1.

To diminish the advantage that might be gained by knowing 3 of the 4 cards, no Late Tie bets are accepted if the hands are tied with a visible score of 5 or greater.

The Commerce Casino in suburban Los Angeles formerly offered a version of 21st Century Baccarat that bore many of the features of blackjack, including the use of an 8-seat table instead of a 14-seat "batwing" table common in baccarat salons. Three 2-card hands were dealt: one belonged to the player/dealer, one was a community hand for all players in seats 1–4, and the third was a community hand for all players in seats 5–8. As at Ocean's Eleven, one player/dealer card was dealt face down.

Players who stood behind the table could make *backline* bets on either hand. This wagering option permits players who are not seated at the table to place their own bets on the hands of seated players. In card rooms offering this choice, the player in the chair has control of the hand and makes all decisions concerning play.

Once the two community hands were dealt, players had the following options:

- Stand, drawing no additional cards. This option was required on a hand of 7–9, and optional on a 5 or 6.

- Hit, drawing one additional card with no additional wager. All hands from 0–4 must hit. Players have the option to hit or stand on a 5 or 6.

- Double down, drawing one more card and doubling the initial bet. This is not as restrictive as its blackjack counterpart, since no baccarat hand will draw more than one card. Essentially, this represents a chance to double without giving anything up when a player judges the circumstances to be especially favorable. A player may double down on any hand totaling 6 or less.

- Surrender, forfeiting half the wager and exiting play for that hand. Like doubling down, surrender is available when the player hand is 0–6.

With two player hands in action, the player/dealer's choice to draw a third card or not cannot simply be keyed to what the player hands do. Player/dealers still hit on 5 or less and stand on 6–9. The "natural 9" rule is not played here; a 3-card 9 pushes a 2-card 9. There is no Tie bet; the player/dealer wins ties on 0 and 1 while ties from 2–9 push.

At the Hustler Casino in Gardena, an alternate version of 21st Century Baccarat is played with an 8-deck shoe containing 8 jokers. Jokers are wild: a hand containing one joker is scored as a 9; a hand with 2 jokers is called a natural 9 and ranked higher than any other hand summing to 9.

The probability of a natural 9 from the top of a fresh shoe is

$$\frac{\binom{8}{2}}{\binom{424}{2}} = \frac{8 \cdot 7}{424 \cdot 423} \approx 3.122 \times 10^{-4},$$

about once in 3203 hands. Counting jokers may indicate a change in this probability, but since there is no bonus for a natural 9, tracking jokers—though easy, since a full shoe contains only 8—brings no significant edge. A joker-rich shoe has just as much chance of delivering a natural 9 to the player/dealer as to any one gambler.

Each player is dealt their own 2-card hand and faces off directly against the player/dealer, like in blackjack. All hands hit on 0–4 and stand on 7–9. Players must also stand on 6 but have the option of hitting or standing when holding a 5. The player/dealer may choose to hit or stand on both 5s and 6s.

4.4 Side Bets

Baccarat is very, very boring.—Stanley Roberts, in [105].

Playing baccarat has been likened to wagering on a coin toss: a simple betting choice between 2 near-equally likely outcomes without any role for strategy in a gambler's decisions. While whether or not a game is boring is surely a subjective matter, it is undeniable that baccarat can be played with very little activity, since only two cards are initially dealt to each of two hands and there is no place for the gambler to influence gameplay directly. High-limit baccarat salons may allow the highest bettor on Player or Banker to turn over the cards, but this is merely for show; the players do not affect the game by their actions.

One way to enliven a table game is to introduce a new betting option; given the simplicity of baccarat, there is considerable room for new ideas. Many of these ideas are intended to increase the traffic at baccarat tables by drawing in new gamblers, because many high rollers resist any baccarat wager other than Banker, Player, and Tie.

Natural 9 and Natural 8

Two early baccarat side bets were *Natural 9* and *Natural 8*. These were simple wagers that could be made on either the Banker hand or the Player hand, as the gambler might choose. Each of the 4 bets paid 9–1 if the chosen hand had the indicated sum in its first 2 cards. From an 8-deck shoe, the probability of a natural 9, regardless of the hand, is $P(9) = .0949$, giving the casino a 5.10% advantage on Natural 9. For Natural 8, the probability is $P(8) = .0945$ and the HA is 5.47% [147].

Edward O. Thorp and William Walden assessed these bets mathematically and found some shoe compositions that could result in a player advantage on these bets. Since the bets pay 9–1, the player has the edge if

$$(9) \cdot P(9) + (-1) \cdot (1 - P(9)) > 0,$$

or

$$P(9) > \frac{1}{10},$$

where $P(9)$ denotes the probability of drawing a 9 on the chosen hand. The same inequality holds for $P(8)$ and a bet on Natural 8 [147]. The values given above for $P(8)$ and $P(9)$ are very close to .1, suggesting that as the shoe composition changes, and with it these two probabilities, the advantage on these bets may shift from the casino to the gambler.

Consider an 8-deck shoe. Let n be the number of cards yet to be dealt and t be the number of 9s remaining in the shoe. The probability of a 2-card 9 as a function of n and t is then

$$P(9\,|\,n,t) = \frac{2 \cdot (n-t) \cdot (32n + 351t - 32)}{1149n \cdot (n-1)}.$$

A similar formula for $P(8\,|\,n,t)$ can be derived, tracking 8s instead of 9s, of course:

$$P(8\,|\,n,t) = \frac{(n-t) \cdot (127n + 1405t - 127)}{2298n \cdot (n-1)}.$$

We then have

$$|P(9\,|\,n,t) - P(8\,|\,n,t)| \leqslant \frac{1}{2298},$$

so strategies for betting Banker 9 and Banker 8 are essentially the same [147]. Moreover, a shoe composition favoring 9s or 8s makes no distinction between Player and Banker hands, so we can reasonably consider these 4 bets as being equivalent from a card-counting standpoint, excepting only the need to track 8s for Natural 8 and 9s for Natural 9.

In 1966, Thorp led a team of players who took this knowledge to Nevada and tested their counting and betting schemes in live casino play. The team won $100 per hour for seven nights in one casino and $1000 per hour for two hours in another before being barred from both. Nevada casinos responded quickly by discontinuing both side bets [145, p. 180].

On a purely academic level, a plausible counting system for Natural 9 and Natural 8 focuses on the 9s and 8s, respectively, since the most common Natural 9 consists of a 9 and a 10-count card. The count is simple: For Natural 8, 8s count –12 and other cards count +1; for Natural 9, 9s count –12 and other cards +1. These were tested with billion-shoe computer simulations long after Thorp and Walden's work [69]. The betting strategy for these counts follows.

- For Natural 9, when the true count is +7 or greater, make the Natural 9 bet on both Player and Banker hands.

- For Natural 8, make both bets when the true count is +8 or greater.

Though these bets, in their original form, were effectively eliminated by casinos after Thorp and Walden published their results, game parameters can be changed. One effective way to thwart baccarat card counters is to move the shuffle point up, from 14 cards to 52 cards in an 8-deck shoe [69]. It is much more likely that high counts occur very late in the shoe; cutting off one full deck means that those favorable configurations are shuffled away instead of being played out.

Another option would be changing the payoff odds. If Natural 8 and Natural 9 were paid at 8–1 instead of 9–1, the HA would increase. For Natural 9, the HA rises from 5.10% to 14.59%, and for Natural 8, from 5.47% to 14.95% [69]. Taken together, the dual strategies of changing the shuffle point and cutting the payoff change Natural 8 and Natural 9 to a point where card counting is effectively useless in securing a player advantage.

WINSUIT

Stanley Roberts, who called baccarat "very, very boring" above, proposed a baccarat side bet in *WIN Magazine* in 1990. *WINSUIT* was designed by Roberts in part to address this perception. WINSUIT was a side bet made, as the name suggests, on a suit: in this case the suit of the fifth card dealt to a baccarat hand, assuming there is one. If the hand ends after four cards, WINSUIT bets lose. If a fifth card is required, whether it is dealt to the Player or Banker hand, WINSUIT bets on that card's suit win, at odds (for 1) equal to the rank of the card. Face cards pay off at 10 for 1; aces merely push by paying 1 for 1 [105].

The key to making WINSUIT successful for casinos is the hands that stop at 4 cards, which happens either when one hand has a natural 8 or 9, or when both hands have an initial total of 6 or 7. This gives the house its edge; the exact magnitude of that edge—balancing the desire to attract players with the need for a meaningful return—can be determined by how the payoff odds are set.

In the original proposal for WINSUIT, the chance of a baccarat hand requiring a fifth card was said to be .5941, so the probability of an immediate loss was .4019 [105]. These fifth cards may be assumed to be equally distributed among the 4 suits, so each suit has a win probability of

$$\frac{1}{4} \cdot .5941 \approx .1485.$$

The probability of losing a WINSUIT bet is then

$$.4059 + \frac{3}{4} \cdot .5941 \approx .8515.$$

The chance that a WINSUIT wager on a specific suit wins, .1485, is a probability evenly distributed among the 13 card ranks. Each value—which determines the payoff—comes up $\frac{1}{13}$ of the time, so the probability of each payoff from 0 through 8 is approximately .0114. The probability of a ten-count card and a 9–1 payoff is 4 times this value, or .0456.

The expected value of a WINSUIT wager is then

$$E = \left(\sum_{k=0}^{8} (k) \cdot .0114 \right) + (9) \cdot .0456 + (-1) \cdot .8515 \approx -\$0.0289,$$

giving the casino an edge just under 3%. This is higher than the HAs on the

Player and Banker wagers, but far lower than the edge on the Tie bet, and so it fits in well with the established baccarat bets, as it needed to.

Later analysis would turn up a flaw in these calculations. Computer simulation of card games such as baccarat required considerable programming facility in 1990; in 2014, mathematician Eliot Jacobson released a comprehensive spreadsheet counting every possible baccarat game—all 4,998,398,275,503,360 (nearly 5 quadrillion) of them—and illustrating a number of important calculations [68]. This spreadsheet showed that 1,893,735,611,458,560 of these games would require only 4 cards; thus that the probability that a baccarat hand draws a fifth card is approximately .6211, which is 4.35% higher than originally stated in 1990.

This 4.35% error turns the original WINSUIT bet into one with a player advantage. The new expected value of the proposed bet with payoffs based on the rank of the fifth card, is $0.0153, giving players a 1.53% edge.

However, payoffs can be modified, and there's certainly room to change the pay table for WINSUIT in order to generate a viable side bet. The challenge lies in making the pay table simple enough that a dealer can handle winning payments with a minimum of confusion.

- A simple and reasonable-sounding pay table might be one offering a flat 3–1 payoff to winning bettors. The gambler would be encouraged to think "There are 4 suits, and I've got 1 of them, so the odds are 3–1 against. Sounds fair!"—overlooking the high probability of a hand ending in 4 cards.

  Using the Jacobson probabilities, this version of WINSUIT has a 37.89% HA: too high for baccarat even as it might be comparable to some of the worst casino bets and better than the expectation in many lotteries. Raising the flat payoff to 4–1 gives a 22.36% edge, and offering 5–1 takes the HA closer to reasonable territory, at 6.83%. A 6–1 payoff gives the players an edge of 8.70%.

- A 3.25% HA can be achieved with the following payoff scheme, which is fairly dealer-friendly:

  - 20–1 if the winning card is an ace.
  - 10–1 if the winning card is a face card.
  - 2–1 if the winning card is a 2 through 10.

From a game designer's perspective, it might be useful to know how much total payoff is possible if the desired HA is specified. Since the 13 payoff cards of a winning suit have equal probability, we can solve a simple linear equation for the total payoff X to be distributed across the 13 cards. (In the original proposal, $X = 72$.) If the expected value of a $1 WINSUIT bet is desired to be $Y(X) < 0$, a function of X, we have

$$P(\text{Lose}) = .3789 + \frac{3}{4} \cdot (.6211) = .8447,$$

$$P(\text{Win}) = \frac{.3789}{4 \cdot 13} \approx .0119, \text{ and}$$

$$Y(X) = (-1) \cdot .8447 + (X) \cdot .0119.$$

Table 4.14 shows the WINSUIT house advantage for integer values of X from 60 to 70.

TABLE 4.14: WINSUIT: House advantage for a given total payoff X

X	HA
60	12.80%
61	11.61%
62	10.41%
63	9.22%
64	8.02%
65	6.83%
66	5.64%
67	4.44%
68	3.25%
69	2.05%
70	0.86%

Example 4.17. If we distribute 64 units among 13 card ranks, Table 4.14 shows that the HA is 8.02%, regardless of how the payoffs are allocated among the cards. One simple way to implement WINSUIT under these conditions would be to pay 16–1 on fifth-card aces and 4–1 on other all winning bets. ∎

Baccarat World

Casino baccarat culture allows players to track the outcome of each game dealt from a shoe and look for patterns in an effort to guide their wagering. As baccarat has been shown to be essentially immune to card counting, any patterns that a gambler might identify are almost certainly meaningless. The *World* side bet, approved for play in Washington state casinos, might appeal to a player seeking to capitalize on a run of three winning hands for either the Banker or Player hand [57].

The World bet is made by choosing a side, Player or Banker, and wins if that side wins three straight hands. Unlike in standard baccarat, a tie hand causes all World bets to lose. Variable pay tables allow Washington casinos to customize the HA as management might wish.

On a given hand of baccarat, the 3 outcomes have the probabilities in Table 4.15. Since successive baccarat hands are effectively independent, it

TABLE 4.15: Baccarat hand probabilities [8]

Result	Probability
Banker wins	.4584
Player wins	.4461
Tie	.0955

follows that the probability of winning a World bet on Banker is

$$(.4584)^3 \approx .0963 \approx \frac{1}{10.4}$$

and the probability of winning a World bet on Player is

$$(.4461)^3 \approx .0888 \approx \frac{1}{11.3}.$$

If the Banker World bet pays off at X to 1, the resulting expectation is

$$E = (X) \cdot .0963 + (-1) \cdot .9037.$$

The approved pay tables for the World bet in Washington allow X to be 8, 8.5, or 9; this allows the HA to range from 3.70–13.33%. For the slightly less likely bet on Player, a 10–1 payoff in addition to 8, 8.5, and 9 is permitted. HAs then range from 2.32% to 20.08%.

Of course, these lower odds and higher house edge can be masked by advertising payoffs as "x for 1" instead of "x to 1".

Fabulous 4 Baccarat

At the Marina Bay Sands Casino, *Fabulous 4 Baccarat* adds a number of side bets centered around the number 4 [111].

- The *Precious Pair* bet may be made on either the Player or Banker hand, and pays off 30–1 if the chosen hand is a pair of 4♦s, 15–1 on any other pair of 4s, 12–1 on any other diamond pair (such as 9♦ 9♦) and 9–1 on any other pair.

  Since Fabulous 4 Baccarat may be dealt out of a 4- to 10-deck shoe, we use the infinite deck approximation for convenience. The probability of a pair of 4♦s is

  $$\left(\frac{1}{52}\right)^2 = \frac{1}{2704},$$

so a 30–1 payoff is a boon to the casino. The probability of any other pair of 4s is

$$\left(\frac{1}{13}\right)^2 - \left(\frac{1}{52}\right)^2 = \frac{15}{2704} \approx .0055.$$

The probability of a diamond pair, of a rank other than 4, is

$$\frac{12}{52} \cdot \frac{1}{52} = \frac{3}{676} \approx .0044.$$

Finally, the probability of any other pair, not a pair of 4s or diamonds, is

$$12 \cdot \left(\frac{4^2 - 1}{52}\right)^2 = \frac{45}{676}.$$

Here, the 12 counts the number of possible ranks other than 4, and the factor

$$\left(\frac{4^2 - 1}{52}\right)^2 = \frac{15}{52^2}$$

computes the probability of drawing two cards of this rank that are not both diamonds.

The probability of winning the Precious Pair bet is the sum of these 4 probabilities, or $\frac{1}{13}$. This probability may also be derived by recognizing that Precious Pair pays off if the selected hand comprises a pair, of any sort, and that the payoff odds simply vary according to the type of pair that's dealt.

- *Fabulous 4* bets, on either Player or Banker, win if the chosen hand wins with a total of 4. A winning Player bet pays 50–1 and a winning Banker bet pays 25–1.

Baccarat Bonus

The Gardens Casino in Hawaiian Gardens, California offers a two-handed California version of baccarat which incorporates an optional Baccarat Bonus wager based on the four cards dealt to the player and the player/dealer. Table 4.16 shows the pay table.

TABLE 4.16: Gardens Casino Baccarat Bonus pay table [62]

4-Card Hand	Payoff
Pair in both hands	40–1
Flush	10–1
Pair in either hand	3–1

With a 6-deck shoe, the probabilities of these winning outcomes are:

$$P(\text{Pair in both hands}) = \frac{13 \cdot \binom{24}{2}}{\binom{312}{2}} \cdot \frac{12 \cdot \binom{24}{2} + \binom{22}{2}}{\binom{310}{2}} \approx .0055,$$

$$P(\text{Flush}) = \frac{4 \cdot \binom{78}{4}}{\binom{312}{4}} \approx 0147,$$

and

$$P(\text{Pair in either hand}) = 2 \cdot \frac{13 \cdot \binom{24}{2}}{\binom{312}{2}} - P(\text{Pair in both hands})$$
$$\approx .1424.$$

The total probability of a win on this bet is just over 16%. The expectation of the Baccarat Bonus bet is then

$$E = (40) \cdot .0055 + (10) \cdot .0147 + (3) \cdot .1424 + (-1) \cdot .8374 \approx -\$.0432,$$

which gives the casino a 4.32% edge.

Side Bets Based On Hand Totals

A number of baccarat side bets are resolved by looking at the final total of the Player and Banker hands. For convenience, Table 4.17 shows the probabilities of reaching the various hand totals in 2 or 3 cards, in an 8-deck game. Certain 2-card hands are only possible if the other hand is a natural 8 or 9; for example, a 2-card 0 in either hand.

Example 4.18. The event that both hands tie at 6 can arise with 2 or 3 cards in either hand. Enumeration of all hands that can be dealt from an 8-deck shoe gives the figures shown in Table 4.18.

Adding gives 96,170,001,308,416 ways for two hands to tie at 6. Dividing by the 4,998,398,275,503,360 possible hands gives

$$P(\text{6-6 Tie}) = \frac{96,170,001,308,416}{4,998,398,275,503,360} \approx .0192.$$

∎

Table 4.19, derived through a count of all possible baccarat hands dealt from an 8-deck shoe, collects the probability of every possible Player vs. Banker outcome.

TABLE 4.17: Probability distribution for 8-deck baccarat hand values in 2 and 3 cards [119]

Total	Player		Banker	
	2 Cards	3 Cards	2 Cards	3 Cards
0	.0279	.0661	.0279	.0608
1	.0180	.0565	.0180	.0513
2	.0179	.0564	.0179	.0511
3	.0180	.0565	.0225	.0502
4	.0179	.0564	.0496	.0439
5	.0180	.0565	.0588	.0419
6	.0945	.0387	.0855	.0355
7	.0949	.0388	.0949	.0335
8	.0945	.0387	.0945	.0335
9	.0949	.0388	.0949	.0335

TABLE 4.18: Number of 8-deck baccarat hands tying at 6, by number of Player and Banker cards [119]

Player cards	Banker cards	Hands
2	2	44,487,098,110,464
2	3	22,532,336,357,376
3	2	17,217,820,643,328
3	3	11,932,746,197,248

TABLE 4.19: Probability distribution for 8-deck baccarat outcomes [119]

Banker	Player										Total
	0	1	2	3	4	5	6	7	8	9	
0	0.0058	0.0049	0.0048	0.0048	0.0049	0.0050	0.0115	0.0117	0.0176	0.0177	0.0888
1	0.0049	0.0041	0.0040	0.0039	0.0041	0.0042	0.0098	0.0100	0.0121	0.0121	0.0693
2	0.0048	0.0041	0.0040	0.0039	0.0041	0.0042	0.0098	0.0100	0.0121	0.0121	0.0691
3	0.0054	0.0046	0.0046	0.0045	0.0041	0.0042	0.0098	0.0100	0.0129	0.0127	0.0728
4	0.0092	0.0081	0.0077	0.0076	0.0073	0.0069	0.0103	0.0100	0.0129	0.0134	0.0934
5	0.0092	0.0082	0.0086	0.0090	0.0084	0.0079	0.0114	0.0111	0.0135	0.0135	0.1007
6	0.0097	0.0081	0.0086	0.0090	0.0092	0.0093	0.0192	0.0190	0.0145	0.0145	0.1211
7	0.0108	0.0092	0.0091	0.0090	0.0092	0.0093	0.0202	0.0204	0.0156	0.0156	0.1284
8	0.0170	0.0116	0.0114	0.0114	0.0115	0.0117	0.0156	0.0158	0.0110	0.0110	0.1280
9	0.0171	0.0116	0.0115	0.0114	0.0116	0.0117	0.0156	0.0159	0.0111	0.0110	0.1284
Total	0.0940	0.0745	0.0743	0.0745	0.0743	0.0745	0.1332	0.1337	0.1332	0.1337	1.0000

ZooBac

Some side bets bear the names of animals from the Chinese zodiac, in an effort to appeal to Asian gamblers. *ZooBac* is another California game, from the Bay 101 Casino in San Jose, that includes 4 extra side bets: Rabbit, Tiger, Monkey, and Zoo.

- The *Rabbit* bet wins when the Player hand wins with a 3-card total of 7. It pays 25–1 unless the Banker hand is 0, in which case it pays 75–1.

- The *Tiger* bet wins on a winning Player total of 8 in 3 cards. Its payoffs are the same as Rabbit: 75–1 when the Banker hand is 0 and 25–1 on any other Banker total.

- *Monkey* pays 150–1 if the two hands tie at 0.

- All three animal wagers may be bet at once with the *Zoo* bet. This bet wins if any of the Rabbit, Tiger, and Monkey bets win. Note that only one of these 3 bets can win on a single hand. The pay table for Zoo is shown in Table 4.20.

TABLE 4.20: ZooBac: Zoo bet pay table

Result	Payoff
Rabbit wins with Banker 0	30–1
Rabbit wins with Banker 1–6	10–1
Tiger wins with Banker 0	30–1
Tiger wins with Banker 1–7	10–1
Monkey wins	30–1

The necessary probabilities to assess this collection of side bets are found in Table 4.17. A full collection of the 205 different possible hand outcomes, with their probabilities, is found in [119]; this is the source for many of the probabilities quoted in this section.

Example 4.19. For the Monkey bet, the tie at 0 must necessarily be with two 3-card hands, since a 2-card Player or Banker hand of 0 will draw another card. From Table 4.2, we see that the probability of winning this bet is .0058, and so the expectation is

$$(150) \cdot .0058 + (-1) \cdot .9942 \approx -.1242,$$

so this bet gives the house a 12.42% edge. Since the casino charges a $1 collection fee on this bet, a $1 Monkey wager is ill-advised, as that drives the HA to 112.32%. On a $100 wager with the same $1 collection fee, the HA is 13.42%. ∎

These wagers are all susceptible to card-counting techniques.

- The Rabbit bet, which carries a 9.37% HA, can be attacked with the following count system [67]:

Card	Value
A	+1
2, 3, 4, 5, 7	−1
6, T	0
8, 9	+2

The cards which benefit the Rabbit bettor most when removed from the deck are aces, 8s, and 9s. Given the surplus of 10s in a deck, removing 8s and 9s decreases the chance of a natural 8 or 9—while this is good for a gambler making the Player wager, it's not going to win a Rabbit bet. The index number for the rabbit bet is +5: if the true count exceeds 5, Rabbit holds an edge for the bettor [67].

- For the Tiger bet, where the target is a winning 3-card 8, the following count is used:

Card	Value
A, 2	+1
3, 4, T	0
5, 6, 7, 8	−1
9	+2

and the index number is +5, just as it is for Rabbit [67].

- Counting the Monkey bet calls for this system:

Card	Value
A, 2, 3, 4	+2
5, 6, 8, 9	1
7	0
T	−3

The Monkey bet is favored when this count is greater than 7 [67]. Since Monkey wins on a 3-card 0–0 tie and a hand of 0 consisting of 3 10s is the most common 3-card 0, the 10s are the most important card in this scheme.

- A viable Zoo count needs to consider all of the important cards for Rabbit, Tiger, and Monkey. It is perhaps easiest to see that 9s on the table improve a player's chances of a winning Zoo wager: the 9 has the highest positive count in all three separate count systems since 9s produce more natural 9s, and those eliminate the chance of getting a winning 3-card hand.

Card	Value
A	+2
2, 3, 4, 6	0
5, 7, T	−1
8	+1
9	+3

Looking at the counting system shows that aces are also important to the Zoo bet, and getting them out of the shoe improves the player's chances when betting Zoo. Aces matter for this wager because Zoo has 3 winning conditions: winning 3-card Player 7, winning 3-card Banker 8, and 3-card 0–0 tie, and none of these totals can be reached with an ace as the third card [67].

- A 3-card Player 7 with an ace as the third card requires drawing an ace to a two-card 6, and Player stands on a 2-card 6.

- Similarly, Banker cannot draw to a 2-card 7, so it cannot pull the ace that results in a 3-card 8.

- To reach 0 with a third-card ace necessitates drawing to a natural 9, which is forbidden by the rules of baccarat.

Accordingly, we see that aces dealt to the table improve the chance that one of the individual bets pays off, since they cannot then be dealt as third cards to either hand. If the hand of interest moves to draw a third card, it's better that the chance of drawing an ace is diminished, since the side bet cannot then win.

The index number for the Zoo count is +5.

Example 4.20. If 186 cards have been dealt from an 8-deck shoe in the following amounts of each rank, find the counts for the 4 ZooBac wagers.

Card	A	2	3	4	5	6	7	8	8	T
Number dealt	16	13	12	17	11	16	16	15	16	54

The running counts for the 4 wagers are

- Rabbit: +9.

- Tiger: +3.

- Monkey: +12.

- Zoo: +14.

Approximately 3½ decks have been dealt, so we divide by 4½ and round appropriately to convert these to true counts.

- Rabbit: +2.

- Tiger: +0.
- Monkey: +2.
- Zoo: +3.

Comparing the true counts with the index numbers shows that none of the 4 bets should be made. ∎

Golden Frog Baccarat

Golden Frog Baccarat, a game from the Bay 101 Casino and the Commerce Casino, is a rich source of side bets based on precise hand compositions. Several bets other than Player, Banker, and Tie are defined. So long as a gambler is making a Banker or Player wager, there is no additional collection fee collected on these bets. No animal names are involved; these bets state their winning criteria up front.

- A bet that is exactly what it says is the *3 Card 9 Over a 3 Card 1* bet. If one of the two hands wins with a 3-card 9 against a 3-card 1, this bet pays off at 150–1. In making this bet, the gambler need not pick the winning hand; the bet wins whether the Player or Banker hand wins the hand.

  The probability of winning this bet is .0057 $\approx \frac{1}{175}$, making this a slightly less unfavorable wager than the Monkey bet [119]. Since no collection fee is charged, a \$1 bet is not immediately contraindicated by the collection fee. Its expected value is

  $$E = (150) \cdot .0057 + (-1) \cdot .9943 = -.1393,$$

  so the card room has a 13.93% HA.

- Continuing the "truth in advertising" wagers is *Natural 9 Over 2-Card 7*. Once again, all that matters is that the winning and losing hands hit the designated values. In this wager, the triggering event has probability .0180 $\approx \frac{1}{55.5}$. [119]. The payoff is 50–1, giving the card room an edge of 8.11%.

- The *8 Over 6* bet does not specify number of cards, so it can be won by 2- or 3-card hands in either position, so long as the winning hand is an 8 and the losing hand equals 6. Reference to Table 4.2 gives

  $$P(8 \text{ Over } 6) = .0156 + .0145 = .0301.$$

  Since 8 Over 6 pays 25–1, the HA on this bet is 21.74%.

Total Shot

At the Capitol Casino in Sacramento, California, one baccarat variation is *Supreme Baccarat*, a local name for Nepal Baccarat. Supreme Baccarat comes with the *Total Shot* side bet. Total Shot pays off based on the sum of the two final hands, without regard to the number of cards they contain. If the sum is 18, the maximum possible, Total Shot pays 40–1; if that sum is 17, the payoff is 20–1. The wager loses if the sum is less than 17.

A sum of 18 can be made either with two 2-card hands or two 3-card hands. From Table 4.17, we see that this probability is

$$P(18) = (.0949)^2 + .0388 \cdot .0335 = .0110.$$

The probability of a sum of 17 similarly compares only hands with the same number of cards. Since this sum must consist of a 9 and an 8, a 2-card 8 or 9 would end the round before the other hand draws a third card. This means that

$$P(17) = .0221,$$

again by reference to Table 4.17.

The expectation is

$$E = (40) \cdot P(18) + (20) \cdot P(17) + (-1) \cdot [1 - P(18) - P(17)] = -.0849.$$

While the casino holds an 8.49% advantage, this side bet can be attacked by counting cards. The count focuses on 8s and 9s, since these cards combine with the abundant 10s and face cards to form the needed totals of 8 and 9. Here's the counting scheme to use:

Card	Value
A, 2, 3, 10, J, Q, K	+1
4, 5, 6, 7	0
8	−1
9	−6

When the true count exceeds +5, which occurs roughly $\frac{1}{7}$ of the time, the player making a Total Shot bet enjoys approximately a 9.8% edge [67].

Flex Action

In the *Flex Action* side bet, offered as part of Supreme Baccarat at the Aviator Casino in Delano, California, gamblers have the opportunity to make a baccarat version of a craps Place bet on the Banker hand. This bet can extend over multiple hands before it is resolved.

Place bets (page 114) allow the gambler to bet that a selected number: 4, 5, 6, 8, 9, or 10, will be rolled before a 7 is rolled. In baccarat, Flex Action lets the player pick his or her own total for the Banker hand: 0, 1, 2, 3, 5, 6, or 7; the bet wins if the Banker hand hits that total before a Banker hand totals 9. Once a bet is made, the only numbers that matter are the selected total and 9. If the Banker hand totals 4 or 8, the bet continues without either side losing; these numbers are not available for player wagers. A single player may make up to 7 Flex Action bets at a time, covering as many numbers as desired.

Table 4.21 shows the payoff odds and Banker hand probabilities for the Flex Action bet.

TABLE 4.21: Flex Action bet: Probabilities and payoffs [47]

Banker hand	Probability	Payoff
0	.0888	7–5
1	.0693	7–5
2	.0691	7–5
3	.0728	7–5
4	.0934	N/A
5	.1007	1–1
6	.1211	1–1
7	.1284	1–1
8	.1280	N/A
9	.1284	Lose

N/A indicates No Action.

Suppose that the gambler has chosen the number k, and let $P(k)$ be the probability that the Banker hand totals k. The probability that a Flex Action bet on this number wins is

$$\frac{P(k)}{P(k) + P(9)} = \frac{P(k)}{P(k) + .1284}.$$

The best number for a Flex Action bet is 7, where the probability of winning is vanishingly less than ½: .49998. This bet has a HA of only .0035%. Since this wager does not require a collection fee and may be made without placing a Player or Banker bet with its collection fee, it is essentially a bet on a coin toss.

The highest HA comes with a Flex Action bet on 2. The probability of winning this bet is

$$\frac{P(2)}{P(2) + P(9)} = \frac{.0691}{.0691 + .1284} \approx .3499.$$

With the given payoff odds of 7–5, the HA is 16.03%. Changing the payoff to 8–5 reduces the house edge to 9.03%. At 9–5 odds, it's only 2.04%.

4.5 Exercises

Solutions begin on page 329.

4.1. The expected values for baccarat Player and Banker bets computed on page 196 are the expected values *per resolved hand,* with tied hands removed from consideration. An alternate way to compute the advantages is to consider the HA *per dealt hand,* where there are 3 outcomes to consider. The PDF for baccarat hands is shown in Table 4.22.

TABLE 4.22: Baccarat: PDF with ties included

Outcome	Probability
Player wins	.4461
Banker wins	.4584
Tie	.0955

Compute the expected values of the Player and Banker bets using this model.

4.2. As a promotion, some online casinos offer "Bonus Baccarat Zero Mondays". For a few hours in the afternoon and early evening on Mondays, baccarat is dealt with the payoffs shown in Table 4.23.

TABLE 4.23: Bonus Baccarat Zero pay table [11]

Wager	Payoff
Player	1.027–1
Banker	.9725–1
Tie	9.5–1

The bet range on Bonus Baccarat Zero is $1–50. It's clear that this is better for gamblers than standard baccarat, as every bet has a higher payoff. Since the payoffs are all electronic, the decimal payoffs pose no problem to a dealer. Using the probabilities given on page 195, find the HA of each bet.

Dragon Tiger

Dragon Tiger is a one-card version of baccarat that is popular in Asia and on offer at the Lucky Ruby Border Casino in southern Cambodia on the Vietnam border. Dragon Tiger is dealt from a 8-deck shoe, and instead of Banker and Player, the two one-card hands are denoted Dragon and Tiger. Players bet on which card will be higher, with aces ranking low and kings high. In the event of a tie, the house takes half of all bets, so there is no need to charge a commission on any winning wagers.

4.3. What is the probability of the two hands tying?

4.4. Find the HA on a hand of Dragon Tiger dealt from the top of a shoe.

4.5. A separate Tie bet pays 8–1 if the two cards match in rank, just as in baccarat. How does the HA of this bet compare to the house edge on the Tie bet in baccarat, which is 14.36%?

Fa Fa Fabulous 4 Baccarat

Another alternate deck is used in *Fa Fa Fabulous 4 Baccarat*, also a variation from the Marina Bay Sands. This game is played using 4–10 decks of 65 cards each. Each card appears in 5 suits, dubbed "elements": Fire, Gold, Earth, Wood, and Water. Each suit contains 13 cards, ace through 10 and cards representing the 3 Lucky Immortals: Fuk, Luk, and Sau. These are the gods of wealth, good fortune, and longevity, and replace the jack, queen, and king in a standard deck. The Immortals count zero when dealt, just as face cards do.

4.6. In a 6-deck shoe, find the probability that the Player hand is dealt 3 Immortals, for a hand totaling 0.

4.7. How accurate is the infinite deck approximation when it's used to estimate the probability of dealing 3 Immortals from a 6-deck shoe?

4.8. Fa Fa Fabulous 4 Baccarat extends the Precious Pair bet from Fabulous 4 Baccarat, with the Gold suit playing the role of diamonds in the original bet. This Precious Pair bet pays off according to Table 4.24.

TABLE 4.24: Fa Fa Fabulous 4 Baccarat: Precious Pair pay table

Hand	Payoff
Pair of Gold 4s	30–1
Any other pair of 4s	15–1
Another Gold pair	12–1
Any other pair	9–1

Using the infinite deck approximation, compare the probabilities of the events in Table 4.24 with the corresponding events in Fabulous 4 Baccarat's Precious Pair bet (page 222).

4.9. The Tie bet at Fa Fa Fabulous Baccarat pays 8–1, but sweetens the deal a bit with the *Tie on Element 8s* bonus. A player making a Tie bet wins

800–1 if the Player and Banker hands tie at 8 with a pair of 4s each. Consider a game dealt from the top of an n-deck shoe. Find the probability $P(n)$ of a Tie on Element 8s as a function of n.

Side Bets

4.10. Use Table 4.2 to compute the house advantage of the Fabulous 4 bet in Fabulous 4 Baccarat, on both Player and Banker hands.

4.11. Find the house advantage of a \$5 Flex Action bet on 0, under the assumption that a 50¢ commission is charged on the bet.

4.12. Suppose that the World bet (page 221) was extended to alow players to bet on Tie, so the bet would win if the next 3 hands ended in ties.

a. Find the probability of winning this bet.

b. If this bet pays 1000–1, find the house advantage.

c. Find a payoff that gives a HA as close as possible to 4%.

Exercises 4.13–4.19 cover side bets which have been offered in California baccarat variations. When evaluating the expected value or house advantage of these wagers, assume that no extra collection fee is charged on these additional bets, as is usual practice in California card rooms provided that the gambler has made, and paid a collection fee on, a Banker, Player, or Tie bet.

4.13. The *Dragon 7* baccarat side bet pays 40–1 if the Banker hand wins with a total of 7 in 3 cards.

a. Find the probability that the Dragon 7 bet wins.

b. What is the HA on Dragon 7?

4.14. A common baccarat side bet similar to Dragon 7 wins if the Player hand wins with a 3-card total of 8, paying 25–1. In EZ Baccarat, this bet is called *Panda 8*. In Golden Frog Baccarat, this proposition is known as the *Koi 8* bet. A California variation simply called *California Baccarat* calls this bet *China Bear*, and another California game uses *Ox 8* for this bet.

a. Find the probability that this bet, no matter the animal, wins.

b. What is the HA?

4.15. *Monkey Baccarat* is a California variation which has been seen at the Commerce Casino and the Morongo Casino in Cabazon. The game is essentially EZ Baccarat with some additional side bets. Among them is the *Grand Monkey Bonus* bet. Monkey Baccarat designates the face cards as "monkey cards"; Grand Monkey Bonus pays 6000–1 at the Commerce Casino, and 5000–1 at the Morongo Casino, if the hand requires 6 cards and all 6 are monkey cards. Note that 10s are not monkey cards, even though they have the same value as face cards.

Find the probability of winning the Grand Monkey Bonus bet if the hand is dealt from the top of

a. A single deck.

b. A 6-deck shoe.

c. An 8-deck shoe.

4.16. The Grand Monkey Bonus bet is highly countable, since a surplus of face cards will favor getting 6 of them in 6 cards. The Morongo limited Grand Monkey Bonus wagers to $1, so the opportunity to raise one's bet with a favorable count doesn't exist, but a good counting system can indicate when this bet is worth making. In an 8-deck shoe, consider a count system where aces through 10s count +1 and face cards count −1. Keeping a running count with these values is a simple matter of counting the surplus or deficit of monkey cards over non-monkey cards. Convert the running count to a true count by dividing by the number of decks remaining, as always. If 2 decks remain in the shoe and the count shows that 38 monkey cards are left, up from the 24 that would be expected, find the probability of winning the Grand Monkey Bonus bet.

4.17. The Commerce Casino also was home to *Bai Cao Monkey 9*, another baccarat variation that deals individual 2-card hands to each player. Players receive a third card unless their initial hand is a natural 9; the player/dealer takes a third card if its total is 1–4. Since multiple Player hands are dealt, only the initial value of the player/dealer hand is used to determine if that hand receives a third card.

Bai Cao Monkey 9 takes its name from its optional bonus bet, the *Monkey 9* bonus bet. The Monkey 9 bet wins if the player/dealer hand wins with a 3-card total of 9. The payoff on Monkey 9 is 40–1 unless the winning hand consists of a 2-card total of 4 followed by a 5 on the third card, which pays 30–1. Using an infinite deck approximation, find the probability of a 2-card 4 followed by a 5 on the third card.

4.18. Another side bet in Bai Cao Monkey 9 is the *9 Bonus* bet, which players may make on their own hands. The 9 Bonus is paid if the player's hand totals 9, with Table 4.25 as the pay table.

Using an infinite deck approximation, find the probability of drawing the following hands:

a. 3–3–3.

b. 2–3–4.

c. 0–0–9.

d. Any other total of 9 with a pair.

TABLE 4.25: 9 Bonus pay table

Player hand	Payoff
3–3–3	200-1
2–3–4	40–1
0–0–9	15–1
Total of 9 with a pair (other than 0–0–9)	10–1
Any other 3-card 9	5–1
Natural 9	1–1

4.19. The *One Up* bonus bet is associated with Fortune 7 Baccarat, at the Crystal Park Casino in Compton, California. Rather than looking at the values of the two hands, One Up looks at the difference between the two hands, paying off if that difference is 1. If the player hand beats the player/banker hand 1–0, One Up pays 30–1; if the player hand wins by any other 1-point margin, the bet pays 9–1.

a. Find the probability that the player scores a 1–0 win.

b. Find the probability that the player wins by 1 and both hands consist of 2 cards.

c. Find the probability that the player wins by 1, other than winning 1–0, and both hands contain 3 cards.

Too many side bets at a live table game run a risk of slowing down the game, and make it more likely that there will be errors in gameplay or payoffs. With an electronic version of a table game, this is far less of a concern. Unity Technologies has developed an electronic baccarat table that offers 38 different side bets, beyond Player, Banker, and Tie. Some of these extra bets are bets we have seen already, including variations on the Dragon 7 and Panda 8 bets described above. Exercises 4.20-4.23 examine several additional betting options.

4.20. Eighteen of these bets are Straight Up wagers on the exact value of a winning Player or Banker hand, which pay off according to Table 4.26.

Use Table 4.19 to find the probability of winning and the house edge for each wager. Which number carries the lowest house edge for Player and for Banker?

4.21. Players may also bet that either the Banker or Player hand contains a pair. These two bets pay off at 11.5–1. Find the expected value of the Pair bet if the game is dealt from a fresh 8-deck shoe.

TABLE 4.26: Baccarat: Straight Up pay table

Hand value	Player	Banker
1	188–1	191–1
2	104–1	104–1
3	73–1	63–1
4	53–1	28–1
5	37–1	21–1
6	14–1	16–1
7	10.5–1	11–1
8	7.5–1	8–1
9	6.5–1	7–1

4.22. The Majestic Match side bet may be made on either hand, Player or Banker. Majestic Match pays off 25–1 if the first two cards dealt to that hand are a suited king and queen, and 2.5–1 if the hand starts with any other 2 suited cards. Find the probability of each winning outcome, again assuming a fresh 8-deck shoe.

4.23. Unity Technologies' game expands the baccarat Tie bet to 10 wagers that allow players to specify the total at which the hands tie. Table 4.27 gives the payoff odds.

TABLE 4.27: Baccarat: Individual tie bet pay table

Hand value	Payoff odds
0	160–1
1	226–1
2	232–1
3	208–1
4	128–1
5	117–1
6	48–1
7	45–1
8	84–1
9	84–1

The house advantages on these bets are all between 5 and 7%. Which bet has the highest HA?

Card Counting

4.24. Seven decks have been dealt from an 8-deck shoe, in the following quantities:

Card	A	2	3	4	5	6	7	8	9	T
Number dealt	29	28	26	27	28	30	28	29	30	109

Compute the true counts for the four ZooBac wagers. Which, if any, of these bets is favored in the remainder of the shoe?

4.25. Apply the combined count to the card distribution in Exercise 4.24. Which bet is favored: Player or Banker? Does the favored bet hold an edge over the casino?

Chapter 5

Blackjack

Blackjack is the most popular table game played with cards, and uses anywhere from one to eight decks of cards shuffled together. When more than 2 decks are used, the cards are dealt from a *shoe*: a box that holds the cards and allows them to be removed one at a time. A short-lived variation, Blackjack II, dealt at Harrah's Marina in Atlantic City, New Jersey used 16 decks.

5.1 Blackjack Basics

In a hand of blackjack, two cards are dealt to each player and two—one face up (the *upcard*) and the other face down (the *hole card*)—to the dealer. Players' cards are customarily dealt face down from the dealer's hand in single- or double-deck games, and face up from a shoe in games using four or more decks. Each card from 2 through 10 counts its face value, face cards count 10, and an ace may be counted as either 1 or 11, at the player's discretion. In describing blackjack, it is common to use the shorthand "ten" or "ten-count card" to refer to any card—10, jack, queen, or king—counting as 10.

A player dealt a two-card total of 21 consisting of an ace and a ten-count card—called a *natural* or *blackjack*—wins immediately until the dealer also has a natural. Naturals pay either 3 to 2 or 6 to 5, depending on the rules of the casino. Some electronic blackjack games pay only even money on naturals. The alternate name "blackjack" for a two-card total of 21 derives from the earliest days of the game, when a 21 consisting of the ace and jack of spades qualified for a bonus payoff.

Example 5.1. What is the probability of a natural?

The probability depends on the number of decks in play and on the cards which have been dealt earlier in the shoe. For convenience, we shall begin by computing the probability of a natural dealt from the top of an n-deck shoe. The shoe contains $4n$ aces and $16n$ 10s; since the two cards comprising a natural may appear in either order, we have

$$P(\text{Natural}) = \frac{4n}{52n} \cdot \frac{16n}{52n-1} + \frac{16n}{52n} \cdot \frac{4n}{52n-1} = \frac{32n}{13(52n-1)}.$$

As the number of decks n increases, this probability decreases. If $n = 1$, the chance of a natural is .0483, approximately 1 in 21; with 6 decks in use, we have $P(\text{Natural}) \approx .0475$.

As a fraction, this value is very close to $\dfrac{1}{21}$, so approximately 1 blackjack hand in 21—in the long run—is a natural. ∎

Using the infinite deck approximation, we have

$$P(\text{Natural}) = \frac{4}{52} \cdot \frac{16}{52} + \frac{16}{52} \cdot \frac{4}{52} = \frac{8}{169} \approx .0473,$$

which is reasonably close to the 6-deck probability computed above.

Though it is sometimes described that way, the object of blackjack is not to get a hand of cards totaling as close to 21 as possible without going over. Blackjack players play against the dealer's hand, and the player's object is to get a hand that is *closer to 21 than the dealer's hand* without going over. If the object of blackjack was simply to get as close to 21 as possible but not to exceed 21, then stopping on a hand of 12, as blackjack basic strategy sometimes directs, would be a bad idea. However, there are times when standing on a low hand in hopes that the dealer's hand will bust is the best possible player choice.

If the dealer does not have a natural, then each player in turn has the opportunity to "hit" their hand and draw additional cards in an effort to bring their total closer to 21 without going over. If a player's hand exceeds 21, this is called *busting* or *breaking*, and the bet is lost and collected at once.

A hand containing an ace counted as 11 is called a *soft* hand, because it cannot go over 21 by drawing one card—if a hand counting an ace as 11 goes over 21 upon drawing a card, the ace may simply be revalued at 1. A hand with a total of 12 to 16 without any aces, or with all aces counted as 1, is called a *hard* or *stiff* hand, because drawing a single card risks busting the hand. *Pat* hands are hands ranging from 17–21: these hands will usually not take a second card unless they are also soft hands totaling 17 or 18.

Once all player hands are settled, the dealer exposes his hole card. If the dealer's hand is 16 or less, he must take additional cards until the hand totals at least 17. More recently, a popular rule change in many casinos requires dealers to hit a soft 17 hand. When the dealer's hand is complete—either by busting or reaching a total of 17 or higher—the hand is compared to those of all players who have not yet busted.

Player hands that are closer to 21 than the dealer's are paid off at 1 to 1; if the player and the dealer have the same total, the hand is called a *push*, or tie, and no money changes hands. If the dealer's completed hand is closer to 21 than the player's, the player loses and his or her wager is collected.

Depending on the casino, players may be offered several options during play to make additional bets that offer the chance of winning more money (or, in the case of surrender, losing less money). These are not options available to the dealer.

- If the player's first two cards have the same rank—as in a pair of 8s or aces—they may be **split** to form the first card of two separate hands. The player must match his bet on the new hand, and the two hands are played out separately. Some casinos allow players to split two 10-count cards, such as a jack and queen, and a standard casino rule allows the player to draw only one additional card to each hand after splitting aces. If a third card matching the first two is drawn to a split hand, many casinos allow that hand to be split again, although some do not allow resplitting of aces. Most casinos have a limit on the number of times a given hand may be split: a maximum of four separate hands is common.

- The player has the option to **double down**—to double his or her initial bet after the first two cards are dealt. This represents a chance for the player to get more money in play upon receiving a good initial hand, but this opportunity comes at a cost: only one additional card may be drawn to a doubled hand. A player who doubles down on a 10 and draws a 3 cannot draw again in an attempt to prove her 13. Candidates for double-down hands are hands totaling 9, 10, or 11, as well as certain soft hands. Casinos may place restrictions on which hands may be doubled; some, for example, restrict doubles to 10s and 11s. Some casinos do not allow players to double down after splitting pairs. It is also possible to "double down for less": to increase the bet by less than the full amount originally wagered—this is only recommended if you absolutely cannot afford to double your bet, as doubling for less means taking less than full advantage of a situation where you have the edge over the house.

  On a practical level, this means that a gambler should never stake all of his or her chips on a blackjack hand. Always hold back enough funds so that you can double down or split a pair if called for.

- If the dealer's face-up card is an ace, players have the opportunity to make an **insurance** bet. This is a separate bet of up to half their initial wager and pays 2 to 1 if the dealer has a natural. Of course, if the dealer does have 21, the main hand loses (unless the player also has a natural), and this is the reason for the name—the player is "insuring" the main hand against a dealer 21.

  Some players routinely make the insurance bet when holding their own natural, in an effort to make sure that they win something for their strong hand rather than being tied. This is sometimes called the *even-money option*, because the net effect is that the player wins the amount of her original wager regardless of the dealer's hand. A player making a $1 bet on the hand makes an additional 50¢ insurance bet. If the dealer holds a natural, the original bet pushes and the insurance bet pays 2–1, for a net profit of $1. If the dealer's hand is not a natural, the insurance bet loses while the original bet pays 3–2, making the net profit $1 again.

- Some casinos offer a **surrender** option, in which a player may elect not to play out his hand while forfeiting only 50% of his initial wager. This might be something worth considering when the chances of beating the dealer are small—for example, when the player's hand is 16 and the dealer's upcard is a 10.

Surrender comes in two versions: *early surrender*, where the option is available when the dealer's upcard is a 10 or ace and before the dealer checks his or her hole card for a possible natural, and the far more common and slightly less player-friendly *late surrender*, which is only offered after the check for a natural is complete.

Expected Value and House Advantage

In craps and roulette, the house advantage is achieved by paying off winning bets at less than true odds. The HA for blackjack is somewhat more complicated to compute, since it depends on the exact rules of the game being played and on the cards that have been dealt previously. The casino derives its edge on the blackjack tables from the fact that the players must play out their hands first, and if they bust, they lose even if the dealer subsequently busts with a higher total. The blackjack rule that "ties are a push and the player neither wins nor loses" applies *only* to ties at 21 or less—if a player and dealer tie with hands of 22, the player's chips are already in the dealer's rack before the dealer busts, and they aren't returned. (Some blackjack variations played in California card rooms are exceptions to this rule; see Section 5.5.)

The house advantage for blackjack is also affected by player decisions, because there is an element of skill to blackjack play. One player option arises from thinking that since the dealer must draw on all 16s and stand on all 17s, this is how they should play out their hands. This is called the "mimic the dealer" strategy, although, as we shall see, it is far from optimal for players. Since the probability of a dealer bust is about 28.77%, a player opting to mimic the dealer by drawing on 16, standing on all 17s, and eschewing doubling down and splitting pairs has the same probability of busting [114]. The probability of a double bust is then approximately $.2877^2 = .0828$, and this is the casino's initial edge over a mimicking player.

From this, we must subtract an amount to account for the 3–2 payoff on naturals, which is half the probability that the player has a natural and the dealer does not. We multiply this probability by ½ to reflect the extra payout of half the player's wager. This quantity is approximately .0259, so the blackjack HA against a mimicking player is 5.68% [41]. A player who properly exercises the game options listed above by using basic strategy as described in section 5.2 can reduce the HA by about 5.50%, which makes a single-deck game nearly an even proposition.

Casinos may combine rules as they wish, which produces similar-looking blackjack games with very different house advantages. Table 5.1 lists the effect on the house edge of various rule changes [14, 56, 61].

Notice that the house edge increases as the number of decks in use goes up, but the effect levels out pretty quickly. Whatever additional advantage a casino

TABLE 5.1: Effects of rule changes on blackjack house advantage [14, 56, 61]. Negative values favor the player.

Rule change	Effect on HA
Two decks	+.32%
Four decks	+.48%
Six decks	+.54%
Eight decks	+.57%
Continuous shuffling machine used	+.30%
No double downs on soft hands	+.13%
Double down only on 10 or 11	+.26%
No double downs on 9	+.13%
No double downs on 10	+.52%
Double downs only allowed on 11	+.78%
No double downs allowed	+1.60%
Double down on two or more cards	−.24%
Double down after pair splitting	−.14%
No nonace pair resplitting	+.03%
No splitting of aces	+.18%
Pair splitting forbidden	+ .40%
Resplitting aces allowed	−.06%
Draw more than one card to split aces	−.14%
Dealer wins ties	+9.00%
Naturals pay 6 to 5	+1.39%
Naturals pay even money	+2.32%
Naturals pay 2 to 1	−2.32%
Six cards under 21 automatically wins	−.15%
Dealer hits soft 17	+.20%
Late surrender	−.06%
Early surrender	−.62%

may gain from using 12 or more decks is outweighed by the inconvenience of handling and shuffling that many cards. Time spent shuffling is, after all, time not spent dealing hands and collecting the HA, on the average, from every hand dealt. A continuous shuffling machine (CSM) cuts down on shuffle time and allows more hands per hour by immediately recycling discards into the shoe and reshuffling them. This deals the game from an effectively infinite shoe.

These percentages may be added to determine the house edge for a blackjack game with a specified set of rules. One particularly common set of rules is known as *Las Vegas Strip Rules*, though these rules are neither restricted to the Strip nor universal on that street.

Las Vegas Strip Rules

- The dealer must draw to 16 and stand on all 17s.

- Players may double down on any two cards.

- Players may split any pairs, and resplit up to 4 hands, except for aces.

- Aces may not be resplit. Only 1 card may be drawn to split aces.

- Doubling down after splitting a pair is not permitted.

When played according to these rules, the casino holds an advantage over a basic strategy player based on the number of decks in use: from .40% on a single-deck game to .97% on an 8-decker.

Before the 1976 legalization of casinos in Atlantic City, New Jersey, the only American state with legal blackjack games was Nevada. Casinos in northern Nevada settled on their own set of rules over time, the so-called *Northern Nevada Rules*. These differ from Las Vegas Strip rules in two ways:

- The dealer must draw to soft 17 or lower, and must stand on hard 17 and higher, instead of standing on all 17s.

- Players may double down only on two-card totals of 10 and 11, rather than any two cards.

Table 5.1 shows that these changes raise the house advantage by .46% relative to Las Vegas Strip rules for a game with the same number of decks.

A particularly onerous and common rule change reduces the payoff on a natural from the traditional 3–2 to 6–5. On a $10 bet, this changes the payoff on a natural from $15 to $12. While this may seem insignificant due to the scarcity of naturals, Table 5.1 shows that this rule raises the HA by 1.39%—more than a threefold addition to the HA in a 1-deck game. This rule change finds its origins in *Super Fun Blackjack*, a 1999 single-deck variation on blackjack invented by Howard Grossman and designed in part as a defense against card counters. Super Fun Blackjack paid either even money or 6–5 on naturals, but offered a number of game options that might have appeared to make up the difference. At the Orleans Casino, the following options were available [8]:

- Doubling down is permitted on hands of two *or more* cards, including after hitting or splitting pairs. Players could double down, for example, on a three-card total of 11, which is not allowed under standard blackjack rules.

- Pairs, including aces, may be resplit up to three times, resulting in as many as four hands in play at a time. Some casinos allowed multiple-card draws to split aces.

- Hands may be surrendered at any point, including after drawing one or more cards, if the dealer does not have a natural.

- A hand totaling 20 or less with six or more cards pays even money regardless of the dealer's hand, although if the hand has been doubled, the doubled amount is not an instant winner. The doubled amount does remain in play and is paid off if the hand beats the dealer.

- A 21 with five or more cards pays 2 to 1 instantly, with the same restriction as for six-card 20s.

- Player naturals are instant winners even if the dealer also has a natural. Naturals with both cards in diamonds pay 2 to 1, an exception to the decreased payoff for other naturals.

Despite the appearance of a more player-friendly game, reducing the payoff on naturals raised the HA by so much that the new rules still left casino operators with a healthy edge. The game could not be effectively attacked by players without mastering a specialized strategy for Super Fun 21 itself that took the different rules into account [131].

Super Fun Blackjack was protected by patent, but the simple idea of changing the payoff on a natural was not. Casino operators were quick to see the potential for increased income at blackjack tables by offering 6–5 on naturals without any of the compensating rules in Super Fun Blackjack. Such payoffs have become very common at low-minimum games; players who wish to compete for a 3–2 payout on naturals must often move to higher-stakes tables, perhaps from a $5 minimum game to a $25 minimum.

At the other end of the payoff spectrum, Binion's Horseshoe Casino in downtown Las Vegas (now simply Binion's) regularly offered a holiday special of a 2–1 payoff on naturals in the few days before Christmas each year. Because this gave players such a big edge, 2.32% more than with a 3–2 payoff, this goodwill promotion was restricted to bets of $5 or less.

To estimate the effect of changing from 3–2 to 6–5 on naturals, assume a fixed $10 wager across 63 hands, roughly an hour's play at a reasonable pace. With the probability of a natural approximately 1 in 21, the player would expect 3 naturals over this span. The total win on these 3 hands would be $36 instead of $45, so $9 would be retained by the casino. Dividing by the $630 wagered shows that the casino has held 1.42% of the player's wagers, which is very close to the value of 1.39% shown in Table 5.1. Similar calculations confirm that the 2–1 payoff at Binion's Horseshoe gives the $5 bettor approximately the 2.32% edge shown in that table.

Example 5.2. Figure 5.1 shows a hand-held single-deck electronic blackjack game with the following deviations from Las Vegas Strip rules:

- No pair splitting allowed (HA + .4%)

- Doubling down only permitted on 10s and 11s (HA + .26%)

- The player has the chance to push a dealt dealer natural by drawing out to 21, as that natural is not revealed until the player has completed

FIGURE 5.1: Simple handheld electronic blackjack game.

her hand. This unusual rule was seen in live casino play at the Hide-Away Casino in West Wendover, Nevada in the mid-1980s. Peter Griffin, author of *The Theory of Blackjack* [56], estimated that this rule gives the player an additional .5% edge by turning a sure loser into a possible push [134].

Adding up the cumulative effects of these rule changes results in a game with a house edge of .56%—the rule about dealer naturals more than counters the base house edge.   ∎

5.2   Basic Strategy

A novice blackjack player can reduce the house advantage by using *basic strategy*, which is a set of rules for blackjack decisions to be followed by players. Based on millions of simulated hands, basic strategy gives a player the best mathematical advice on how to play a hand, given the composition of the hand and the dealer's upcard. Calculations about the HA typically assume that the player is using basic strategy correctly; deviations from these instructions will work to increase the casino's long-term edge.

In the early days of blackjack, the dealer's play was not prescribed: the first two cards to each player were dealt face down and the dealer chose how to play his hand in reaction to the faceup cards drawn to players' hands [43]. Players had the option of bluffing by standing on a low hand in order to encourage the dealer to draw and risk busting. When the dealer had a choice of options, no one player strategy could prevail. Once the dealer's actions—draw to 16 or less, stand on 17 or more—were standardized and choice was removed, the possibility of a "best" way for a gambler to play her hand emerged.

The formal history of basic strategy begins with "The Optimum Strategy in Blackjack," a paper by Roger Baldwin, Wilbert Cantey, Herbert Maisel, and James McDermott that was published in the September 1956 issue of the *Journal of the American Statistical Association* [3]. The four authors were mathematicians serving in the Army at the Aberdeen Proving Grounds in Maryland, and their casual blackjack games inspired the foursome to analyze the game mathematically. This paper was the first to suggest that a set of rules for player action, which has now evolved into basic strategy, would lead to the best possible return for players. For the first time, the idea that the player's position in blackjack could be improved by skillful play was backed up with a rigorous mathematical argument. Among other revelations, this paper was the first to advocate that a pair of 8s should always be split and that standing on a 12 against a dealer 4, 5, or 6 is mathematically the best play.

Given the player's cards and the dealer's upcard, basic strategy gives the best choice for the player: when to hit, stand, double down, or split pairs for the best long-term results. The key word here is *long-term*: basic strategy, for example, tells a player to hit rather than stand when holding hard 15 against a dealer 8. Note that the only way a player 15 can win is if the dealer busts, since a completed dealer hand will always be 17 or greater. While hitting a 15 will certainly bust some hands—more often than not, in fact: the chance of busting a hard 15 is $\frac{7}{13} > .5$—computer simulation indicates that this action is better for the player, in the sense that, over a lifetime of gambling, you will win more (or lose less, which is effectively the same thing) by hitting and taking the chance of busting than by standing and hoping the dealer will bust.

The goals of basic strategy are twofold:

1. To get the most money in play, via doubling down and splitting pairs, when conditions favor the player, and to minimize exposure by forgoing these options when conditions are less favorable. A weak dealer upcard means that players should double down and split pairs more frequently when those options are available. When the dealer's hand is strong—a ten or ace showing, for example—then basic strategy players take more risks as they try to compete against a hand that is more likely to be close to 21, and don't invest additional money under the less favorable circumstances.

2. To make optimal decisions about hitting and standing on stiff hands. If the dealer's upcard indicates that his hand is likely to bust—say, a 4, 5, or 6—then players should not risk busting their own hands and let the dealer assume the risk of exceeding 21.

Table 5.2 contains the complete basic strategy for a 6-deck game. "T" stands for any 10-count card.

TABLE 5.2: Basic strategy for six-deck blackjack [61]

Player Hand		Dealer's upcard									
		2	3	4	5	6	7	8	9	T	A
	17+	S	S	S	S	S	S	S	S	S	S
	16	S	S	S	S	S	H	H	H	H	H
	15	S	S	S	S	S	H	H	H	H	H
	14	S	S	S	S	S	H	H	H	H	H
No pair	13	S	S	S	S	S	H	H	H	H	H
or ace	12	H	H	S	S	S	H	H	H	H	H
	11	D	D	D	D	D	D	D	D	D	H
	10	D	D	D	D	D	D	D	D	H	H
	9	H	D	D	D	D	D	H	H	H	H
	8−	H	H	H	H	H	H	H	H	H	H
	A/9	S	S	S	S	S	S	S	S	S	S
	A/8	S	S	S	S	S	S	S	S	S	S
	A/7	S	DS	DS	DS	DS	S	S	H	H	H
Soft hand,	A/6	H	D	D	D	D	H	H	H	H	H
no pair	A/5	H	H	D	D	D	H	H	H	H	H
	A/4	H	H	D	D	D	H	H	H	H	H
	A/3	H	H	H	D	D	H	H	H	H	H
	A/2	H	H	H	D	D	H	H	H	H	H
	A/A	SP	SP	SP	SP	SP	SP	SP	SP	SP	SP
	T/T	S	S	S	S	S	S	S	S	S	S
	9/9	SP	SP	SP	SP	SP	S	SP	SP	S	S
	8/8	SP	SP	SP	SP	SP	SP	SP	SP	SP	SP
Pairs	7/7	SP	SP	SP	SP	SP	SP	H	H	H	H
	6/6	H	SP	SP	SP	SP	H	H	H	H	H
	5/5	D	D	D	D	D	D	D	D	H	H
	4/4	H	H	H	H	H	H	H	H	H	H
	3/3	H	H	SP	SP	SP	SP	H	H	H	H
	2/2	H	H	SP	SP	SP	SP	H	H	H	H

Never take insurance.
Key: D: Double down. DS: Double down if permitted, otherwise stand.
H: Hit. S: Stand. SP: Split.

Summary of Table 5.2

- If the dealer is showing a 7 through ace, hit any hard hand less than 17.
- If the dealer's upcard is 2–6, stand on any hard hand that's 12 or greater, except hit hard 12 against a 2 or 3.
- Double down on 11 except when the dealer shows an ace, on 10 except against a dealer 10 or ace, and on 9 if the dealer shows a 3–7.
- Never stand on a soft hand less than 19, except stand on soft 18 against a dealer 2, 7, or 8.
- If the casino permits doubling down on any first 2 cards, double soft 17 and 18 against a 3–6, soft 15 and 16 against 4–6, and soft 13 and 14 against 5 or 6.

- Always split aces and 8s.

- Never split 4s, 5s, or 10s—indeed, a pair of 5s should be doubled unless the dealer's upcard is a ten or ace.

- Split 9s against any dealer upcard except 7, 10, or ace. Split 7s against 2–7, 6s against 3–6, and 2s or 3s against 4–7.

Example 5.3. Basic strategy calls for players to stand on a hard 12 or higher against a dealer 6. This being so, casinos were able to promote the "dealer hits soft 17" rule change as a second chance for players to win, as there was a chance that the dealer would turn over an ace to go with the 6 and then bust while hitting a winning 17.

This was not mathematically true. Hitting soft 17 gives the dealer a good chance to improve 17 to a higher hand that not only beats player 12–16s, but some higher hands as well. ∎

Example 5.4. A less obvious bit of basic strategy is the admonition never to split 4s. This represents the understanding that a hand of 8 is moderately strong, with probability $\frac{6}{13}$ of improving to a pat hand after a one-card draw. A hand starting with 4, by contrast, is weak.

If the casino permits doubling down after splitting pairs, though, basic strategy advises players to split 4s against a dealer 5 or 6 [131]. In this situation, the player will adopt a "do not bust" strategy for drawing since the dealer's bust probability is high, but can gain an additional edge by doubling if one of the 4s receives a 5, 6, 7, or ace as its second card. The chance of drawing one of these cards is $\frac{4}{13}$ under the infinite deck approximation. ∎

If surrender is an option, blackjack experts have, through the repeated simulation mentioned above, identified certain hands of 15 or 16 that should be surrendered, and these are listed in Table 5.3. Note that hand composition

TABLE 5.3: Strategy for surrender [102]

If dealer shows	Surrender
Ace	10/6.
10	10/6, 10/5, 9/7, 9/6, 7/7.

can be as important as the total: a pair of 8s should be split, not called a 16 and surrendered.

Local Promotions

Over the course of blackjack's history, many casinos and game designers have tried local variations on the basic rules in an effort to promote their casino, cleverly gain an extra edge over the players, or just to add some excitement to this traditional casino favorite. Since these are decidedly nonstandard rules, basic strategy must often be adjusted to make the best use of these modifications.

Doubling Down

- Two casinos in Moscow, the Crystal Casino and the Golden Palace, extended the double down option, allowing players to triple their original bet by adding an additional betting unit after doubling down and receiving a third card. Players would then receive one more card, for a total of 4 in hand. One place where this rule would be useful is when doubling a 9 and receiving a deuce for a total of 11.

  While the triple down rule added approximately .31% to the player's edge when used properly, a more punitive casino rule could more than negate the advantage. Russian casinos at the time, early 2001, extracted a 35% tax on player's winnings [54]. At some casinos, this tax was collected when a player cashed out his chips, so after buying in for $500 and losing $250, a player redeeming $250 in chips would only receive $162.50, disregarding that this was not money won. Unlike in the USA, players could not use their accumulated gambling losses to offset the amount of their winnings subject to tax. While the Crystal and Golden Palace did not charge this tax (behavior which was questionable from a legal perspective), casino gambling in Russia was in this way very player-unfriendly.

- At Hart's Palace in upstate New York in the late 1980s, blackjack rules included the "hard double": a player option to double down on hard 12–17. Ordinarily, this would be an ill-advised move, but the sweetener for players was that the doubled bet pushed if both player and dealer busted. Revised basic strategy with the hard double is to double hard 12 against a dealer 4, 5, or 6 [128]. The risk of the player busting is at its lowest when hitting a hard 12, while the dealer's bust probability is highest with one of those cards up.

  The probability of a player busting when hitting hard 12, using the infinite deck approximation, is $\frac{4}{13} \approx 30.77\%$, as the hand only busts when receiving a ten. The dealer, by contrast, busts over 40% of the time when starting with a 4, 5, or 6 [131]. Doubling like this against a weak dealer upcard is a move that gets more money on the table when conditions favor the player, and the chance of recovering from a busted 22 when the dealer also busts is high, justifying the risk.

Insurance

- In 1985, the Tropicana Casino in Atlantic City faced an unusual puzzle at its $3 minimum blackjack tables. A standard insurance bet of 50% of this minimum wager would be $1.50, but the casino did not stock fractional chips valued at less than $1. Their solution was to allow $1 insurance bets but to pay them off at the $3 that would accrue from a winning $1.50 bet unless the player was insuring their own natural [33]. In the latter case, the payoff was the traditional "even money" 2 for 1 return.

The Tropicana paid a small price for its simplicity. The optimal player strategy was to bet the minimum and insure every hand except for dealt naturals, which gave an expectation on the insurance bet, using the infinite deck approximation, of

$$E = (3) \cdot \frac{16}{52} + (-1) \cdot \frac{36}{52} = \frac{12}{52}.$$

This translates into a 23.1% edge on the insurance bet. The opportunity to make an insurance bet arises on 1 hand in 13; the player has a natural on $\frac{1}{21}$ of these hands, so the net advantage accruing to a player from this rule using correct strategy was

$$\frac{12}{52} \cdot \frac{1}{13} \cdot \left(1 - \frac{1}{21}\right) \approx .0169,$$

or just under 1.7%.

As basic strategy players routinely skip the insurance bet, this may have had only limited practical significance.

Jokers

- In 1989, just prior to the enormous boom in themed casino resorts on the Las Vegas Strip, the Sands Casino (now the Venetian) launched a blackjack promotion where 3 wild jokers were inserted into a 6-deck shoe. Jokers dealt to the dealer were burned: discarded without counting. This opportunity did not require deep mathematical analysis to exploit. Since the value of a joker had to be declared before the next card was dealt and could not exceed 11, the following algorithm gave the best strategy for assigning a value to a joker [160]:

  1. If possible, make the hand 21. This is especially valuable when hitting a stiff hand and receiving a joker.

  2. If possible, bring the hand to a total of 20.

  3. Either make the hand 11 or call the joker an ace, retaining some flexibility to count the joker as either 1 or 11 later in the hand.

  As a result, this promotion, and its 1.8% increased player advantage, lasted only days [28].

- The Kewadin Casino in Sault Sainte Marie, Michigan ran a joker promotion in late 1997. For 8 hours daily, two jokers were added to each 4-deck shoe. Any winning hand with a joker paid double the usual payoff: 2–1 on a win, 3–1 on a natural, and 4–1 on a doubled hand [19]. The joker carried no numerical value in the hand; it served more as a coupon for a doubled payoff. Two jokers per shoe used in this fashion gave the players an addition 1.3% edge, so it was no surprise that advantage players converged on northern Michigan to make the most of this promotion, which started at 4:00 A.M. An observant casino manager could not have failed to notice the unusual patron influx at such a late hour and suspect that something was not quite right with the joker promotion.

- The El Condado in Lima, Peru briefly offered a bonus joker in its 6-deck games. The joker replaced one ace of spades in the shoe and counted as an ace for the purpose of valuing the hand, but a hand containing a joker paid off an extra unit. A natural with a joker thus paid 5–2, an winning hand paid 2–1, and a push paid even money [34]. Since the joker was a replacement for the $A\spadesuit$, no change in player strategy was necessary.

Naturals

- At the Continental Casino in Las Vegas (now the Silver Sevens), a lucrative September 1992 promotion offered players an average 1240% hourly return for their $5 wager. The deal was simple and required no additional wager: A player dealt consecutive naturals received a $500 bonus, regardless of the size of their initial wager [32].

  That was it.

  Also attractive was the fact that this promotion did not require a change in player strategy. Such a lucrative player deal could not last forever, and over the ensuing days, the bonus was lowered to $200, then $100, then $25. At that final level, a $5 bettor could still expect an average bonus of $2.25 per hour while risking no additional funds.

  With the infinite deck approximation, the probability of consecutive naturals is
  $$2 \cdot \frac{4}{52} \cdot \frac{16}{52} \approx \frac{1}{446},$$
  so the expected return from the $500 bonus on every hand was approximately $1.12, since it required no extra player investment. For a $5 wager, this gave players a 22.4% bump in their edge.

- On bets of $5 to $100, the El Rancho Casino in Las Vegas, open from 1982–1992, paid 8–5 on naturals. This rule required only that players bet in multiples of $5 for full effect; fractional payoffs were rounded down, in the casino's favor. This corresponds to a $16 payoff on a $10 natural, up from the $15 payoff that 3–2 would yield. This rule gave players a net .15% edge at double-deck games and pulled 4-deck games essentially even [29]. As with the Continental's promotion, this required no change in basic strategy because it was based on naturals.

Splitting

- No matter how it's dealt, an initial hand of 16 is not good for a blackjack player. The Hide-Away Casino went one step beyond splitting a pair of 8s when it allowed players to split any 2-card total of 16: 10/6 and 9/7 as well as 8/8 [22].

Example 5.5. Starting with a 10 gives the player a 13% edge over the house on that hand, so turning a 10/6 into two hands, one starting with 6 and one starting with 10, turned a bad hand (an initial 6 is the worst card to see in your hand, even as it's welcome as the dealer's upcard) into a bad hand and a great hand. ■

Splitting 8s against all dealer upcards is a defensive measure: an attempt to salvage two mediocre hands out of one bad hand. The opportunity to split all 16s extended this salvage operation to other initial hands, but with the difference that following the proper strategy for splitting 16s could garner the player a 1.7% boost to his or her advantage, making this rule nearly non-viable unless it was balanced by some player-unfriendly rules [161]. Proper strategy was to split all 16s.

Surrender

- A number of casinos in Macau offered a "reverse surrender" option. If a player drew out to a 5-card hand totaling under 21, he or she could opt to collect a win of half their wager and end the hand before the dealer's hand was complete [12]. This rule is especially advantageous when the dealer does not take a hole card until all of the players have played out their hands. This is sometimes called "European style", and was common in Macau at the time. In a 4-deck game, this rule offers the player a .70% advantage, and in a single-deck game, the player edge is still impressive: .54% [56].

Example 5.6. If you have drawn out to a 5-card 16 and the dealer is showing a 10, reverse surrendering allows you to win 50% of your wager in a situation where you are the decided underdog and risk losing it all: potentially a 150% turnaround. The dealer could draw an ace for a natural, or could beat your 16 in a number of other ways, but the reverse surrender rule gets you half a win [56].

Note that basic strategy calls for hitting 16 against a dealer 10, so there's more risk ahead for the basic strategy player. ■

Basic strategy needs to change to accommodate both this new rule and the dealer taking his hole card later. A bit more risk is taken in an attempt to reach a 5-card unbusted hand in unfavorable circumstances and benefit from the reverse surrender option. The following changes to Table 5.2 are recommended for 4-card hands [56]:

- Hit soft 19 and below against any upcard other than 7 or 8.
- Hit hard 15 and less against a 2.
- Hit hard 14 and less against a 3 or 4. Hit T/2/A/A against a dealer 5. This is the only 4-card 14 which should be hit with the dealer showing a 5; it is estimated that this last opportunity arises about once in 35,000,000 hands.
- Hit hard 13 and less against a 5 or 6.

Additionally, hit all soft 18s against an ace, all 3-card soft 18s against an 8, and all 3-card 12s against a 4.

- For several years in the late 1980s, the Silver City Casino in Las Vegas offered "September Surrender", a promotion in which "surrender" meant the more desirable "early surrender". As per Table 5.1, this gave single-deck players a positive edge using basic strategy alone [28].

- The Free Ride promotion at the Klondike Casino in Las Vegas (Figure 5.2) was an ill-fated variation on surrender that lasted for 3 weeks in 1996.

FIGURE 5.2: The Klondike Hotel and Casino on the Las Vegas Strip, in 2001 [153].

Players who were dealt a natural were, in addition to a 3–2 payoff, given a lammer (Figure 4.2, page 196) imprinted with the amount of their bet.

On a future hand, the player could redeem the lammer before the dealer checked her hole card to declare "Free Ride!" and call any two-card hand a push, effectively surrendering for the full value of the bet rather than half. The value of the lammer was the value of the initial bet, so it was to a player's advantage to *flat-bet* the game: wager a constant amount per hand.

Example 5.7. If a player betting $25 drew a natural, her lammer would be marked "25". On a future $100 bet, the lammer could only be used to save $25; the remaining $75 would have to be played out. ∎

One extra condition was that a player could hold only one lammer at a time: a second natural would not garner an additional lammer if the player still held one from a previous natural [135]. This meant that holding onto a lammer ran the risk of "wasting" a natural, and so strategy calculations focused on finding circumstances to make the best use of a lammer quickly, balancing expected return against efficient use.

For a short while, the promotion attracted only slight attention, perhaps in part due to the Klondike's location at the far south end of the Las Vegas Strip and its generally unglamorous architecture. That changed on June 23, 1996. Blackjack author Stanford Wong alerted the audience at a seminar at the Tropicana, just up Las Vegas Boulevard from the Klondike, that this rule carried approximately a 1% player advantage using basic strategy alone. He later estimated that a player using the surrender strategy detailed above in Table 5.3—for this was, in effect, a variation on the early surrender option—gained 1.5% over the casino.

This was too great an advantage to keep quiet, and Wong notified subscribers to his *Current Blackjack Newsletter* of this opportunity via fax and email. In very short order, the Klondike was packed with gamblers making table maximum wagers of $100 at every spot on the 5 blackjack tables [135]. Casino officials were puzzled by their accumulating losses; some attributing the unusual player success to a giant team of players that had the ability to steer losing cards to the dealer by selective hitting or standing. Wong's mathematics told a different story.

Free Ride was discontinued at 1:45 A.M. on June 25 [99]. After the fact, Edward Thorp devised an ideal strategy as an academic exercise. Balancing the desire to use a lammer for maximum benefit with the competing wish not to have a lammer expire on a second natural, he derived the following equation for the overall player gain g from using a lammer on a given hand [146]:

$$g = b \cdot \frac{pe}{b + p},$$

where

b = The frequency that lammers are issued.

e = The expected gain when a lammer is used.

p = The probability that a lammer is used on a given hand.

What does it take to maximize g? It's easy to see that increasing e leads to an increase in g; finding the best possible value for e requires assessing the conditions that contribute to the amount saved by using the lammer. This comes down to the value of the hand as a function of the player's initial 2 cards and the dealer's upcard.

This suggests that Wong's initial estimate that using the early surrender strategy was a sound first approximation to an ideal strategy, and almost certainly the best available given his limited time to act.

A further approximation leaves out the effect of the player's hand and looks solely at the dealer's upcard, using it to determine e. Thorp's analysis concluded that the best simple strategy was simply to use a lammer if the dealer's upcard was an ace, except when the player held a natural. The player edge with this approach is 1.25% [146].

More detailed analysis improves this edge only slightly: to 1.43% if the lammer is used according to Table 5.4. Wong's quick estimate of a 1.5%

TABLE 5.4: Optimal strategy for the Klondike's Free Ride promotion [146]

Dealer upcard	Take the **Free Ride** on
Ace	Hard 4–6, Hard 12–17, Soft 16–17
10	Hard 12–17
9	Hard 14–17
8	Hard 15–16, except 8/8

player advantage was extremely close to ideal.

It should be noted that most of these hard hands are ones where the basic strategy player is hitting and risking busting against a strong dealer upcard, so invoking Free Ride on these hands turns a probable loss into a push. Playing a lammer on 8/8 against an ace, 10 or 9 instead of splitting the pair is settling for a push when splitting would be a defensive measure aimed at cutting one's losses. Cutting losses down to 0 is a good deal.

5.3 Card Counting

Alone among common casino games, blackjack is not subject to mathematical independence from hand to hand. Until the deck or shoe is shuffled, a card that has been dealt cannot appear again. (Successive baccarat hands may depend on the previously-dealt cards, but we have seen that this is nothing that can be exploited by counting cards. For all practical purposes, we may consider baccarat as a game of independent trials.) Each hand in blackjack depends in a meaningful way on the cards that have been dealt since the last shuffle, and as a result, probability calculations for blackjack must take this into account. The infinite deck approximation described above assumes independence in its calculations and sacrifices a small amount of accuracy for ease of computation. Since we are looking specifically at how previous cards affect probabilities, we must discard the assumption of independence and thus the infinite deck simplification.

Recall that the probability of a natural from the top of a fresh single deck is

$$2 \cdot \frac{4}{52} \cdot \frac{16}{51} \approx .0483 \approx \frac{1}{21}.$$

If one ace is removed from the deck, the new probability of a natural is

$$2 \cdot \frac{3}{51} \cdot \frac{16}{50} \approx .0376 \approx \frac{1}{27},$$

a decrease of 22%. By contrast, if a single deuce is removed from the deck, the chance of a natural rises, to

$$2 \cdot \frac{4}{51} \cdot \frac{16}{50} \approx .0502 \approx \frac{1}{20}.$$

This change is smaller, but it illustrates the significance of deck composition as cards are removed from the deck in the course of play. If 7 cards, none of them aces or 10s, are removed, the probability of a natural is

$$2 \cdot \frac{4}{45} \cdot \frac{16}{44} \approx .0646 \approx \frac{1}{15},$$

so in a single-deck game, it doesn't take much to shift probabilities up—or down. If those 7 removed cards are all 10s, we have

$$2 \cdot \frac{4}{45} \cdot \frac{9}{44} \approx .0364 \approx \frac{1}{28}$$

as the probability of a natural.

Early History

Gambling lore tells of advantage players who successfully won at blackjack before the value of card counting was well-known. Early attempts at card counting were called "casing the deck". This was a time when casino blackjack was routinely dealt from a single deck down to the last card. A card counter who identified a favorable set of cards near the end of a deck would be inclined to raise his bets to take full advantage of the surplus of high cards—10s, face cards, and aces—that remain. This action, however, could easily be foiled by the dealer who shuffled the deck whenever a suspected deck caser significantly raised his bet [158]. Properly camouflaging their activity to avoid detection for as long as possible remains a challenge for card counters. While card counting is not illegal, casinos are free to restrict bet ranges or even to bar players from gambling, so long as they violate no laws in doing so—and card counters are not a protected class in the eyes of the law.

The High-Low Count

There is nothing mathematically complicated about card counting; all that is required for the most commonly used counting system is the ability to add and subtract 1, and to divide. To do so under casino conditions while avoiding detection is, as portrayed in the movie *21*, considerably more challenging. It is not necessary, nor is it desirable, to try and track every card in a deck or shoe; all that most card-counting systems do is give the player a sense of the balance between high and low cards in the remaining undealt portion of the deck. This information, combined with the ability to adjust bet sizes in line with the count and the willingness to learn exceptions to basic strategy to take the best advantage of the changing deck composition, makes card counting a tool that can give a blackjack player an advantage over the casino.

A simple but powerful counting system, and arguably the most popular, is the *high-low count*. The idea behind the high-low count is that a deck with a surplus of high cards favors the player, in the sense that the probability of naturals and of drawing a high card on a double down both increase. Of course, the probability of the dealer pulling a natural also increases, but since player naturals are paid 3 to 2 while dealer naturals lose only the amount bet, an increased probability of naturals is still a benefit for the player.

In *Beat The Dealer*, published in 1962, mathematician Edward O. Thorp outlined a strategy for tracking cards as they were dealt and described the mathematics underlying the influence of the count on player decisions. Thorp's original high-low card-counting system simply tracked the ratio of non-10s to 10s in a deck, a ratio that starts at 36/16 in a full deck [144]. This was soon eclipsed by more sophisticated counting methods. In the high-low counting system, the cards are assigned the values shown in Table 5.5.

A complete deck, or shoe of several decks, has initial count 0, since there are exactly as many high cards as low cards in a deck. Cards are counted as

TABLE 5.5: Card values for the Hi-Low count

Cards	Value
Low: 2, 3, 4, 5, 6	+1
Neutral: 7, 8, 9	0
High: 10, J, Q, K, A	−1

they appear: add 1 to the count for every low card dealt and subtract 1 for every high card. The total is called the *running count*, RC.

Example 5.8. At the start of a four-player single-deck game, suppose that the cards dealt are 10/K, 4/5, 9/J, and 7/4, with a 4 for the dealer's upcard. The running count is $(-1 + -1) + (1 + 1) + (0 + -1) + (0 + 1) + 1 = 1$. ∎

For playing purposes, the running count must be converted into the *true count*, TC. This is done by dividing the RC by the number of decks remaining to be dealt. The divisor is estimated from the progress of the hand; a look at the discard rack on the table can help with the estimation. Switching from the RC to the TC gives a per-deck accounting of the excess or deficiency of high cards. A running count of +4, signifying an excess of four high cards over low cards, is a lot more meaningful if only one deck remains than if three decks are left in the shoe.

Example 5.9. Completing the hand above: The players hold hands of 20, 9, 19, and 11. Since the dealer's upcard is a 4, basic strategy dictates that the 20 and 19 stand and that the 11 and 9 double down. The 9 receives the 10♢ for a 19, and the count is now 0. The 11 draws the 2♢ for a 13. While this player may regret the decision to double down due to the low total, basic strategy still calls for her to stand on 13 against a 4. The count is now +1. The dealer's hole card is a 6 (RC +2) and he draws a a 3, a 2, and a jack, busting with 23. The running count is now +3. All 4 players win. Fifteen cards have been dealt, leaving roughly 3/4 of a deck—in general, the number of decks remaining is estimated to the nearest half- or quarter-deck. The TC is then $+3 \div (3/4) = +4$, meaning that high cards currently exceed low cards, making the game more favorable to players. Consequently, slightly larger bets are called for—the players may choose to let their winnings ride into a favorable deck. ∎

There are two uses for the TC. The first is to indicate a point when the player's initial bet should be increased because the cards are favorable and a good hand is more likely, or conversely, when the cards favor the casino and bets should be kept as small as possible. To be more explicit, every increase of 1 in the true count translates into approximately a .5% increase in the player advantage [148]. Since the game starts out with a house edge of about .5%, a count of +1 makes the game essentially even, and if the true count is +4 as in Example 5.9, players have a 1.5% edge over the casino.

The second purpose of the TC is to highlight times in the course of a deck or shoe when the composition of the remaining cards is such that certain deviations from basic strategy are, probabilistically, more favorable to the player. Once again, these true counts, called *index numbers*, have been determined by computer simulation.

Example 5.10. Consider the insurance bet. Since you're basically making a side bet that the dealer has a ten in the hole, a positive count indicating more high cards might indicate that that bet is worth making. Since insurance pays 2 to 1, this bet has a player advantage (independent of the player's hand) if the ratio of 10s to non-10s exceeds 1 : 2. Under the infinite deck assumption, this ratio is 4:9, less than 1:2, which is why basic strategy simply suggests "Never make the insurance bet." Counting cards may reveal deck compositions where the insurance bet is favored. Suppose that on the first round of a one-deck game, 11 cards were dealt: 2 ten-count cards and 9 lower cards. On the second round, the dealer shows an ace and your hand is a pair of 5s. The composition of the deck is now $16 - 2 = 14$ ten-count cards and $36 - 12 = 24$ nontens. Since the ratio of tens to nontens is 14:24, greater than 1:2, an insurance bet has an expectation in your favor. Specifically, the expected value of a 50¢ insurance bet is

$$E = (1) \cdot \frac{14}{24} + (-.50) \cdot \frac{10}{24} = \$\frac{9}{24} > 0,$$

so you should make the insurance bet. ∎

This simple reasoning based on the ratio given does not account for the fact that if the insurance bet wins, your original \$1 bet loses unless you also have a natural, and a careful analysis of game options will have to take that into account.

Translating this into the world of card counting, this means that if the TC is +3 or greater in a 6-deck game, the insurance bet favors the player. This index number takes into account the fact that aces as well as 10s count –1 in the high-low count, hence the true count by itself is not a perfect indicator of the ratio of 10s to non-10s remaining in the shoe.

At the other end of the spectrum, a highly negative true count indicates an abundance of low cards over high ones. In addition to indicating that insurance bets should not be made and wagers should be kept to the table minimum, a negative TC may mean that hitting certain hard hands despite the advice of basic strategy—for example, hitting a hard 12 against a 5 at a true count of –2—is called for. Eighteen deviations from basic strategy for a four-deck game where the dealer stands on all 17s—dubbed the "Illustrious 18" by blackjack experts—are listed in Table 5.6 in decreasing order of their contribution to the player's edge.

Insurance tops the list; the gain from making an insurance bet under favorable shoe composition is itself over one-third of the advantage attained from adjusting strategy with the count [116, p. 67]. The list of indices does not stop at 18—there are hundreds of other changes to basic strategy recommended by

TABLE 5.6: Blackjack's Illustrious 18 [116]

Strategy deviation	If the high-low true count is
Take insurance	$\geqslant +3$
Stand on 16 vs. 10	$\geqslant 0$
Stand on 15 vs. 10	$\geqslant +4$
Split 10s vs. 5	$\geqslant +5$
Split 10s vs. 6	$\geqslant +4$
Double 10 vs. 10	$\geqslant +4$
Stand on 12 vs. 3	$\geqslant +2$
Stand on 12 vs. 2	$\geqslant +3$
Double 11 vs. ace	$\geqslant +1$
Double 9 vs. 2	$\geqslant +1$
Double 10 vs. ace	$\geqslant +4$
Double 9 vs. 7	$\geqslant +3$
Stand on 16 vs. 9	$\geqslant +5$
Hit 13 vs. 2	$\leqslant -1$
Hit 12 vs. 4	$\leqslant 0$
Hit 12 vs. 5	$\leqslant -2$
Hit 12 vs. 6	$\leqslant -1$
Hit 13 vs. 3	$\leqslant -2$

the true count, but they have less effect on the player's ultimate return than any of the Illustrious 18.

A closer analysis of the effect on the player's advantage of removing individual cards from the deck can be found in [56], where theoretical and experimental evidence were combined in the following model: For each of the ten card ranks, define the following variables:

- x_1 = The number of stiff hands (12 to 16) which that card will improve to 17 or higher.

- x_2 = The number of stiff hands which the card will bust.

- x_3 = 1 if the card is a ten-count card, 4 if the card is an ace, and 0 otherwise.

For example, if the card in question is a 5, we have $x_1 = 5$, $x_2 = 0$, and $x_3 = 0$. Simulation of many hands indicated that removing an ace, 9, or 10 decreased the player's edge, removing an 8 had no effect, and removal of any other card increased the player's advantage. The cumulative effect of card removal, denoted by y, was found to be well-modeled by the formula

$$y = .14x_1 - .07x_2 - .18x_3.$$

The values predicted by this formula are all within .1 of the effect determined by simulation, and in most cases are far more accurate than that.

Card Counting and the Surrender Option

Given a two-card blackjack hand, the dealer's upcard, and the specific game rules in force, it is possible through repeated simulation to obtain an accurate estimate of the probability p of winning the hand. At what probability is surrendering the correct choice?

If you surrender, your expectation on a $1 bet is a flat –$.50. If you choose instead to play out the hand, your expectation is

$$E = (1) \cdot p + (-1) \cdot (1 - p) = 2p - 1.$$

This will be greater than –$.50 if $p > .25$, so if you have at least a 25% chance of winning, surrendering is not the optimal play. Turning this around, if your hand has greater than a 75% chance of losing, you should surrender if the option is available.

The challenge here is that p is not a probability that is readily calculable from the three visible cards. The strategy described in Table 5.3 (page 249) encompasses the best basic-strategy surrender advice. When playing a game where surrender is allowed, judicious use of this option in conjunction with the true count can further add to a player's edge. In addition to the recommended surrenders in Table 5.3, the "Fab 4" surrender decisions listed in Table 5.7, again in order of the advantage they carry, are profitable at or above the indicated count [116].

TABLE 5.7: Blackjack's Fab 4 [116]

Surrender	If the high-low true count is
14 vs. 4	$\geq +3$
15 vs. 10	≥ 0
15 vs. 9	$\geq +2$
15 vs. ace	$\geq +1$

It is important to understand the inherent risk here: Part of the appeal to casinos of offering the surrender option is that many players will not use it even when it's the "correct" decision. A blackjack player who surrenders correctly may be a card counter, and permitting this game choice may be part of a casino's strategy for identifying and barring counters.

Side-Counting Aces and Hi-Opt I

The ace can be a tricky card for card counters. The High-Low count classifies aces as high cards, but their ability to count as 1 or 11 means that they have some of the qualities of low cards at the same time. With this in mind, some counting strategies advocate that counters keep a side count of aces as additional information.

Example 5.11. One example of a rule change favoring an ace count comes from the Royal Scandinavian Casino (now the Royal Casino) in Aarhus, Denmark. The Royal Scandinavian once offered a rule variation where an ace dealt to a split 10 qualified as a natural and paid 3–2. With this rule in effect, proper basic strategy called for splitting 10s against a dealer 4, 5, or 6 if an ace count showed that aces were particularly abundant in the cards remaining to be dealt [87].

This strategy requires some knowledge of the proportion of aces remaining in the shoe if it's to be used effectively. ∎

Some counting schemes account for the dual nature of the ace by assigning aces a count of 0 and then adjusting the count, temporarily and only for betting purposes, to account for a surplus or deficiency of aces. One method is *Hi-Opt I*, which is short for "Highly Optimum" (the "I" refers to its standing as the first Hi-Opt system and to the fact that the values assigned to card ranks are 0 and ±1):

Cards	Value
Low: 3, 4, 5, 6	+1
Neutral: A, 2, 7, 8, 9	0
High: 10, J, Q, K	−1

Here the aces and deuces are counted as neutral cards, so the count remains balanced: a full shoe or fresh deck begins with a running count of 0. As with High-Low, Hi-Opt I calls for adjusting the running count to the true count by dividing by the number of decks remaining.

A further adjustment that accounts for the aces calls on the counter to add 1 to the running count for each "extra" ace in the deck beyond 1 per quarter-deck and subtracting 1 for each "missing" ace [56].

Example 5.12. In a 5-hand game dealt from a fresh single deck, the players' hands are 10/J, 4/4, A/9, 9/8, and A/8, with the $A\heartsuit$ as the dealer's upcard. The Hi-Opt I count is $(-1 + -1) + (1 + 1) + (0 + 0) + (0 + 0) + (0 + 0) + 0 = 0$. The corresponding High-Low count would be –3, all due to the aces counting as –1 in High-Low rather than 0 as in Hi-Opt I.

The hand is completed with the second hand drawing a 5 and a king to a pair of 4s, busting at 23 while the running count remains 0. All other players stand. The dealer turns over a 3 and draws a 10 and a 9 to bust. The new running count is 0, but 3 aces have been dealt among 16 cards. Sixteen cards is roughly one-quarter of a deck, so we would have expected 1 ace to appear. The remaining deck is ace-deficient. Since 2 more aces have been dealt than expected, we modify the running count by subtracting 2. For betting purposes, we consider the deck as having a running count of –2, calculate the true count as

$$\text{TC} = \frac{-2}{\frac{3}{4}} = -\frac{8}{3},$$

and bet the minimum, but when the game resumes, we continue the running count from 0. ∎

Once the bets are properly sized in accordance with this modified running count, we resume counting with the unmodified value. In a single-deck game, the 4 aces are easy to count; in an eight-deck shoe, tracking all 32 aces while also maintaining the proper running and true counts is somewhat more challenging.

The ace count with Hi-Opt I is just the next step on the road to more intricate blackjack counting systems. The pinnacle of complexity is reached with Edward Thorp's Ultimate Counting System, which is shown in Table 5.8. While far too complicated to be of any practical use, the values assigned to the various ranks represent the best possible illustration of the relative importance of all 10 ranks.

TABLE 5.8: Edward Thorp's Ultimate Counting System [56]

Rank	A	2	3	4	5	6	7	8	9	T
Value	-9	5	6	8	11	6	4	0	-3	-7

This system shows that the 5 is the most important card for players. This is because the dealer must draw to a 16, and so the absence of 5s diminishes the chance of the dealer making a hand instead of busting.

And the Race was On

Casino reactions to card counting seek an elusive balance between the fact that successful card counters have the potential to win a lot more money than the house advantage says that they should and the equally important fact that far more people think they can count cards successfully than actually can, which leads to increased traffic at blackjack tables and, ultimately, the likelihood of increased casino income from the tables.

Since the first of these facts is easier for casinos to address, we begin by describing some casino countermeasures. One of the first things that casino managers did after *Beat the Dealer* was published was to tighten up the rules of the game and so eliminate the advantages of card counting. On April 1, 1964, the Las Vegas Resort Hotel Association announced that henceforth, double downs would be allowed only on hands counting 10 or 11 and that blackjack players were forbidden to split aces [145, p. 128]. Table 5.1 shows the effect of these changes: the HA with these new rules went up by .44%. Blackjack players—counters and noncounters alike—stayed away from the games with new unfavorable rules, and they were reversed within three weeks.

After that reversal, casinos started looking for other game modifications that might restore their advantage and moved from single-deck hand-dealt games to multideck shoe games. Adding decks, in addition to increasing casino profits because more time is spent dealing the cards than shuffling them, increases the HA. Card counters then responded by developing new guidelines for counting into multideck shoes, which led to the running count and true count, and so the arms race of sorts between players and casinos was on. As casinos found a new rule or game procedure that would improve their position, the card-counting community developed a modification to their tactics that would combat it, and vice versa. Some examples of these modifications follow.

- Card counters developed team play, as popularized in the movie *21*. One team member, the "spotter", kept the count and made only minimum bets. When the count rose sufficiently high, the counter would signal another player to join the game and begin making large bets with a good advantage. When the count dropped, the "big player" left the table in search of another spotter and another hot table. One big player could monitor three or four table simultaneously.

  The roles of the spotter and big player came together in the "gorilla", who was responsible for keeping a count and acting the part of a high roller by making large bets correlated with the count. The gorilla was not intended to evade detection by casino security.

  Casinos responded by forbidding mid-shoe entry, allowing players to enter the game only after a shuffle.

- Counters started experimenting with tiny wearable card-counting computers. In the late 1970s and early 1980s, wearable computers were limited by technical matters: it was difficult to keep thin undetectable wires from breaking and to devise secure ways to output the count to the player. The input to one particularly notable card-counting computer was through toe switches that were prone to operational challenges [131].

  Nevada and New Jersey gaming authorities responded by banning portable computers and any other devices that aided in game analysis. These were the only states with legal blackjack at the time; the ban has been adopted as gaming has expanded to other states.

- Counters developed new ways to influence the fall of the cards, including shuffle tracking and key-card locating. Shuffle trackers look for a particularly rich stack of high cards, follow it through the shuffle, and cut the stack of cards to bring those cards to the top, so that they're immediately in play. Players on a shuffle-tracking team can make big bets off the top of the shoe—a time when the game usually holds no great advantage for either side—and know that they're tapping into a rich vein of high cards.

  Key-card locators attempt to control a single card, either an ace that can be directed to a collaborating player's hand or a ten that they can direct to the dealer's hand in an attempt to bust a stiff hand. When a dealer shuffles a 4- or 6-deck shoe, it is sometimes possible for players to catch a glimpse of the bottom card of the stack. A player skilled with a cut card can cut the stack 52 cards from the end; when these cards are moved to the top, the 52nd card in the shoe is known. Players then make small bets until the key card is due to come out, and then raise their bets in hopes of an ace landing in a collaborator's hand or a ten going to the dealer as a third card, which will bust the hand if the dealer holds a stiff hand. Steering a card to the right hand may involve making unusual plays that go against basic strategy. A player may hit a hand such as a hard 17 unnecessarily while betting a small amount in order to use up cards and ensure that an upcoming ace goes to a collaborating player. This is called "card eating"). Alternately, when betting big and steering a bust card to the dealer, players may forgo drawing to a hand to avoid taking the dealer's bust card. While this is a risky proposition, the possibility of a huge payoff makes it attractive.

- As card-counting strategies were developed in response to and to get ahead of changes in casino procedures, basic strategy techniques were also revised with the aid of better computers. A strong candidate for the least-used piece of basic strategy advice was described in *The Theory of Blackjack* [56]:

In a game with 27 or more decks in use, don't double ace-4 against a dealer 4. Hit that hand instead.

5.4   Variations

Blackjack is effectively in the public domain due to its long history. Game designers seeking to develop the next big casino table game sometimes begin with the basic rules of blackjack and then introduce a new modification that makes the game eligible for patent or trademark protection.

Experto 21

Experto 21 was a blackjack variation that launched at Vegas World (now the Strat) in Las Vegas. A skilled card counter could have had a huge advantage at this game, which was a standard one-deck blackjack game dealt down to the last card. One card was burned at the start, so 50 of the 52 cards came into play. As compensation to the casino for this deep penetration into the deck, naturals paid only even money, but players could double down on any 2 cards and the dealer stood on soft 17. These 3 changes raise the HA by 2.32%, which was a big margin for average players to overcome.

These rules still gave an advanced card counter, one who could track individual card ranks beyond a simple high-low count, the potential for a big win. All that was necessary was the ability to track most of the card ranks and a bankroll that facilitated taking full advantage of playing situations that arise as the cards are dealt. For example, a player dealt a pair of 7s against a dealer 7 should stand rather than split the pair as basic strategy calls for.

Example 5.13. An extreme example of the potential available to a player who could accurately track a full single deck is due to Edward O. Thorp [145]. Suppose that you have determined that the last 5 cards to be dealt are three 8s and two 7s. Your should immediately place as large a bet as you can, borrowing money if necessary—because you are a guaranteed winner.

Your initial hand will be 14, 15, or 16. Stand, because the dealer will also be dealt a hand between 14 and 16. The dealer must draw, and is certain to bust. ∎

Double Exposure

Double Exposure is a blackjack game where both of the dealer's initial 2 cards are dealt face up. It was originally described by gaming mathematician Richard Epstein under the name *Zweikartenspiel* in 1977 [40]. Bob Stupak, owner of the Vegas World Casino, brought Double Exposure to the Las Vegas Strip in 1979. Giving players a look at the dealer's hole card meant that the rules of the game needed considerable additional changes to restore the casino's edge.

- Naturals pay even money, except for $A\spadesuit\ J\spadesuit$, which pays 2–1. Only $\frac{1}{16}$ of all naturals qualify for the 2–1 payoff, though. The effective payoff on naturals is then

$$(1) \cdot \frac{15}{16} + (2) \cdot \frac{1}{16} = 1.0625 \text{ to } 1.$$

The effect on the HA is

$$(2.32\%) \cdot \frac{15}{16} + (-2.32\%) \cdot \frac{1}{16} = 2.03\%,$$

a significant benefit to the house.

- Players may double down on any 2 cards. This gives players a .24% edge over Strip rules.

- A 21 consisting of a suited 6, 7, and 8 pays 2–1.

  In the 5-deck game played at Vegas World, the probability of a suited 6, 7, and 8 is

$$\frac{4 \cdot 5^3}{\binom{260}{3}} \approx \frac{1}{5791},$$

so this feature does not add much to the player's expectation. It is not worth violating Double Exposure basic strategy (see below) in pursuit of this bonus.

- Most significantly, while tied natural hands are won by the player, all other ties are won by the casino—this rule alone adds 9.23% to the casino's edge.

While seeing both dealer cards gives the player valuable additional information, the rule that the house wins ties below 21 forces some plays that no player would make when playing traditional blackjack. For example, when dealt a hard 20 with the dealer also showing 20, the player must hit and hope for an ace—and no basic strategy player or card counter would ever hit a hard 20. A Double Exposure player following standard basic strategy, even if she deviated from the strategy in obvious cases like hitting hard 20 vs. a dealer 20, would hand the house a larger advantage.

Example 5.14. Standard basic strategy calls for players to stand on a pair of 10s and not to split them. In Double Exposure, we have more information about the dealer's hand, and so if the dealer has a hard 13–16 and is guaranteed to take a third card with a good chance of busting, the Double Exposure basic strategy calls for splitting the 10s to take greater advantage of the dealer's higher probability of going over 21. No matter what cards are drawn, a player splitting 10s should stand on the resulting two hands, even if holding hard 12 against hard 13. ∎

It's possible to bring the HA for Double Exposure down near the value for blackjack with basic strategy by following the appropriate strategy for this game. Since the Double Exposure player knows both of the dealer's cards, a sound basic strategy needs to account for the dealer's total as well as its composition. Stiff hands, which will be drawing a third card and have the potential to bust, need to be considered differently from pat hands, which will not draw a third card or can reach at least hard 17 if their third card is a 10.

The full Double Exposure basic strategy is shown in Tables 5.9, when the dealer's hand is pat or potentially pat, and 5.10, when the dealer is holding a stiff hand and the chance of busting increases.

TABLE 5.9: Basic strategy for Double Exposure: Dealer's hand is pat or possibly pat [131, 162]

	Player Hand	A6	7	8	9	10	11	17	18	19	20
						Dealer hand					
Hard hands	20	S	S	S	S	S	S	S	S	S	H
	19	S	S	S	S	S	S	S	S	H	H
	18	S	S	S	S	S	S	S	H	H	H
	17	H	S	S	S	S	S	H	H	H	H
	16	H	H	S	S	S	S	H	H	H	H
	15	H	H	H	H	S	S	H	H	H	H
	14	H	H	H	H	H	S	H	H	H	H
	13	H	H	H	H	H	H	H	H	H	H
	12	H	H	H	H	H	H	H	H	H	H
	11	H	D	D	D	H	H	H	H	H	H
	10	H	D	D	H	H	H	H	H	H	H
	9−	H	H	H	H	H	H	H	H	H	H
Soft hands	A9	S	S	S	S	S	S	S	S	S	H
	A8	S	S	S	S	S	S	S	S	H	H
	A7	S	S	H	H	H	H	S	H	H	H
	A6−	H	H	H	H	H	H	H	H	H	H
Pairs	AA	SP	SP	SP	SP	SP	H	H	H	H	H
	TT	S	S	S	S	S	S	S	S	S	H
	99	S	S	SP	S	S	S	S	SP	H	H
	88	SP	SP	SP	S	S	S	SP	H	H	H
	77	H	H	H	H	H	H	SP	H	H	H
	66	H	H	H	H	H	H	SP	H	H	H
	55	H	D	D	H	H	H	H	H	H	H
	44	H	H	H	H	H	H	H	H	H	H
	33	H	H	H	H	H	H	SP	H	H	H
	22	H	H	H	H	H	H	SP	H	H	H

Key: D: Double down. H: Hit. S: Stand. SP: Split.
Use A6 column only when dealer hits soft 17.

Summary of Table 5.9

- On hard 17 or higher, including pairs of 9s or 10s, hit if you're beaten or tied by the dealer's hand.
- Never double down on a soft hand.
- Double down on 11 only when the dealer's hand is 7, 8, or 9.
- Double down on hard 10, including a pair of 5s, only when the dealer's hand is 7 or 8.
- Split aces only against a dealer 7–10, and against A6 if the dealer hits soft 17.
- Split 9s only against a dealer 8 or 18.
- Split 8s only against dealer 7, 8, or 17, including soft 17 if permitted.
- Split 2s, 3s, 6s, and 7s only against a dealer hard 17.

TABLE 5.10: Basic strategy for Double Exposure: Dealer's hand is stiff [131, 162]

Player Hand		AA	A2	A3	A4	A5	4	5	6	12	13	14	15	16
	13+	S	S	S	S	S	S	S	S	S	S	S	S	S
	12	H	H	S	S	S	S	S	S	S	S	S	S	S
	11	H	D	D	D	D	D	D	D	D	D	D	D	D
	10	H	H	D	D	D	D	D	D	D	D	D	D	D
Hard	9	H	H	H	H	H	H	D	D	D	D	D	D	D
hands	8	H	H	H	H	H	H	H	H	D	D	D	D	D
	7	H	H	H	H	H	H	H	H	H	H	D	D	D
	6	H	H	H	H	H	H	H	H	H	H	D	D	D
	5	H	H	H	H	H	H	H	H	H	H	D	D	D
	A9	S	S	S	S	S	S	S	S	S	D	D	D	D
	A8	S	S	S	S	S	S	S	S	D	D	D	D	D
	A7	H	H	H	H	H	D	D	D	D	D	D	D	D
Soft	A6	H	H	H	H	H	D	D	D	D	D	D	D	D
hands	A5	H	H	H	H	H	H	D	D	D	D	D	D	D
	A4	H	H	H	H	H	H	H	D	D	D	D	D	D
	A3	H	H	H	H	H	H	H	D	D	D	D	D	D
	A2	H	H	H	H	H	H	H	D	D	D	D	D	D
	AA	SP	SP	SP	SP	SP	SP	SP	SP	SP	SP	SP	SP	SP
	TT	S	S	S	S	S	S	S	S	S	SP	SP	SP	SP
	99	S	S	S	SP	SP	SP	SP	SP	SP	SP	SP	SP	SP
	88	S	S	S	S	S	SP	SP	SP	SP	SP	SP	SP	SP
Pairs	77	S	S	S	S	S	S	SP	SP	SP	SP	SP	SP	SP
	66	H	H	S	S	S	SP	SP	SP	SP	SP	SP	SP	SP
	55	H	H	D	D	D	D	D	D	D	D	D	D	D
	44	H	H	H	H	H	H	H	H	SP	SP	SP	SP	SP
	33	H	H	H	H	H	H	H	SP	SP	SP	SP	SP	SP
	22	H	H	H	H	H	H	H	SP	SP	SP	SP	SP	SP

Key: D: Double down. H: Hit. S: Stand. SP: Split.

Summary of Table 5.10

- Double any hard hand from 5–11 against dealer's hard 14–16.
- Stand on hard 13 or higher against any stiff hand.
- Stand on hard 12 except when facing a dealer AA or A2.
- Double 11 against any stiff hand except AA, and double 10, including 55, against any stiff hand except AA and A2.
- Split pairs more frequently, in general:
  - Split 10s against hard 13–16.
  - Split 9s against A4, A5, and any stiff hand without an ace.
  - Stand on 88 against stiff hands with an ace; split 8s otherwise.
  - Split 7s against hard 5 and higher, 6s against hard 4 and higher, and 4s against hard 12 and higher.
  - Split 2s and 3s against hard 6 and higher.

Double Exposure II, a later version of the game, only revealed the dealer's hole card when the upcard was a 10. All other rules remained the same.

Jack Jack

In 2017, Jack Casino in Cincinnati, Ohio (now the Hard Rock Casino Cincinnati) invented and introduced a new version of blackjack called *Jack Jack* that played off the casino's name and modified only the payoffs on natural 21s [81]. Standard blackjack rules pay constant odds of 3–2 or 6–5 on naturals; video blackjack machines often pay 1–1. Jack Jack uses the following pay table for naturals, separating the payoffs by the cards that form the total of 21:

- A natural with two black cards pays 2–1.
- A natural with two red cards pays 3–2.
- A natural with two cards of different colors pays 1–1.

Averaging over all possible naturals, what is the mean payoff for a natural at Jack Jack? If the ace is black, then 50% of naturals are black-black and 50% are mixed. Similarly, if the ace is red, then 50% of naturals are red-red and 50% are mixed. It follows that ¼ of naturals are black-black, ¼ red-red, and ½ black-red in the long run.

The average payoff on a natural is then

$$(2) \cdot .25 + (1.5) \cdot (.25) + (1) \cdot .5 = 1.375$$

to 1. This corresponds to payoff odds of 11–8, which is less than 3–2 (1.5 to 1 or 12 to 8) but greater than 6–5 (1.2 to 1 or 9.6 to 8). The standard payoff on live-game naturals at the casino is 3–2, making this game less lucrative for players.

Since same-color naturals pay more than mixed-color ones, it is plausible that a card-counting scheme tracking high cards by color, perhaps with a separate count for aces, could give the player an extra edge at Jack Jack. Such a system might be too complicated for a single person to execute alone—but a team of two, each counting one color, might be able to pull it off.

Easy Jack

In an effort to retain the excitement and strategy of blackjack while cutting out some of the arithmetic, Matthew Stream developed *Easy Jack*, a variant where players and dealer are dealt 1 card initially, and the player goal is to get closer to 11 than the dealer without going over [4]. Effectively, Easy Jack assumes that all hands start with a 10, and doesn't bother dealing it. Aces count 1 or 11, and an ace as the first card is a natural and wins automatically, paying 3–2. The dealer must draw when holding 6 or less and stand on 7 or higher. Winning hands other than naturals pay even money.

Easy Jack is dealt from an 8-deck shoe, so the infinite deck approximation gives a good estimate to the probabilities involved. Suppose that the dealer's card is a 6. The probability that the dealer will bust with his mandatory draw is $\frac{8}{13} \approx .6154$. Should a player hit a 2–6 against a dealer 6?

If the player stands, she can only win if the dealer busts. Her expectation is then

$$E = (1) \cdot \frac{8}{13} + (-1) \cdot \frac{5}{13} = \$\frac{3}{13} \approx \$.2308,$$

so standing delivers a 23.08% edge. Much like a stiff hand against a dealer 6 in standard blackjack, letting the dealer bust first is a sound strategy here.

As long as the probability of a dealer bust exceeds ½, Easy Jack strategy favors the player standing on 2–6. Consequently, the player should stand whenever the dealer's card is a 5 or 6.

Crown Pontoon

The Crown Casino hosts a New Games Lab on its gaming floor, where new games can be tested in a live casino environment. Among the games offered in the Lab is a blackjack variant called *Crown Pontoon*, which takes its name from "Pontoon", the traditional Australian name for blackjack. This name is derived from an Australian mispronunciation of "vingt et un", French for "21".

Crown Pontoon is played with multiple *Spanish decks*, which are 48-card decks consisting of standard decks with all of the 10s removed. From a card-counting perspective, this starts a round off with a true count of –4 per deck, a considerable disadvantage to the player. Some effort to level the game comes in the form of an optional Pontoon Pandemonium bet, which offers players a bonus opportunity if their next hand is a natural.

With Spanish decks in use, the probability of a natural in the infinite deck approximation drops, from 1 in 21 to

$$2 \cdot \frac{4}{48} \cdot \frac{12}{48} = \frac{1}{24},$$

a decrease of nearly 12%. This deduction, of course, applies to all players, not just those making the Pontoon Pandemonium bet, and the dealer. Players eschewing the extra wager are giving the casino an additional advantage by buying into a game where the HA is increased relative to full-deck blackjack. Players making that bet are putting more money at risk chasing a less-likely outcome.

Example 5.15. The payoff on Pontoon Pandemonium is determined by an electronic device, and ranges from 2–1 to 100–1. The permitted payouts on this bet include 2–1, 3–1, 5–1, 10–1, 20–1, 25–1, 50–1, and 100–1 [92]. The average value of this prize is determined by the device's programming; there is no reason to believe that the various outcomes are all equally likely. Since the removal of four 10s from each deck results in an increased HA of about 2% (absent any other rule changes), what would that average have to be to make up for the missing cards?

Let the average payoff be $x-1$. Assuming that the Pontoon Pandemonium bet is the same size as the player's main wager, we need

$$E = (x) \cdot \frac{1}{24} + (-1) \cdot \frac{23}{24} = .02,$$

or

$$x - 23 = .48.$$

The average payoff must be at least 24–1 if Pontoon Pandemonium is to make up all of the rise in the HA caused by the absence of the 10s. It is extremely unlikely that the game is programmed to permit an average payoff of 24–1. ■

Black Jackpot

A casino patron passing by a Black Jackpot table, which had a field trial in 2018 at the Atlantis Casino in Reno, Nevada, would surely be drawn to the inscription "Dealer Never Wins" printed on the felt.

A seasoned casinogoer, on seeing that inscription, would guess that there was more to the story than the phrase indicated.

Black Jackpot was part of casinos' collective efforts to introduce games and resort attractions that would appeal to millennial gamblers, who were thought—perhaps rightly—to be interested in skill-based games and social gaming. One early sign that the game was different from conventional blackjack or other casino games was that two or more players were necessary to play; no one could play Black Jackpot head-to-head against the dealer.

Other differences became apparent in gameplay:

- Players were restricted to a single fixed wager determined by the casino, perhaps $10.

- The collected wagers were combined into a communal pot.

- The cards were then dealt. A dealer natural meant that the pot went to the house.

- If the dealer did not have a natural, and 3 or more players did, the house contributed 25 times the fixed wager to the pot, and the hand ended with the pot evenly divided among the players with naturals.

- Absent these naturals, the blackjack hand proceeded with the following rules in force:

  - No doubling down.
  - No insurance.
  - No surrender.
  - Players could split pairs up to 3 times. Each split required that the player contribute an additional bet in the amount of the fixed wager to the pot.

- When the hand was complete, players were awarded shares of the pot based on the strength of their hands, as listed in Table 5.11.

TABLE 5.11: Black Jackpot: Player pot share awards

Player's hand	Shares
5 cards without busting	5
Natural 21	4
21, not a natural	3
Win against dealer	2
Push with dealer	1

The pot was then divided among the players in proportion to their shares. Individual shares were rounded down to the nearest 25¢.

- If the dealer beat all player hands without drawing a natural, the entire pot rolled over to the next hand.

Example 5.16. Consider a 5-handed game, with $50 in the pot. The dealer's upcard is the $J\diamondsuit$.

- Player 1 is dealt $6\heartsuit Q\spadesuit$ for a 16. She hits, draws the $A\diamondsuit$, and stands on 17.

- Player 2 has a natural 21: $J\spadesuit A\clubsuit$.

- Player 3 has $10\diamondsuit 6\clubsuit$, also hits her 16, and busts with the $J\heartsuit$.

- Player 4 holds $4\clubsuit A\heartsuit$. He draws the $K\diamondsuit$, making 15, hits again with the $9\heartsuit$, and busts.

- Player 5 has an 11 with $3\diamondsuit 8\spadesuit$, but cannot double down. He hits successively with the $3\clubsuit$, $2\clubsuit$, and $2\spadesuit$. His 18 consists of 5 cards under 21.

The dealer turns over the $5\diamondsuit$ for a 15 and busts with the $8\diamondsuit$. Player 1 gains 2 shares, Player 2 has 4, and Player 5 has 5, so the pot is divided into 11 portions of $4.5454... each. Player 1 collects $9, which is $1 less than her original wager—winning has cost her money. Player 2 receives $18 for a profit of $8, which is less than the $15 or $12 ordinarily paid on a natural with a $10 wager. Player 5 gets $22.50, and so is $2.50 better off than he would have been in a standard blackjack game. These amounts are awarded electronically, so there is no need for anyone to handle low-denomination chips or small change.

The casino has paid out $49.50 of the original $50 pot. The remaining 50¢ is assigned to a side pot, which accumulates over time. After 10 hands, the player who has won the most money over those hands receives a badge, and the player with the most badges after a set number of hands collects the side pot. ∎

The casino's advantage comes from dealer naturals, mitigated somewhat by the large payout when 3 or more players hold naturals. What is the probability that 3 or more players are dealt naturals without a dealer natural, triggering a 25× payout?

For convenience, we shall use the infinite deck approximation, in which the probability of a natural is a constant

$$2 \cdot \frac{4}{52} \cdot \frac{16}{52} = \frac{8}{169},$$

and the probability of not getting a natural is

$$1 - \frac{8}{169} = \frac{161}{169}.$$

If there are n players at the table, the probability of 3 or more naturals without a dealer natural is

$$P_n = \sum_{k=3}^{n} \binom{n}{k} \cdot \left(\frac{8}{169}\right)^k \cdot \left(\frac{161}{169}\right)^{n-k+1}.$$

For $3 \leqslant n \leqslant 7$, Table 5.12 collects the values of P_n. The casino is not

TABLE 5.12: Black Jackpot: Probability of 3 or more player naturals without a dealer natural

n	P_n
3	0.0001
4	0.0004
5	0.0009
6	0.0018
7	0.0031

taking much of a risk in offering this jackpot; even at a full table, the chance of 3 or more player naturals is less than $\frac{1}{3}\%$.

Example 5.17. Standard blackjack rules pay 3–2 for a natural. In Black Jackpot, a natural gives its owner 4 shares of the pot. In Example 5.16, we saw that this might lead to less than a standard blackjack payoff. Of course, if all of the other players lose, the payoff might be better than 3–2. The payoff depends on the number of players at the table, n, and the number of pot shares awarded to players, s.

At a flat bet of $10 per player, the total pot comes to $10n$. Any players dealt a natural are awarded 4 shares of the pot, out of a total of s, where $4 \leqslant s \leqslant 5(n-1) + 4 = 5n - 1$. Their individual share of the pot comes to

$$A(n, s) = 10n \cdot \frac{4}{s} = \frac{40n}{s}.$$

Recognizing that $A(n, s)$ is rounded down to the nearest quarter, we can assess the natural-holder's status by solving some simple inequalities. For convenience, we shall ignore the very small probability of three or more player naturals, which triggers the $250 bonus addition to the pot.

- A 3–2 payoff would translate into a total return of $25 from the pot. The player does at least this well if

$$\frac{40n}{s} \geqslant 25,$$

  or if

$$s \leqslant \frac{8n}{5}.$$

  In a three-player game, the total number of shares would have to be less than 4.8, so a natural pays off at less than 3–2 if either of the other players scores at least a push. Of course, if both other players bust, the player with the natural collects all $30: a 2–1 payoff.

  With 4 players, the threshold is $s \leqslant 6.4$, so the other players can collect no more than 2 shares among them.

- A 6–5 payout on naturals corresponds to a $22 pot share for a natural. This gives

$$s \leqslant \frac{20n}{11}$$

  for the total number of shares. The average player in such a game must receive no more than 2 shares, so the dealer busting would be treacherous for a natural-holder.

- When does a natural lead to a net loss? When

$$\frac{40n}{s} \leqslant 10,$$

  or

$$s \geqslant 4n.$$

  This is unlikely. In a three-handed game, if no other player holds a natural, then there must be at least 8 more shares beyond the natural: either $5 + 3$ or $5 + 5$. If one other player among the three has a natural, the other must score 5 shares.

■

Black Jackpot certainly qualified as a social game, but since winning players were literally taking money from the other players—often their friends—its success was unlikely.

A 50-Card Deck

In blackjack, as in baccarat, a fifth suit means nothing in the play of the game. The change in ranks in moving to Empire Global Gaming's 50-card deck with aces through 10s in five suits or colors has a profound effect. Fewer ten-count cards means fewer naturals, but also fewer bust cards for stiff hands like 12 or 13.

Example 5.18. Let's start with the probability of a blackjack with a fresh single 50-card deck. This probability is about .0483 $\approx \frac{1}{21}$ with a standard deck; we expect that the probability is lower with the new deck. We seek the probability of drawing an ace and a 10 off the top, without regard to order. The probability is

$$p = 2 \cdot \frac{5}{50} \cdot \frac{5}{49} = \frac{1}{49} \approx .0204$$

—less than half the chance in standard blackjack.

If six 50-card decks are used, the probability of a blackjack becomes

$$p = 2 \cdot \frac{30}{300} \cdot \frac{30}{299} = \frac{6}{299} \approx .0201.$$

The difference of .0003 is essentially negligible, as it was for a standard deck when comparing single-deck and 6-deck games (page 239). ∎

The patent application for the 50-card deck contains rules and a basic strategy table for a six-deck blackjack game with the 50-card deck called Five Color Blackjack [15]. The standard rules of Five Color Blackjack are not far removed from ordinary blackjack rules:

- Naturals pay either 2–1 or 3–1, at the casino's discretion. This is the major difference between the two games, and reflects the scarcity of naturals with a 50-card deck as computed above.

- The dealer must hit all 16s and soft 17s, and must stand on hard 17 or higher.

- Pairs may be split, but only once. Only one card is dealt to split aces.

- Ace-10 after a split is considered 21 and not a natural, as is the case in blackjack generally.

- Doubling down is allowed on any two cards except after splitting aces.

- Surrender is not offered.

The basic strategy table for Five Color Blackjack is reproduced in Table 5.13.

There are some stark differences between this strategy and basic strategy for a standard deck:

TABLE 5.13: Basic strategy for six-deck Five-Color Blackjack [15, p. 4]

Player's Hand		Dealer's upcard									
		2	3	4	5	6	7	8	9	T	A
No pair or ace	18+	S	S	S	S	S	S	S	S	S	S
	17	S	S	S	S	S	S	S	S	S	H
	16−	H	H	H	H	H	H	H	H	H	H
Soft hand, no pair	A/9	S	S	S	S	S	S	S	S	S	S
	A/8	S	S	S	S	S	S	S	S	S	S
	A/7	H	H	H	H	H	S	H	H	H	H
	A/6	H	H	H	H	H	H	H	H	H	H
	A/5	H	H	H	H	H	H	H	H	H	H
	A/4	H	H	H	H	H	H	H	H	H	H
	A/3	H	H	H	H	H	H	H	H	H	H
	A/2	H	H	H	H	H	H	H	H	H	H
Pair	A/A	H	H	H	H	H	H	H	H	H	H
	T/T	S	S	S	S	S	S	S	S	S	S
	9/9	S	S	S	S	S	S	S	S	S	SP
	8/8	SP	SP	SP	SP	SP	SP	SP	SP	SP	H
	7/7	H	H	H	H	H	SP	H	H	H	H
	6/6	H	H	H	H	H	H	H	H	H	H
	5/5	H	H	H	H	H	H	H	H	H	H
	4/4	H	H	H	H	H	SP	SP	H	SP	H
	3/3	SP	SP	SP	SP	SP	SP	SP	SP	SP	H
	2/2	H	H	H	SP	SP	SP	SP	SP	SP	H
Never take insurance.											
Key: H: Hit. S: Stand. SP: Split											

- **Strategy for hard hands is simple: Hit hard 16 or lower, stand on hard 17 or higher, *except* hit hard 17 against a dealer ace.** The notion of standing on a hard 12–16 against a weak dealer upcard in the hopes that the dealer will bust is a strategy doomed to failure: with fewer 10s in the deck, *everyone* busts less often. In the long run, hitting a hard 16 will only bust 50% of the time, as opposed to the 61.5% chance of busting a hard 16 with a single hit from a standard deck.

- **Soft hands are also easy to play: Hit soft 18 or lower, except stand on soft 18 against a 7. Stand on soft 19 or higher.** Once again, this strategy is motivated by the increased number of small cards relative to the scarce 10s.

- **Doubling down is never advisable.** A blackjack player typically doubles down hoping to catch a 10 on a 9, 10, or 11 or to get more money on the table with a soft hand against a weak dealer upcard. Since the 50-card deck contains only one quarter of the 10s of a standard deck, there are fewer to be had, and it's better to retain the option of drawing a fourth card. The probability of reaching 21 when doubling down on an 11 is $\frac{4}{13}$ with a standard deck, but only $\frac{1}{10}$ with this deck. As to soft hands, the diminished number of 10s means that the dealer will bust less often with a weak upcard, and so there's less benefit in risking more money on a soft hand.

- **Pairs should be split less often, and differently.** Split 9s against an ace, 8s and 3s against anything *but* an ace, 7s against a 7, 4s against a 7, 8, or 10, and 2s against a 5–10. That's it.

  In particular, contrary to basic strategy against a standard deck or decks, splitting aces is never advised. One thing that's unchanged is that 10s should never be split.

It is instructive to analyze the 50-card deck from a card counter's perspective. Using the high-low count, a fresh 50-card deck has a 15-card excess of low cards over high cards and so can be regarded as a deck with a running count of –15; the true count after dividing by 1 would also be –15. Six decks shuffled together would result in a running count of –90 and a true count of –15 again. Employing standard basic strategy against this deck would effectively hand the casino an additional 7.5%, for a total edge of about 8%. To get the best game, it would be necessary to use the correct basic strategy for 5-Color Blackjack as shown in Table 5.13. Any successful attempt to count cards into this deck would have to use a counting system different from the High-Low count.

Five Color Blackjack also provides for an optional side bet: Color Match 21. This simple wager pays off if the player's first two cards are the same color or suit. In a single-deck game, that event has probability

$$p = \frac{9}{49} \approx .1837,$$

since the first card doesn't matter; all we need to find is the chance that the second card matches the first. Another version of the bet increases the payoff if the player's cards and the dealer's upcard all match in suit, an event with probability

$$q = \frac{9}{49} \cdot \frac{8}{48} = \frac{3}{98} \approx .0306.$$

If the payoff for two matched cards is X and that for matching three cards is Y, we can compute a general formula for the expectation of a $1 side bet and then adjust the parameters X and Y to fine-tune the HA as the casino may desire. Using the probabilities p and q defined above, we have

$$E = X \cdot (p - q) + Y \cdot q + (-1) \cdot (1 - p) = p(X + 1) + q(Y - X) - 1.$$

Substituting the established values of p and q into this expression gives

$$E(X, Y) = \frac{15X + 3Y - 80}{98}.$$

In this formula, it is understood that $X \leqslant Y$ and that $X \geqslant 0$. $X = Y$ corresponds to no additional bonus if all three cards have the same suit; $X = 0$

TABLE 5.14: Color Match 21 expectations [15]

X	Y	$E(X, Y)$	HA
0	25	−0.0510	5.10%
3	7	−0.1429	14.29%
3	8	−0.1122	11.22%
3	9	−0.0816	8.16%
3	10	−0.0510	5.10%
3	11	−0.0204	2.04%
4	4	−0.0816	8.16%
4	5	−0.0510	5.10%
4	6	−0.0204	2.04%

means that the bet only pays off if all three cards match in suit. We have the expected values and house advantages listed in Table 5.14. From this table, we can see that Color Match 21 can be nearly as lucrative for the casino as the Big Red craps bet (HA = 16.67%) or more player-friendly than European roulette (HA = 2.70%), depending on the payoff structure.

Lucky 13s Blackjack

A different variation on blackjack including a nonstandard deck was introduced at the 2014 Global Gaming Expo Asia as a game called *Lucky 13s Blackjack*. Lucky 13s was developed and marketed by the Australian firm Engaging Table Games Pty Ltd, whose credo is that new table games should follow what they call the SPA principle: **S**imple, **P**layable, and **A**wesome [76]. The game was soon placed at two casinos in the United Kingdom and on Royal Caribbean cruise ships.

This game uses a single 64-card deck, with 11s, 12s, and 13s in each of the four suits added [76]. In the play of the game, certain important blackjack rules have been changed in light of the new high-value cards:

- The dealer must stand on hard 16 rather than 17. Soft 16s must be hit.

- A two-card 21 without an ace, such as a 12 and a 9, is a hard 21 and not a natural. It does not qualify for a 3–2 payoff, and it can be tied by a dealer 21.

- Insurance is not offered when the dealer's upcard is an 11.

In light of the new composition of the deck and the fact that a natural 21 still requires an ace and a 10, we expect that the probability of a natural in Lucky 13s Blackjack will drop relative to the probability with a standard deck. Reasoning similar to that used for standard decks shows that this probability is

$$p = 2 \cdot \left(\frac{4}{65}\right) \cdot \left(\frac{16}{64}\right) = \frac{2}{65} \approx .0308.$$

This probability lands in between the .0483 chance of drawing a natural from a single standard deck and the probability .0204 of a blackjack with the 50-card deck computed on page 277.

The 11s, 12s, and 13s also lead to a new basic strategy chart, which is presented here as Table 5.15.

TABLE 5.15: Basic strategy for Lucky 13s Blackjack [76]

Player's Hand		Dealer's upcard											
		2	3	4	5	6	7	8	9	10	A	11	12	13
Hard hands	16+	S	S	S	S	S	S	S	S	S	S	S	S	S
	15	S	S	S	S	H	H	H	S	S	S	S	S	S
	14	S	S	S	S	H	H	H	H	S	S	S	S	S
	13	S	S	S	S	H	H	H	H	S	H	S	S	S
	12	S	S	S	S	H	H	H	H	S	H	S	S	S
	11	D	D	D	D	H	H	H	H	H	H	H	D	D
	10	D	D	D	D	D	D	H	H	H	H	H	D	D
	9	D	D	D	D	D	H	H	H	H	H	H	D	D
	8	D	D	D	D	H	H	H	H	H	H	H	D	D
	7−	H	H	H	H	H	H	H	H	H	H	H	D	D
Soft hands	A/9	S	S	S	S	S	S	S	S	S	S	S	S	S
	A/8	S	S	S	S	S	S	S	S	S	S	S	D	D
	A/7	D	D	D	D	D	D	D	D	D	H	D	D	D
	A/6	D	D	D	D	D	D	H	H	H	H	D	D	D
	A/5	D	D	D	D	D	H	H	H	H	H	H	D	D
	A/4	D	D	D	D	H	H	H	H	H	H	H	D	D
	A/3	D	D	H	H	H	H	H	H	H	H	H	D	D
	A/2	H	H	H	H	H	H	H	H	H	H	H	D	D
Pairs	13/13	SP	SP	SP	SP	SP	SP	SP	B	B	B	SP	SP	SP
	12/12	SP	SP	SP	SP	SP	SP	SP	SP	SP	B	SP	SP	SP
	11/11	SP	SP	SP	SP	SP	SP	SP	SP	SP	SP	SP	SP	SP
	A/A	SP	SP	SP	SP	SP	SP	SP	SP	SP	H	SP	SP	SP
	10/10	S	S	S	S	S	S	S	S	S	S	S	S	S
	9/9	SP	SP	SP	SP	S	S	S	SP	S	S	S	SP	SP
	8/8	SP	SP	SP	SP	SP	SP	SP	SP	SP	SP	SP	SP	SP
	7/7	SP	SP	SP	SP	SP	SP	SP	SP	SP	SP	H	SP	SP
	6/6	SP	SP	SP	SP	SP	SP	H	H	S	H	SP	SP	SP
	5/5	D	D	D	D	D	D	H	H	H	H	H	SP	SP
	4/4	D	D	D	D	H	H	H	H	H	H	H	D	D
	3/3	H	H	H	H	H	H	H	H	H	H	H	SP	SP
	2/2	SP	SP	SP	H	H	H	H	H	H	H	H	SP	SP

Never take insurance.
Key: B: Bust (don't split). D: Double down. H: Hit. S: Stand. SP: Split.

Of particular interest in this table is what to do against a dealer 11, 12, or 13. With these upcards, a dealer two-card bust is a possibility, so this is a time to consider putting more money in play if the risk in doing so is not too great. Toward that end, the following actions are called for:

- **Double down on hard 11 or lower against a 12 or 13, including hands as low as a hard 5.** The slight risk of busting a hard 9, 10, or 11 is more than balanced by the dealer's increased chance of busting, which is $\frac{27}{61} \approx .4426$ from a single deck when the dealer shows a 12 and $\frac{31}{61} \approx .5082$ when her upcard is a 13. The player edge comes since dealer busts when drawing a 10 or face card and the player does not.

- **Double down on any soft hand of 19 or lower against a 12 or 13.** Here, the only risk of busting comes when drawing a 13 to an ace-8 hand.

- **Except for 10s and 4s, split all pairs against a 12 or 13. Split any pair from 6–8, 11–13, or aces against an 11.**

Pairs of 11s, 12s, or 13s should almost always be split, with only the following four exceptions, which are akin to surrendering (an option not available in Lucky 13s Blackjack):

- **Bust a pair of 13s against a dealer 9, 10, or ace.**

- **Bust a pair of 12s against a dealer ace.**

Expanding the standard deck this way introduces the possibility that a two-card hand might bust. For players, this is accommodated by an optional "Protection" side bet, which pays off if the total of the player's first two cards, with aces counted as 1, exceeds 21. A protected hand busting with a total of 22–25 pays off at 5–1; if the hand busts at 26 (a pair of 13s), the payoff is 40–1. If the dealer's upcard is a 13, these payoffs are multiplied by 5, making a 200–1 payoff on a single hand possible [77].

An initial hand consisting of a pair of 11s, 12s, or 13s can be split like any other pair, and so need not automatically lose the main bet. Unless it can be split, a hand which busts in two cards loses the main bet. This may induce players to make the Protection wager as a new form of insurance. On the other hand, given the higher busting probability with a 12 or 13, a player may choose not to split a pair of 12s or 13s against a strong dealer upcard and just take a one-unit loss rather than risking two betting units in an unfavorable situation.

Example 5.19. To determine whether or not the Protection bet is a good wager, we begin by computing the probability that a two-card hand busts. We need to count the following combinations, which we shall call *bust pairs*:

Count	Combinations
22	11/11, 12/10, 13/9
23	12/11, 13/10
24	12/12, 13/11
25	13/12
26	13/13

A bust pair consisting of 13/9, 12/11, 13/11, or 13/12 can occur in 16 different ways. A bust pair including a 10 has 64 possibilities, and a bust pair with two cards of the same rank has 6 possibilities. Adding gives a total of $16 \cdot 4 + 64 \cdot 2 + 6 \cdot 3 = 210$ possible bust pairs. The probability of a bust pair from the top of a fresh deck is then

$$p = \frac{210}{\binom{64}{2}} = \frac{210}{2016} \approx .1042.$$

■

The expectation for the Protection bet depends on the dealer's upcard, which in turn depends on the composition of the bust pair. In a single-deck game, if the bust pair includes x 13s, the probability that the dealer shows a 13 is

$$P(x) = \frac{4-x}{62}.$$

92 bust pairs contain no 13s, 112 bust pairs contain one 13, and 6 contain two. These last 6 are the bust pairs that receive a 200–1 payoff if the upcard is a 13.

We proceed one case at a time. Each of the following six probabilities, labeled p_1 through p_6, is the product of the probability of drawing a bust pair with x 13s and a factor measuring the chance that the dealer either does or does not have a 13, given the composition of the first two cards. The latter is either $P(x)$ defined above or $1 - P(x)$. For a bust pair with no 13s and no dealer 13, the probability is

$$p_1 = \frac{92}{2016} \cdot \frac{58}{62} \approx .0427.$$

The probability of a bust pair with no 13s and a dealer 13 is

$$p_2 = \frac{92}{2016} \cdot \frac{4}{62} \approx .0029.$$

For bust pairs with a single 13, we have

$$p_3 = \frac{112}{2016} \cdot \frac{59}{62} \approx .0529$$

as the probability without a dealer 13 and

$$p_4 = \frac{112}{2016} \cdot \frac{3}{62} \approx .0027$$

for the probability with a dealer 13.

The corresponding probability of two 13s without a dealer 13 is

$$p_5 = \frac{6}{2016} \cdot \frac{60}{62} \approx .0029;$$

the probability, then, of two 13s with a dealer 13 is

$$p_6 = \frac{6}{2016} \cdot \frac{2}{62} \approx 9.601 \times 10^{-5}.$$

Finally, the probability that the Protection bet loses because the original two-card hand doesn't bust is

$$p_7 = \frac{1806}{2016} \approx .8958.$$

This probability need not incorporate the dealer's upcard.

Assembling all of these factors gives an expectation for a \$1 Protection bet of

$$E = 5 \cdot (p_1 + p_3) + 25 \cdot (p_2 + p_4) + 40 \cdot p_5 + 200 \cdot p_6 + (-1) \cdot p_7 \approx -\$.1428.$$

The house advantage is 14.28%. This is protection you can do without.

If the game is played with a four-deck shoe, the house edge drops to 8.45%—somewhat more reasonable, but still high. In the infinite deck approximation, the house edge only decreases to 6.45%.

If you have made the Protection bet, busting a pair of 12s or 13s still leaves you with a net profit. If not, you are in the long run cutting your losses by giving up one bet instead of doubling your stakes with a high chance of busting in two cards against a strong dealer upcard. Given an initial pair of 13s against a dealer ace, the probability of busting after splitting the pair is

$$\frac{2 + 4 + 4 + 16 + 4}{49} \approx .6122,$$

where the sum in the numerator counts down the potential bust cards from the two remaining 13s in the deck to the four 9s.

5.5   California Games

Beginning in 1989, a number of game designers have tried to develop blackjack variations that conformed to California laws and could then be offered in the state's card rooms. Many of these games have certain general features in common, though the details vary from game to game:

- The target number is not 21. This is an easy way to separate these games from traditional blackjack.

    Example 5.20. California Aces and LA Blackjack have a target of 22. Pure 21.5 Blackjack uses 21.5, and Blackjack Elite's top hand is 31. ∎

- Some games have a "no bust" rule where hands totaling over the target number remain in play. "Closer to 22" as a winning condition can then mean "closer, above, or below".

- These rule changes are accompanied by different rules for the dealer's hand. Standing on 17 when the target is 22 is often suboptimal, so some games require the dealer to stand on hard 18 and draw to soft 18 or lower.

- There may be stipulations on when players can or cannot hit their hands, as well.

Example 5.21. In Hot Action Blackjack, a game where the best hand is a pair of jokers totaling 24, players must stand on a soft 19–21 unless they choose to double down. In practice, this is not much of a restriction, since standing on these hands is usually the best decision, but this illustrates the importance of carefully checking the rules when playing a blackjack variant in California. LA Blackjack requires players to stand on a hard 18 or soft 19 or higher, but they must hit all soft 18s. This is due to a rule change involving ties at 18; see page 293. ∎

- Many games add one or more jokers to a deck, whose values depend on the game.
- As in California's baccarat variations, players are charged a per-hand collection fee to play the game, since the card room has no stake in the action and so cannot count on a house advantage to make a profit. This fee is set by the card room; commonly, the collection fee is 1% of a player's wager with a minimum of $1.

Example 5.22. Some casinos vary the collection fee with the table limits. The Capitol Casino in Sacramento uses several schedules that link player collection fees to the table limits. One fee schedule for tables with $5 minima and $500 maxima is shown in Table 5.16. Collection fees

TABLE 5.16: Capitol Casino player per-hand fees

Wager	Fee
$5–$100	$0.50
$105–$300	$1.00
$305–$500	$2.00

are based only on a player's initial wager; extra bets made to split pairs or double a hand do not lead to increased fees, nor does surrendering (when available) entitle the player to a refund of half the collection fee. ∎

Collection fees need to be taken into account when assessing the house advantage of California card games. In addition to affecting the expected value of a single wager, collection fees are frequently so large, as a percentage of the player's bet, as to make card counting useless. It's typically not possible for the small edge attained from counting cards to offset the effect of the extra charge [136].

- The rotating player/banker, as described on page 214 for baccarat, is a constant element of these games. Since the player/banker has an advantage over the other players, there is a small charge levied by the casino on each player/banker in turn, just as players are charged a per-hand fee.

Example 5.23. The Capitol Casino links the player/banker charge to the total of the players' bets, which is dubbed the "total table action." For the $5–$500 tables in Table 5.16, the player/banker fee is detailed in Table 5.17.

TABLE 5.17: Capitol Casino player/banker fees

Total Table Action	Fee
$5–$50	$0.50
$51–$200	$1.00
≥ $201	$2.00

∎

Player/bankers are required by law to hold that position for no more than 2 hands; the opportunity to bank then passes to the next player in turn willing to accept it. If no other player is willing to bank, the table may be required to close.

This arrangement may pose a challenge if the players together want to wager more than the player/banker is prepared to cover. To counter this, many card rooms have "corporations" on site: groups of investors who will step up and bankroll a game fully when the player/banker cannot [164].

- In the other direction, a feature of some California card rooms is *kum-kum* betting, where several players contribute, usually equally, to a single bet in the same betting circle. This cooperative wager is made by mutual consent of the players involved, and may be used in an effort to reach a benchmark such as $100 and lessen the effect of the per-hand charge.

Example 5.24. Four players, all wishing to make an individual $25 bet, will each pay a $1 fee, 4% of their wager. If they combine their wagers into a single $100 kum-kum bet, the charge is still $1—only 1% of the combined wager. They each have the same potential to double their money, but the card room extracts less money from each of them: 25¢ apiece rather than $1. ∎

How this quartet chooses to play their hand if a decision needs to be made is not a matter that California gaming regulations cover.

Kum-kum bettors usually divide their collective winnings equally; if the amounts contributed to the wager are not equal, they split the winnings in proportion to the amount that they bet.

- In card rooms that operate without corporations, it is possible that the player/banker will run out of money before paying off all winning hands. At this point, no further hands receive action: winning hands receive no payoff and losing hands' wagers are not collected. This rule would place particular importance on where a player sits at the table; seats earlier in the game rotation would have smaller probability of receiving no money after winning. To counteract this, many games with player/bankers use a small token called an *action button* to identify the first player to receive cards and to be paid. The position of the action button may be determined randomly by dice or by using the dealer's hole card to identify a position on the table that receives first action.

Example 5.25. Artichoke Joe's Casino in San Bruno has the player/dealer roll 3 standard dice to determine the location of the action button. On a standard 8-position blackjack table, with position #1 corresponding to the player/dealer, Table 5.18 is used to place the button. While these outcomes are not equally likely, the rotation of the

TABLE 5.18: Action button table at Artichoke Joe's Casino

Rolls	Ways to roll	Position
9, 17	28	1 (PD)
10, 18	28	2
3, 11	28	3
4, 12	28	4
5, 13	27	5
6, 14	25	6
7, 15	25	7
8, 16	27	8

player/dealer position around the table means that, in the long run, each player is approximately equally likely to receive the action button. An equal division of the 216 rolls of 3 dice into 8 subsets of 27 rolls each is easily shown to be impossible unless the exact composition of the die roll is taken into account rather than just the sum. ∎

- Backline betting, where players who are not seated at the table may place their own bets on the hands of seated players, may be permitted at the card room's discretion. As in baccarat, the seated player in the chair has control of the hand and makes all decisions concerning play.

With these changes to the rules, which vary among California card rooms, to speak of one basic strategy is incorrect. Every variation on blackjack offered in California is effectively its own game and has its own optimal strategy. The challenge to players is that developing the correct strategy for any one game is scarcely worth the work it takes [131]. As a result, players are often forced improvise strategies based on standard blackjack, which are frequently suboptimal in light of California rules.

California Blackjack

The first in a long sequence of modified blackjack games for California was *California Blackjack*, also called California No-Bust Blackjack. In an effort to create a game that retained much of the flavor of blackjack while conforming to state law, Roger Wisted invented this game in 1989 [156]. California Blackjack has a target of 22 instead of 21 and uses 6 decks of cards with a joker added to each deck. These jokers function as wild cards: a joker together with any other card other than an ace or another joker results in a hand counting 21. A joker dealt with an ace or a second joker gives a hand of 22, as does a pair of aces. If the dealer's faceup card is a joker, the hand ends, and all players holding hands under 21 lose. If the dealer has a joker in the hole, all players not holding 21s or 22s lose only their original bets, so additional wagers from doubling down or splitting pairs are returned [66].

The goal of California Blackjack is to draw a hand closer to 22 than the player/dealer, so a natural 21 (in the traditional blackjack sense) is not quite an automatic winner. California Blackjack hands do not bust if they exceed 22—hence the alternate name. However, if two hands such as 21 and 23 push at the same distance from 22, the hand lower than 22 wins. A 23 would win against a hand totaling 20 or lower.

A natural in California Blackjack is any hand totaling 22: either two aces, two jokers, or an ace and a joker. The probability of drawing a natural from a fresh six-deck shoe is

$$\frac{\binom{30}{2}}{\binom{318}{2}} = \frac{435}{50,403} \approx \frac{1}{116},$$

which is less than one-fifth of the chance of a natural 21 in blackjack. Naturals are automatic winners unless the dealer also has a natural, but pay only even money unless the player has a pair of jokers and the dealer does not, in which case the player is paid 2–1.

The new rules in California Blackjack call for a new version of basic strategy for this game; there are situations where standard basic strategy is incorrect. Some of these situations involve the no-bust rule: standing on a low hand against an unfavorable dealer upcard in hopes that the dealer will bust, as we often do when following traditional basic strategy, is inadvisable. Dealers must hit all 16s as well as soft 17s and cannot hit a hard 17 or higher, so completed dealer hands range from 17–26. There is absolutely no value in standing on a hand below 17.

This also affects doubling down: since doubling down restricts the player to a one-card draw, a hand without a very good chance of reaching at least 18 in one card should not be doubled.

Example 5.26. a. The hand $A\heartsuit$ $3\spadesuit$ counts either 4 or 14, and should certainly draw a third card since it cannot go over 22 by drawing.

If the third card is the 10♣, then the hand is a 14 or a 24. A player 24 can win only if the dealer draws to a 25 or 26, but the no-bust rule means that a 14 cannot win against *any* final dealer hand—in no circumstance would the dealer draw out to a 31 or higher that would lose to a 14. Therefore, hitting a 14 cannot leave you worse off since the maximum value of the hit hand is 24, so the best choice is to read this hand as a 14 and hit. We need not consider the dealer upcard when making this decision.

If you hit and draw the 4♣, you have a hard 18. An 18 loses to dealer hands totaling 19–25, but hitting hard 18 is roughly the same as hitting a hard 17 in blackjack. The probability of improving to a 19–22 is only $\frac{100}{315} \approx .3175$, less than $\frac{1}{3}$. Stand.

b. With a starting hand of 3♡ 7♡ and the dealer showing the 9♠, you might consider doubling down, just as you might with a 10 against a 9 in the regular game. The lowest player hand that can win a hand of California Blackjack is 18, which wins only against a dealer 17 or 26. The probability of doubling on 3♡ 7♡ against a 9 and getting a guaranteed losing hand by drawing a 2–6 is $\frac{119}{315} \approx .3778$. This represents a 37.78% chance of immediately losing 2 betting units. Another possibility, which has probability $\frac{23}{315} \approx .0730$, is drawing a 7 for 17, which can do no better than tie the dealer.

If you double and draw an A♣, you have 21. If the dealer turns over the 7♠ for 16, she must draw again. A subsequent draw of the 4♠ stops the dealer at 20 and seals your win.

This is, of course, the best-case scenario. You could draw the 4♢ and be stuck with a guaranteed losing hand of 14. You need to retain the possibility of opting for a fourth card by not doubling down.

c. If you have been dealt 9♡ 9♡ against a dealer 8♡, basic blackjack strategy calls on you to split the pair. This choice remains the correct one in California Blackjack. As it stands, your 18 loses to any dealer hand from 19–25; by splitting, you have two chances to improve a 9 to something better than an 18. The probability of reaching 19–21 in a one-card draw is $\frac{126}{315} = .4000$; moreover, you also have a chance to draw another card if the second card gives a low total. If you draw a third 9 for another 18, you would resplit, and it's not possible to hit 22 from 9 in one card.

d. Some things don't change. Consider the initial hand K♡ A♣. This hand would be scored as a 21, but would not qualify as a natural, since the goal is to reach 22. Standing is the obvious and best strategy.

■

California Blackjack charges gamblers a collection fee of 1% of their wagers, rounded—of course—up to the next dollar. With this custom in place, it's necessary to wager in $100 increments to minimize the effect of the rake on the HA. In the simple case where player and dealer have an equal chance of winning, a \$25 wager with a \$1 rake has an expected return of $-\$\frac{1}{2}$, which corresponds to a 2% house advantage on what appears to be an even-up wager.

California Aces

California Aces was another blackjack-derived game designed by Roger Wisted. It premiered in the early 1990s at the Bicycle Hotel & Casino in Bell Gardens, California [84]. The game was played with a single standard deck plus 4 jokers, the so-called "California aces", which counted 1 or 11, just like aces. Beside this revaluing of jokers, California Aces was very similar to California Blackjack:

- The object was to get a total closer to 22 than the dealer, rather than 21. A "natural 22" in two cards was an automatic winner, though it paid even money, not 3–2 as in blackjack.

- Busted hands remained impossible: "closer to 22" applied to hands either above or below 22.

With the jokers added to the deck, a natural 22 could be had with two jokers, two aces, or one of each card. The probability of a natural dealt from the top of a fresh 56-card deck was then

$$\frac{\binom{8}{2}}{\binom{56}{2}} = \frac{28}{1540} = \frac{1}{55},$$

about double the chance of a natural in California Blackjack, but still less than the chance of a natural in traditional blackjack.

Two versions of the game were proposed: a poker-style game where all players competed against each other, and a "Beat the Banker" game where all players faced off against a single designated player/banker. The poker version of California Aces allowed 3 distinct levels of wagering to every player. Level 1 was a mandatory \$5 bet; a player wishing to risk no more could stop there and make no further bets. Level 2 bets ranged from \$10 to \$50 and Level 3 bets were from \$25–\$500. The Level 3 prize was available only to players making level 3 bets, and similarly for Level 2, while everyone at the table was eligible to win the Level 1 jackpot [84].

Example 5.27. Suppose that 5 players: Terry, Dale, Robin, Sandy, and Chris, are seated at a California Aces table. Prior to the deal, they make the following wagers:

Player	Level 1	Level 2	Level 3
Terry	$5	$25	—
Dale	$5	—	—
Robin	$5	—	$300
Sandy	$5	$25	$300
Chris	$5	—	—

Everyone is included in the Level 1 pool. Only Terry and Sandy are eligible for Level 2, and only Robin and Sandy are competing at Level 3.

Each player is then dealt two cards, face down.

Player	Terry	Dale	Robin	Sandy	Chris
Cards	5♠ 5♡	8♡ Joker	4♠ Joker	4♡ 2♠	6♡ 7♣
Total	10	9/19	5/15	6	13

Since there are no natural 22s, players now have the option to draw additional cards, which are dealt face up.

- Terry draws the K♠ and stops with a hand of 20.

- Dale elects to stand on 19.

- Robin draws the 8♣, resulting in a hand of 13 or 23. Since 23 is not a bust hand in California Aces, and beats every hand except a 21 or 22, calling the hand 23 and standing is the correct decision.

- Sandy draws the 10♣, and hits the resulting 16. This is a carryover from regular blackjack; without much knowledge about opponents' hands, 16 calls for a hit, especially against 4 opponents rather than a lone dealer. The third card is the K♣. At 26, this hand exceeds 22, and so by rule may draw no further cards.

- Since each player's initial hand is dealt face down, there is no meaningful advantage to drawing last, as there can be in a face-up game for a card counter in the last position. Chris draws the 10♠, for a total of 23, and like Sandy, must stay on a total above 22.

The three levels are resolved from the top down.

- At Level 3, Robin's 23 beats Sandy's 26 and wins $600. Chris could have split this pot with Robin, but chose not to participate at Level 3.

- At level 2, Terry's 20 beats Sandy's 26. Terry collects $50.

- All 5 players participate in the Level 1 reveal. Chris and Robin tie at 23 and divide the pot, each receiving $12.50.

■

In the Beat the Banker version of California Aces, the player/banker places a wager which is covered by the bets from the other players. If the designated banker wagers $50, for example, the other players' bets cannot add up to more than $50. Once the bets are made, play proceeds like blackjack. One player/banker card is dealt face up as an aid to other players in playing their hands. The banker plays last, and must draw on any 17 while standing on 18 or higher, including a hand over 22. This means that the dealer's completed hand will always be in the range 18–27.

Example 5.28. These rules could easily be adjusted to govern a traditional house-banked version of California Aces. Binding the player/dealer to hit all 17s or lower and stand on all 18s or higher would allow all player cards to be dealt face up. Suppose the five players from Example 5.27 try their luck against a casino-employed dealer. The first round of cards is dealt.

Player	Terry	Dale	Robin	Sandy	Chris
Cards	$7\diamondsuit Q\clubsuit$	$J\diamondsuit 2\heartsuit$	$K\heartsuit 6\spadesuit$	$6\clubsuit 10\clubsuit$	$10\heartsuit J\heartsuit$
Total	17	12	16	16	20

The dealer's upcard is the $5\clubsuit$, and so their hand is not a natural 22. It is certain that the dealer will be drawing at least one additional card, and probably stands a good chance of going well over 22. In blackjack, a dealer 5 would be cause for standing on a 12 or higher rather than risking busting, but since busting is not a threat at California Aces and a 12 can never win for a player or the player/banker, a different basic strategy should be employed here.

- Terry's hard 17 would be an obvious standing hand in blackjack, but it will only win at California Aces if the dealer draws to 27. Hitting this hand, a move perhaps motivated by the surplus of 10s in the first 11 exposed cards (the high-low running count here is –6, in a deck starting at –4 if jokers carry the same –1 count as aces) yields the $9\diamondsuit$ and a final hand of 26. Terry is actually no worse off, as a 26 loses to any completed dealer hand less than 27.

- Dale should hit 12 against a 5, and draws the $9\clubsuit$ for a good-looking 21. In blackjack, this would push at worst, but in California Aces, the dealer might draw out to a 22 which would defeat it.

- Robin faces a situation similar to Terry, in that 16 only wins if the dealer's hand exceeds 27, which cannot happen. Hitting is the clear choice, which adds the $2\spadesuit$ for an 18. Hitting again will improve the hand if the fourth card is is a 7 or less, including a joker; since 14 cards have been revealed and only 6 are 7s or lower, the probability of improving is $\frac{26}{40} = 65\%$. Robin hits, and draws the $3\diamondsuit$ for a 21.

- Sandy faces the same choice as Robin, and hits with the $A\spadesuit$, bringing the hand to 17. This hand could be valued at 27, but 17 is better, as it beats a dealer 27 while a player 27 only pushes against a dealer 27. Just as Robin hit an 18 with a 65% chance of improving, Sandy's chance of improving with an 8 or less can be calculated to be $\frac{28}{38} \approx 73.7\%$, which justifies a hit. The fourth card is the 7♡, giving a 24. This hand will win provided the dealer's hand is 18, 19, or anything above 24.

- Chris has it easy—hitting a 20 is as bad an idea in California Aces as it would be in blackjack.

The dealer's hole card is then revealed: the 9♠. The dealer must hit this 14, draws the 4♠ for an 18, and then must stop. Everyone except Terry collects even money on their wagers. ■

One immediate conclusion shown by California Blackjack and California Aces is that small game modifications can lead to a very different game for players.

Postscript: Roger Wisted started a winery in Santa Barbara, California with the proceeds from licensing his two blackjack games. His winery is named Blackjack Ranch.

LA Blackjack

LA Blackjack is similar to California Aces, using 6 standard decks and 2 jokers per deck, so a shoe consists of 324 cards. Jokers count as 2 or 12 in LA Blackjack instead of 1 or 11 as in California Aces [5]. The goal remains to get a hand close to 22; hands are ranked in the following order:

1. A *natural 22* consisting of a pair of aces. As in California Aces, naturals pay only even money. A pair of jokers is not a natural 22; it is a hand of either 4 or 14.

2. A two-card 22 comprised of a joker and a ten-count card.

3. Any 22 formed by 3 or more cards.

4. Hands not equal to 22 are ranked by their distance from 22, with the provision that a hand under 22 beats any hand over 22, so a 14 beats a 23, unlike in California Aces. This is a significant change from California Aces that affects playing strategy: for example, a player 14 can win against any finished player/dealer hand that exceeds 22.

5. Hands of the same value push, *except* for 18s. Ties at 18 are won by the house; this is the casino's advantage over the players. In standard blackjack, the probability of the dealer and a player tying at 18 is approximately 1.77%; this is a good approximation to that probability in LA Blackjack [115].

Unlike in blackjack, both the player and the player/banker are restricted in their play options. Player/bankers must stand on a hard 18 or soft 19 or higher—this is common practice in California games with a target of 22. Players have these same restrictions, but in addition must hit soft 18, although a player may stand on a soft 13 through 17. This last rule gives players a second chance to avoid the "ties on 18 lose" rule. There is also no opportunity for a player to double down or split pairs.

Example 5.29. Consider the cards of Example 5.28 considered as a hand of LA Blackjack.

Player	Terry	Dale	Robin	Sandy	Chris
Cards	7◇Q♣	J◇2♡	K♡6♠	6♣10♣	10♡J♡
Total	17	12	16	16	20

The dealer's upcard remains the 5♣.

- Terry's hard 17 would be an obvious standing hand in blackjack. Unlike in California Blackjack, standing on 17 here will beat any dealer hand over 22. Since the dealer will be drawing and appears to have a good chance of going over 22, standing is the better option.

- Dale should still hit 12 against a 5, and draws the 9◇ for another 21. This hand rates just as highly in LA Blackjack as in California Aces—the only way it loses is if the dealer draws to 22.

- Robin and Sandy both hold hard 16s. While busting may be impossible, a good strategy in LA Blackjack is to think in terms of a hand over 22 busting, since any such hand can be beaten by a finished hand under 22. This is a case where "let the dealer go over 22 first" is sound thinking, and so both players stand.

- Chris should still stand on 20—some things never change as the game rules and deck compositions vary.

The dealer's hole card is the 9♠. The dealer must hit this 14 and draws the 9♣ for a 23. This time, everyone wins. ∎

The player fee at LA Blackjack is $1 for each $100 or fraction thereof wagered, so the fee on a $101 bet would be $2, which has some implications for smart bet sizing [5]. This charge adds to the house advantage and makes it more important for a player to learn and follow a basic strategy in order to cut the house edge down as far as possible. The player/banker pays $2 for the privilege of covering all player wagers.

LA Blackjack has been completely analyzed, and the simple basic strategy shown in Table 5.19 has been developed for hard hands. Soft hands should follow this strategy when the table dictates standing. One thing is clear from the table: Optimal play against the rules of LA Blackjack, especially the fact

TABLE 5.19: Basic strategy for LA Blackjack [5]

Player's Hand	Dealer's upcard										
	J	A	2	3	4	5	6	7	8	9	T
18+	S	S	S	S	S	S	S	S	S	S	S
17	H	H	H	S	S	S	S	S	H	H	H
16	H	H	H	S	S	S	S	H	H	H	H
15–	H	H	H	H	H	H	H	H	H	H	H
Key: J: Joker. A: Ace. T: Ten-count card. H: Hit. S: Stand.											

that doubling down and splitting pairs are not permitted, means that players have very few decisions to make, which means that the chance of the HA rising due to incorrect choices is small. The HA using the infinite deck approximation is 1.33%—but of course, in California, this advantage rests with the banker, not the card room [5]. This may explain why the per-hand player fees are on the high side and the banker is charged $2 per hand: to make sure that the card room profits from LA Blackjack.

Card counting LA Blackjack has been shown to be ineffective: while the tens remain important, the per-hand fees are too high, adding 1% to the player disadvantage. The probability of a favorable shoe, no matter which cards are tracked, is estimated to be less than 1 chance in 1 million [5].

Hawaiian Blackjack

Hawaii is one of two American states, along with Utah, with no legal gambling. The Hawaiian Gardens card room (now the Gardens Casino) and its hometown south of Los Angeles borrowed the state's name for their own and for their variation on blackjack, which launched in 2000.

Hawaiian Blackjack is an extension of LA Blackjack with a target of 22 and a shoe of 4–8 decks with 4 jokers added per deck for a total of 224–448 cards. Jokers may be counted as 2 or 12, and aces count as 1 or 11. The highest-ranking hand is a natural 22 and consists of a pair of aces. In Hawaiian Blackjack, a natural is different from a blackjack, which is a hand of 21 that consists of a joker and a ten-count card.

We can confirm that a Natural 22 outranks a blackjack by computing their respective probabilities. In a six-deck shoe, we have

$$P(\text{Natural 22}) = \frac{\binom{24}{2}}{\binom{336}{2}} = \frac{276}{56,280} = \frac{23}{4690} \approx .0049,$$

and

$$P(\text{Blackjack}) = 2 \cdot \frac{24}{336} \cdot \frac{96}{335} = \frac{96}{2345} \approx .0409.$$

In the general case of an n-deck shoe, we have

$$\frac{P(\text{Blackjack})}{P(\text{Natural } 22)} = \frac{32n}{4n-1},$$

and so the ratio of the two probabilities approaches 8 as n increases.

The rules of Hawaiian Blackjack have the following differences from standard blackjack [62, p. 139–145]:

- The player/dealer must draw to any soft 18 or lower, and must stand on hard 18 or higher.

- Other players can draw to any hard 21 or less, but may not hit a 22. This eliminates the possibility of doubling down on a pair of aces to get more money in play.

- Pair splitting is permitted, but players may only draw one card to each hand when splitting jokers. Since two aces make up a natural 22, a player would not split aces.

- Players may double down on any 2-card hand except a natural 22 or a blackjack.

- Surrender of an initial 2-card hand is allowed, but the 1% collection fee is not refunded, in whole or in part, on a surrendered hand.

- A completed hand under 22 wins against any hand over 22. If both hands go over 22, the hand closer to 22 wins.

- Ties are handled as follows:

  - If both hands tie at 22, natural 22s beat any other 22. Two natural 22s push, as do two non-natural 22s.

  - Ties over 22 push.

  - Ties under 22 push *unless* both hands total 18, in which case the player/dealer wins, as in LA Blackjack.

    This rule is somewhat balanced by the requirement that the player/dealer hit a soft 18, which gives a player who stood on 18 a chance not to lose.

Example 5.30. Assume that all hands are dealt from a fresh 4-deck (224 cards) shoe.

a. The player holds 4♡ 10♣ for a 14 against a dealer 9♣. She hits and draws the 10◇ for a 24. This hand will win if the dealer draws out to a 25 or 26. However, the dealer's hole card is a joker, bringing his hand to 21, which wins.

b. A player holds 10♡ 9◇, and the dealer shows a joker. Standing on this good hand in the face of a strong dealer upcard is the correct play. The dealer turns over the A♡, for a hand totaling 3 or 13. Given that this hand is composed of the two "best" cards, it is odd that this is not an encouraging total. The dealer must hit, drawing the 7◇ and winning the hand with a 20. Since a soft 20 is higher than a hard 18, the dealer may not draw further.

c. The player holds 9♡ 6♣, and the dealer's upcard is the 8♡. Especially in light of the no-bust rule, hitting 15 is correct, and the player draws the 3♡ for an 18. Players are free to hit hard 18s if they wish, but the risk of going over 21 is high. Facing a dealer 8, which has great potential to turn into 18, 19, or 20 when the hole card is revealed, a case could be made for hitting. Examine the probabilities shown in Table 5.20, where we assume that the player will draw at most one more card.

TABLE 5.20: Hawaiian Blackjack: Possible results when hitting 9/6/3 vs. 8

Hand total	Ways to reach	Probability
19	16	.0727
20	32	.1455
21	15	.0682
22	16	.0727
23	16	.0727
24	15	.0682
25	16	.0727
26	15	.0682
27	15	.0682
28	64	.2909

- The probability of improving to 22 or less is .3591. Reaching this total, of course, is no guarantee of winning.
- The probability of a final total of 23–27, which can still beat or tie some completed dealer hands, is .3500.
- The probability of drawing to a total of 28, which is a guaranteed loss, is .2909.

The decision is perhaps not any clearer for knowing these figures, although the 29% probability of drawing and exceeding 27 is a moderately strong argument for standing. Suppose that our player, experienced in the ways of regular blackjack and thus loath to hit a hard 18, stands. The dealer turns over the $J\heartsuit$, and the hands push.

■

No-Bust 21st Century Blackjack Second Chance

The Bicycle Casino is home to *No-Bust 21st Century Blackjack Second Chance*, or "No-Bust" for short. This game uses 8 decks, each with a wild joker added, so a full shoe contains 424 cards. A hand with one joker is automatically valued as a hard 21. A two-joker hand is called a "Natural 22" and is the highest-ranking hand. As in several other California blackjack games, some player actions are specified:

- Players must stand on soft or hard 20 or 21 as well as a Natural 22, except that a pair of 10s may be split.

- Players must hit when holding 11 or less. In practice, this rule should not affect players, as hitting an 11 or lower can only improve a hand.

- Players may choose to hit or stand on hands totaling 12–19.

European rules apply to the player/dealer, who receives only an upcard at the initial deal and does not receive a second card until the players have played out their hands. If that upcard is a joker, then player hands are locked and may draw no additional cards. At this point, the player/dealer's hand is either 21 or 22, so the smart course for any player holding a 20 or lower would be to surrender—but while surrendering is permitted in this game, it is not allowed when the player/dealer shows a joker.

The dealer's second card determines where the action button is placed. Aces through 7s direct the button to position 1–7 at the table, respectively. If the card is higher than a 7, the count resumes at position #1: an 8 puts the button at position 1, a 9 at #2, and so on until a joker moves the action button to position #7.

Player/dealers must hit until their hand is a hard 17 or higher. Natural 22s win against all other hands. For hands below 22, the hand closer to 22 wins. If one hand is greater than 22 and the other is less than 22, the hand below 22 wins. If both hands are over 22:

- The player/dealer wins if his hand is closer to 22.

- If the player is closer to 22, she loses *unless* the player/dealer has a 3-card total of hard 26, in which case the hand is a push.

- Ties above 22 are wins for the player/dealer.

In short, the rules give the players a powerful incentive not to go over 22.

For players, a Natural 22 pays 2–1, a blackjack (2 cards totaling 21) pays 6–5, and all other winning hands pay even money.

The Second Chance option noted in the title of the game allows players, once per hand, to surrender half their wager and replace a card they have drawn. There are certain circumstances where this option should be used; an obvious one is when the draw puts the hand over 22.

Example 5.31. Players betting $100 and dealt 7♣ 4♡ would reasonably double down, increasing their wager to $200. If the double down card is the 5♢, using the second chance option would allow the player to forfeit $100 of the doubled bet and discard the 5♢ in pursuit of a higher 3-card hand. This option may only be exercised once in this hand. ■

When should doubled hands invoke the second chance option? Since half of a doubled bet is being surrendered, it would seem reasonable to take a second chance when the probability of raising the total exceeds 50%.

If the double down card raises an 11 to 21 or 22, we would not use the second chance option. Unlike in regular blackjack, drawing an ace to 11 in this game gives 22, not 12. Using the infinite deck approximation and disregarding the dealer's upcard momentarily, we note that the chance of raising an 11 to 21 or 22 is $\frac{21}{53}$. This is 39.9% of the deck, and they are guaranteed to improve a doubled three-card hand that is under 21. Since 39.9% is less than 50%, it is unwise to break up a 20 in pursuit of a 21 or 22.

That leaves hands totaling 12–19. For example, if dealt a 4 and a 7, and receiving a 5 as the third card for a 16, trading in the 5 results in a higher total if the new card is a 6 or higher. There are

$$32 + 31 + 32 + 32 + 128 + 32 + 8 = 295$$

cards remaining in the shoe, so the probability of improving the hand with a second chance draw is $\frac{295}{421} \approx .7007$. Exchanging the 5 is favored.

The exact probabilities depend in a small way on the 2 cards making up the initial 11, but the rule is simply stated:

Use the second chance option when doubling down on an 11
if the 3-card hand is 18 or less.

The probability of raising a 3-card 18 to a 19–22 is 54.87% or 55.11%: the lower value when the 11 is made up of a 2–9 or 3–8 and the higher value otherwise.

Incorporating the dealer upcard into the count changes the probabilities, but does not raise or lower any of them past 50%, so the conclusion is the same.

Example 5.32. Because No-Bust hands do not bust at 22, it is reasonable to consider doubling down on a hard 12. When holding 10♢ 2♠ and then doubling down, a player should take the second chance if the third card is a 6 or lower, corresponding to a total of 13–18. The probability of improving an 18 is 54.87%, and the chance of improving a 19 is 47.27%. ■

Hot Action Blackjack

Hot Action Blackjack, at the California Grand Casino in Contra Costa, combines many elements of other California blackjack variants. The game is played with multiple decks, each augmented by 1–4 jokers which count 2 or 12. The best hand is a Super Natural 24, which consists of a pair of jokers, but it is only available to players. If the player/dealer's initial hand is a pair of jokers, it counts as a hand of 4.

The next-best hand is a Natural 22, or a pair of suited aces. Natural 21s rank third; these are standard blackjack 21s consisting of an ace and a ten. Any of these 3 hands ends the game early for the player. Super Natural 24s pay 4–1, Natural 22s pay 2–1, and Natural 21s pay 6–5. Absent any natural hands on either side, play proceeds as usual for blackjack. Hands do not bust if they go over 21, although any completed hand under 21 beats any hand over 22. Hands over 22 are ranked in order of their increasing distance from 22. Ties on 22 push, giving players a chance to rescue a busted hand of 22 if the dealer also reaches 22.

Since jokers count 2 unless they're dealt as part of a Super Natural 24, a Hot Action Blackjack shoe is richer in low cards than a standard deck, and this works to the advantage of the player/dealer.

The probabilities of the various natural hands depend on the number of decks in use and the number of jokers added per deck. Since the game is typically dealt from a CSM, assuming a full shoe is computationally correct. If 6 decks are used and 3 jokers are added to each deck, the probability of a Super Natural 24 is

$$\frac{\binom{18}{2}}{\binom{330}{2}} = \frac{18 \cdot 17}{330 \cdot 329} \approx \frac{1}{355}.$$

Even allowing for the rule that players win ties at 24, this event is wildly underpaid at 4–1.

From a 6-deck shoe, the probability of a Natural 22 is

$$4 \cdot \frac{\binom{6}{2}}{\binom{330}{2}} = \frac{4 \cdot 6 \cdot 5}{330 \cdot 329} \approx \frac{1}{905},$$

making this hand over twice as rare as a Super Natural 24.

The probability of a Natural 21 from a 6-deck shoe is

$$2 \cdot \frac{24}{330} \cdot \frac{96}{329} \approx \frac{1}{47}.$$

Adding 3 additional 2s per deck cuts the probability of a 2-card 21 by more than half.

Table 5.21 lists the California blackjack games we have considered with some game parameters.

TABLE 5.21: California blackjack variations compared

Game	Target	Jokers/deck	Joker value
California Blackjack	22	1	Wild
California Aces	22	4	1 or 11
LA Blackjack	22	2	2 or 12
Hawaiian Blackjack	22	4	2 or 12
No-Bust	22	1	Wild
Hot Action Blackjack	24	1–4	2 or 12

5.6 Side Bets

Field Bet

In the 1980s, some casinos experimented with a variation of the craps Field bet (page 111) on their blackjack tables. One version of this bet paid even money if the player's first 2 cards totaled 12–16, with a 2–1 bonus for a pair of aces or a pair of 8s [78, p. 207].

We will assess the Field bet with the infinite deck approximation. The probability of a single card having rank A–9 is $\frac{1}{13}$ and the probability of drawing a 10 is $\frac{4}{13}$. A hand totaling 12 can be drawn as A-A, 2-T, 3-9, ..., T-2. The probability of T-2 or 2-T is $\frac{4}{169}$, and the probability of any other 2-card 12 is $\frac{1}{169}$. Adding up over all ways to draw a 12 gives $P(12) = \frac{16}{169}$; we may perform similar analysis on sums of 13, 14, 15, and 16 to give Table 5.22.

TABLE 5.22: Blackjack Field Bet: Win probabilities using the infinite deck approximation

Hand Total	Probability
12	$\frac{16}{169}$
13	$\frac{16}{169}$
14	$\frac{15}{169}$
15	$\frac{14}{169}$
16	$\frac{13}{169}$

The probability of winning the Field bet is then $\frac{74}{169} \approx .4379$. To compute the expectation, we remove the probabilities of A/A and 8/8 from this sum, since they pay 2–1. We have

$$E = (1) \cdot \frac{72}{169} + (2) \cdot \frac{2}{169} + (-1) \cdot \frac{95}{169} = -\frac{19}{169} \approx -.1124.$$

The house advantage is 11.24%, so the best strategy for this bet is not to play it.

21 + 3

21 + 3 is a simple blackjack side bet based on the player's two cards and the dealer's upcard. If these three cards, taken together, form a flush, straight, 3 of a kind, or straight flush, the side bet pays off at 9–1. How the cards are distributed between the two hands does not matter, and the 9–1 payoff is the same regardless of the final three-card hand.

It seems reasonable that this bet is dependent on the number of decks in play; let's test that assumption. Consider first a fresh single deck. There are 48 possible straight flushes: since aces count both high and low, any card except a king may be the lowest card in a straight flush, so

$$P(\text{Straight flush}) = \frac{48}{\binom{52}{3}} = \frac{48}{22,100}.$$

The probability of three of a kind is then

$$P(3 \text{ of a kind}) = \frac{13 \cdot \binom{4}{3}}{22,100} = \frac{52}{22,100}.$$

Continuing, we have

$$P(\text{Straight}) = \frac{48 \cdot 4 \cdot 4 - 48}{22,100} = \frac{720}{22,100}$$

and

$$P(\text{Flush}) = \frac{4 \cdot \binom{13}{3} - 48}{22,100} = \frac{1096}{22,100}.$$

The total probability of winning this bet is then

$$P(\text{Win}) = \frac{48 + 52 + 720 + 1096}{22,100} = \frac{1916}{22,100} \approx .086697,$$

and the expectation of a \$1 bet is

$$E = (9) \cdot \frac{1916}{22,100} + (-1) \cdot \frac{20,184}{22,100} = -\frac{3540}{22,100} \approx -.1330,$$

corresponding to a house edge of about 13.30%.

Thinking more generally, let n denote the number of decks in use. We have the following:

$$P(\text{Straight flush}) = \frac{48n \cdot n \cdot n}{\binom{52n}{3}} = \frac{48n^3}{\binom{52n}{3}},$$

$$P(\text{Three of a kind}) = \frac{13 \cdot \binom{4n}{3}}{\binom{52n}{3}},$$

$$P(\text{Straight}) = \frac{48n \cdot 4n \cdot 4n - 48n^3}{\binom{52n}{3}} = \frac{720n^3}{\binom{52n}{3}},$$

$$P(\text{Flush}) = \frac{4 \cdot \binom{13n}{3} - 48n^3}{\binom{52n}{3}}.$$

Adding these four probabilities gives, after much simplification,

$$P(\text{Win}) = \frac{52 - 663n + 3485n^2}{26 - 2028n + 35,152n^2}.$$

Starting with a fresh deck or shoe, the probabilities of a winning three-card hand, as well as the corresponding expectations, are listed in Table 5.23.

TABLE 5.23: 21+3 Side Bet: Win Probabilities and Expectations

# of Decks	P(Win)	Expectation
1	.0867	−$.1330
2	.0927	−$.0726
4	.0959	−$.0410
6	.0970	−$.0303
8	.0975	−$.0249

While the probability of winning, and hence the expectation, rises with the number of decks in use, letting $n \to \infty$ gives a limiting upper bound p for the probability of winning:

$$p = \frac{3485}{35,152} \approx .0991.$$

The corresponding limit for the expectation is

$$E = 9p - (1 - p) = -\frac{302}{35,152} \approx -\$.0086.$$

The house advantage approaches 0.86%.

Changing the payoff odds to 8–1 increases the HA to 21.97% for a single-deck game, and 10.77% in the infinite-deck limit.

Lucky Ladies

One of the most successful blackjack side bets, if success is measured by the number of casinos putting it on the gaming floor, is *Lucky Ladies*. A gambler wins this bet when his first 2 cards total 20. There are several pay tables in use that pay different odds for different hands that add up to 20. One pay table is shown in Table 5.24.

TABLE 5.24: Lucky Ladies pay table [67]

Hand	Payoff
Pair of $Q\heartsuit$ with dealer natural	1000–1
Pair of $Q\heartsuit$ with no dealer natural	200–1
Matched 20 (2 identical cards)	25–1
Suited 20	10–1
Unsuited 20	4–1

For a game dealt from a 4-deck shoe, the probability of the top payoff is

$$\frac{\binom{4}{2}}{\binom{208}{2}} \cdot \left[2 \cdot \frac{16}{206} \cdot \frac{62}{205} \right] \approx \frac{1}{76,372}.$$

Here, the first factor is the probability of drawing two $Q\heartsuit$s; the term in brackets is the probability of a dealer natural given that 2 queens have been removed from the shoe.

To compute the probability of a matched 20, we need to exclude the chance that the matched pair is two $Q\heartsuit$s; this leaves 15 possible cards. The probability is

$$15 \cdot \frac{\binom{4}{2}}{\binom{208}{2}} = \frac{90}{21,528} = \frac{1}{239.2} \approx .0042.$$

Lucky Ladies has a high house advantage, anywhere from 5.19–30.05%, depending on the exact pay table and number of decks in use [124]. The

24.94% HA quoted in Table 2.15 is for a 2-deck game using table 5.24 as its pay table.

The bet is so susceptible to card counting that a careful gambler could be very successful by sticking to basic strategy on the main hand and using a Lucky Ladies counting scheme to make the side bet when it's advantageous [67]. The applicable count focuses on the 10s; for a 6-deck shoe, the following count should be used:

Card	Value
A	0
2–9	+1
T	–2

At a true count of +8, Lucky Ladies carries a player edge.

This count system is not far from the High-Low count; using that system instead gives an acceptable—though lower—edge with a true count of +7 in a double-deck game.

An additional enhancement for 2-deck games, keeping in mind the increased payoffs for a player $Q\heartsuit\ Q\heartsuit$, counts the first $Q\heartsuit$ as –4 and the second as 0 [67]. This count does considerably better than the 10-count described above.

Super 31

Super 31 is unusual among side bets in that it is resolved only after the main game is complete. Most blackjack side bets are settled on the initial deal and thus out of the way when the business of hitting, standing, and so forth is considered by the players.

As a first approximation to basic strategy, new blackjack players occasionally are told to assume that the dealer's hole card is a 10 in playing their hands. While this has its limitations—if the dealer's upcard is a queen and you're holding a hard 19, hitting under the assumption that the dealer holds a 20 is unwise—this is not an unreasonable introduction to the decisions that a blackjack player faces and how the dealer's upcard must be taken into account in making those decisions.

The *Super 31* blackjack side bet expands on that assumption, asserting on its Web site that many players automatically assume that the next card off the deck or out of the shoe will be a 10 [141]. This bet is made after a player is dealt a natural and is a wager that the next card dealt to that hand—after the 3–2 natural payoff is made—will be a 10.

This side bet can be interpreted in two ways: as a way for a player to make more money on a dealt natural—even one belonging to another player—or as a way for a casino to get a shot at reclaiming the 3–2 winnings paid out on a natural. Since we are accustomed to the notion that casinos don't offer any bet that doesn't have something in it for them, we initially are inclined toward the second interpretation.

A player receiving a dealt natural has the option of investing all or part of his total winnings, including his initial bet, on the Super 31 bet. Any other player at the table when the natural is dealt can join in with a Super 31 bet up to the table minimum. Following the completion of the round, including the dealer's hand, a third card is dealt to the natural. If this card is a ten-count card, bringing the total to 31, the Super 31 bet pays off at 2–1 [141]. A $10 initial wager can quickly generate a $65 profit from a natural if this bet wins: $15 from the initial 3–2 payoff and $50 more if the entire $25 is parlayed into the Super 31 bet and a 10 is drawn.

As with standard blackjack, this bet is susceptible to exploitation by card counters. If the count is highly positive, signifying an excess of 10s, this bet can be quite profitable—indeed, since a skilled card counter would already have raised his or her bets on a high count, Super 31 provides another opportunity to leverage favorable deck composition into a profit.

We shall consider Super 31 from the non-counter's perspective. First, take the case of a single player head-to-head against the dealer in a single-deck game. On a $1 initial bet, the payoff if the player is dealt a natural but does not make the Super 31 bet is $1.50. The expected value of the Super 31 bet depends on the dealer's hand, which will contain either 0, 1, or 2 10s. Since there is only one player, the dealer's hand will end after its initial two cards are dealt, because the player has a natural.

Let x denote the number of 10s in the dealer's hand. We have $0 \leqslant x \leqslant 2$, and the expectation on a Super 31 bet with the entire $2.50 parlayed, as a function of x, is then

$$E(x) = (5) \cdot \left(\frac{15 - x}{48} \right) + (-2.5) \cdot \left(\frac{33 + x}{48} \right) = \frac{-7.5x - 7.5}{48},$$

which is tabulated below.

x	$E(x)$
0	–$0.16
1	–$0.31
2	–$0.47

While the expectation is negative regardless of the dealer's holdings, this represents something of a a worst-case scenario for the head-to-head player, in that a larger shoe would increase the number of available 10s (while simultaneously, of course, increasing the number of non-10s), and the dealer would still not be able to tie up more than 2 of them. For an n-deck shoe, the expectation, as a function of both n and x, becomes

$$E(n, x) = (5) \cdot \left(\frac{16n - 1 - x}{52n - 4} \right) + (-2.5) \cdot \left(\frac{36n - 3 + x}{52n - 4} \right),$$

or

$$E(n, x) = \frac{-10n - 7.5x + 2.5}{52n - 4}.$$

TABLE 5.25: Expectation for the Super 31 side bet

| $E(n,x)$ | | n | | |
	2	4	6	8
0	−$0.18	−$0.18	−$0.19	−$0.19
x 1	−$0.25	−$0.22	−$0.21	−$0.21
2	−$0.33	−$0.26	−$0.24	−$0.22

Values of $E(n,x)$ for common values of n are found in Table 5.25. Inspection of this table leads to the following observations:

- The expectation remains negative for all combinations of x and n, and the HA runs from 18–33% of the original wager (7–13% of the $2.50 put at risk with the Super 31 bet).

- If $x = 0$, indicating that the dealer's hand contains no 10s, the expectation slowly decreases with increasing n. This is, of course, still the best situation for the player.

- For $x = 1$ or $x = 2$, the expectation slowly increases as n increases and if $n \to \infty$, we see by taking a limit that the expectation approaches −$.1923.

Things can get better—or worse—for the player at a fuller table, depending on the number of 10s in the other players' hands.

Progressive 21

The *Progressive 21* side bet premiered in 6-deck games in Atlantic City in 1993 [7]. As the name suggests, part of the pay table is a progressive jackpot. The $1 wager paid off on the number of 5s dealt to a player's hand in accordance with Table 5.26.

TABLE 5.26: Progressive 21 pay table [7]

Player's Hand	Payoff
Four suited 5s	$200,000 minimum jackpot
Four unsuited 5s	$750 minimum jackpot
Three suited 5s	$500 minimum jackpot
Three unsuited 5s	$$100
Two suited 5s	$50
Two unsuited 5s	$10
One 5	$1

The 3 jackpots in Table 5.26 were separate pots of money; the top jackpot was paid as a series of 20 annual payments, just as many Powerball or Mega Millions lottery jackpots are paid.

We note immediately that pursuing the top Progressive 21 prizes requires going against basic strategy in a number of circumstances. Most significantly, a player dealt a pair of 5s should double down unless the dealer shows a 10 or ace, but reaching the payoffs for four 5s requires skipping the favored double down and retaining the chance to draw a 4th card. Additionally, a player who holds three 5s, suited or unsuited, should stand on the 15 against a dealer 2–6, not hit in pursuit of another 5.

This conflict of strategies can make for a desirable—to the casino—side bet. A player making a $25 main bet and a $1 Progressive 21 side bet cannot, in general, simultaneously act in the best interests of both wagers. Either a basic strategy choice requires abandoning a run at the top progressive prize or acting to chase the progressive jackpots means working against the main bet.

Over/Under

One hallmark of a good blackjack side bet is its ability to be resolved before players start drawing cards, and *Over/Under*, which was launched at Caesars Tahoe in Stateline, Nevada (now the Montbleu) in 1988, fits that description [132]. The player may wager that his or her initial 2-card hand will be either over 13 or under 13, with aces counted as 1 for the purpose of this bet. On a dealt 13, both bets lose. Over/Under does not involve the dealer's upcard, which makes it easier to administer. Once the wager is settled, the blackjack hands are played out against the dealer as usual.

Table 5.27 shows the distribution of the sum of a player's 2 cards using the infinite deck approximation. Using this information, we can compute the house advantages for Under and Over.

$$E(\text{Under}) = (1) \cdot .4497 + (-1) \cdot .5503 = -.1006,$$

for a HA of 10.06%, and

$$E(\text{Over}) = (1) \cdot .4675 + (-1) \cdot .5325 = -.0650,$$

giving a HA of 6.50%.

Example 5.33. A possible wager that could be added to Over/Under is a "13" bet, that the player's hand will total exactly 13. If paid at x to 1, this bet has expected value

$$E = (x) \cdot .0828 + (-1) \cdot (.9172) = .0828x - .9172.$$

A fair game results when $.0828x - .9172 = 0$, or when $x \approx 11.077$. A 10–1 payoff would give a house advantage of 8.92%, which falls nicely between the HAs of Over and Under. ∎

TABLE 5.27: Sum of a player's first 2 cards: Aces count as 1, infinite deck approximation in force

Hand	Probability	Hand	Probability
2	.0059	14	.0769
3	.0118	15	.0710
4	.0178	16	.0651
5	.0237	17	.0592
6	.0296	18	.0533
7	.0355	19	.0473
8	.0414	20	.0947
9	.0473		
10	.0533		
11	.0947		
12	.0888		
Under:	.4497	**Over:**	.4675
Probability of 13: .0828			

The infinite deck approximation conceals, perhaps, the fact that Over/Under is susceptible to card counting. An early study using the High-Low count in a 6-deck game showed that the Over bet had a player advantage at a true count of +5 and the Under bet shifted to favor gamblers when the true count reached –8 [132]. This was a case of adapting an existing count system to a different wager, and while it was adequate for identifying a player edge, a specialized counting system tailored to Over/Under was soon developed as the game spread more widely from northern Nevada.

Of particular significance in the Over/Under count was proper treatment of aces, which count as low cards for Over/Under but are counted with the 10s in High-Low. The Over/Under count uses the following values [137]:

Cards	Value
Low: A, 2, 3, 4	+1
Neutral: 5, 6, 7, 8, 9	0
High: 10, J, Q, K	–1.

With the Over/Under count, players have the advantage on the Over bet when the true count is +3 or greater, and have the edge on Under at –4 or less [137].

Red/Black

The *Red/Black* side bet was a favorite at the Four Queens Casino in downtown Las Vegas. Red/Black was a bet on the color of the dealer's upcard [162]. The bet paid off at even money, with the provision that if the upcard was a deuce of the color bet, the wager pushed.

Computing the expected value from the top of an n-deck shoe is simple:

$$E = (1) \cdot \frac{24n}{52n} + (-1) \cdot \frac{26n}{52n} = -\frac{1}{26} \approx -.0385,$$

which is independent of the number of decks.

Red/Black is easily seen to be susceptible to its own card-counting scheme based simply on colors. If we count -1 for each red card that appears and $+1$ for every black card, a surplus of red cards would be indicated by a positive count; a negative count would suggest more black cards than red remain. As with High/Low, we would divide this running count by the number of decks remaining to get a per-deck assessment of the balance between black and red cards. An advanced counter might consider a side count for deuces.

Example 5.34. Suppose that 2 decks remain and the true count is $+7$, indicating that the Red bet is favored. The 2 decks include 14 more red cards than black cards, so there are 59 red cards and 45 black cards. Even without side-counting deuces, we might reasonably assume that 5 red deuces remain: 4 from the 2 decks and a fifth as one of the 14 surplus red cards. The expected value of a bet on Red is then

$$E = (1) \cdot \left(\frac{54}{104} \right) + (0) \cdot \left(\frac{5}{104} \right) + (-1) \cdot \left(\frac{45}{104} \right) = \frac{9}{104} > 0,$$

a positive value indicating a player advantage. ∎

More generally, suppose that the true count is k and n decks remain. The undealt portion of the shoe then contains $(26 + k)n$ red cards and $(26 - k)n$ black cards. Assuming that the card ranks are evenly distributed for simplicity, there are $2n + \frac{kn}{13}$ red deuces and $2n - \frac{kn}{13}$ black deuces left. For a Red bet, we have

$$E(k, n) = \frac{(26 + k)n - 2n - \dfrac{kn}{13}}{52n} - \frac{(26 - k)n}{52n} = \frac{25k - 26}{52},$$

a result independent of n which is positive if $k > 1$. Similarly, a bet on Black has expectation

$$E(k) = \frac{-25k - 26}{52}$$

and is positive whenever $k < -1$.

Flip Card Blackjack

Flip Card is a blackjack side bet invented in 2015 by University of Nevada, Las Vegas student Aron Kock and marketed by Big Bet Gaming, a gaming company founded by two former executives with Shuffle Master now affiliated with the UNLV Center for Gaming Innovation [48]. This wager pays a bonus when the player has a natural and loses otherwise. The twist is that the amount of the bonus depends on the dealer's upcard. If the upcard is not an ace, the bet pays off at twice the upcard value to 1. A player natural with a dealer ace is paid off in accordance with Table 5.28.

TABLE 5.28: Pay table for Flip Card blackjack side bet with dealer ace [49]

Player's Blackjack	Dealer's Upcard	Payoff
Suited	Ace of same suit	500–1
Same color	Ace of matching color	100–1
Suited	Ace of different suit	80–1
Same color	Ace of other color	40–1
Mixed color	Any ace	25–1

The Flip Card bet wins on any player blackjack, even if the main wager is a push when the dealer also has a natural. Let's consider the Flip Card bet from the top of the shoe in a six-deck game. The probability of winning this bet with a natural is

$$p = 2 \cdot \frac{24}{312} \cdot \frac{96}{311} = \frac{192}{4043} \approx .0474 \approx \frac{1}{21},$$

and so the probability of an immediate loss is approximately .9525.

If the dealer's upcard is not an ace, the payoff is twice the value of the upcard, and this contributes

$$p \cdot \sum_{x=2}^{10} 2x \cdot P(x)$$

to the expectation, where $P(x)$ is the probability that the dealer's upcard is x:

$$P(x) = \begin{cases} \dfrac{24}{310}, & x = 2, \ldots, 9 \\[2ex] \dfrac{63}{310}, & x = 10 \end{cases}.$$

For the purposes of assessing the payoff with a player natural and dealer ace, we subdivide the probability p into three cases: $p = p_1 + p_2 + p_3$, where

- $p_1 = P(\text{Suited blackjack}) = \dfrac{24}{312} \cdot \dfrac{24}{311} \approx .0059,$

- $p_2 = P(\text{Same-color unsuited blackjack}) = \dfrac{24}{312} \cdot \dfrac{24}{311} \approx .0059,$

- $p_3 = P(\text{Mixed color blackjack}) = \dfrac{24}{312} \cdot \dfrac{48}{311} \approx .0119.$

The payoffs in the first two cases vary, depending on how the dealer's ace matches or doesn't match the player's natural.

For a suited blackjack, there are 5 aces of the same suit as the player's, and 310 cards remaining after the natural is dealt, hence the probability that the dealer's ace is the same suit as the player's blackjack is $\frac{5}{310}$ and the probability that its suit is different is $\frac{18}{310}$.

If the player's natural is made up of two cards of the same color but different suits, the probability that the dealer's ace matches it in color is $\frac{11}{310}$ and the chance that it's of the other color is $\frac{12}{310}$.

Combining these probabilities with the payoffs in Table 5.28 gives an expected value of

$$E = p \cdot \sum_{x=2}^{10} 2x \cdot P(x) + p_1 \cdot \left(500 \cdot \frac{5}{310} + 80 \cdot \frac{18}{310} \right)$$
$$+ p_2 \cdot \left(100 \cdot \frac{11}{310} + 40 \cdot \frac{12}{310} \right) + p_3 \cdot \left(25 \cdot \frac{23}{310} \right) + (-1) \cdot .9525,$$

which evaluates to $E = -\$.3088$—a house advantage of 30.88%.

Pick 'Em Blackjack

Instant 18 is a blackjack side bet where a player is assigned a hand of 18 without the need to deal cards [8]. In 2014, the developers of the Instant 18 blackjack side bet debuted *Pick 'Em Blackjack* at the Raving's Cutting Edge Table Games Conference, where it was chosen as Best New Game [125]. Pick 'Em Blackjack expands Instant 18 to instant 17 and 19 bets; as with Instant 18, a player may make a bet and be assigned a card-free hand totaling that value. Players may choose to make as many as four separate bets: three fixed wagers on 17, 18, or 19, or a standard blackjack hand which is dealt and played normally. These fixed bets pay off at 3–2 on 17, 1–1 on 18, and 1–2 on 19, reflecting the relative strength of each hand against the dealer.

How lucrative are these bets? Assuming a single-deck game, the probability distribution for the final value of the dealer's hand may be found in Table 5.29.

TABLE 5.29: Probability distribution for dealer's hand in single-deck blackjack where the dealer stands on soft 17 [61, p. 127]

Hand	Probability
17	.1458
18	.1381
19	.1348
20	.1758
21	.1219
Bust	.2836

We can then derive the following probability distribution for the three bets:

Hand	P(Win)	P(Push)	P(Lose)
17	.2836	.1458	.5706
18	.4294	.1381	.4325
19	.5675	.1348	.2977

For a wager that pays off at x to 1, the expected value of a $1 bet is

$$E(x) = (x) \cdot P(\text{Win}) - P(\text{Lose}).$$

We find that the 17 bet has an expectation of –$.1452, corresponding to a 14.52% HA, which is comparable to that of the baccarat Tie bet. The 18 and 19 bets fare better, with the 18 bet nearly a fair bet: its HA is .31%. Under the assumptions of a single-deck game where the dealer stands on soft 17, you are almost better off just making the Instant 18 bet and forgoing the standard blackjack hand, as this value is close to the HA for the ordinary game under those conditions. The 19 bet is a winner more often than not; this edge is tempered by the 1–2 payoff ($x = \frac{1}{2}$), but the HA is a low 1.40%.

Example 5.35. As is often the case with blackjack, changing game conditions can have a profound effect on the HA. We examine first the effect of the dealer hitting soft 17. Going into this analysis, we expect that the house edge will increase, as this change is known to favor the casino over the players.

Table 5.29 assumes that the dealer stands on soft 17; if the dealer hits soft 17, we can replace that table with Table 5.30.

TABLE 5.30: Probability distribution for dealer's hand in single-deck blackjack where the dealer hits soft 17 [122]

Hand	Probability
17	.1344
18	.1397
19	.1368
20	.1779
21	.1238
Bust	.2875

A comparison of Tables 5.29 and 5.30 shows a decreased probability of a final dealer count of 17, with the difference between the two distributed among the other possible outcomes including a bust, all of which have increased probabilities. This can be used to generate a second probability distribution for Pick 'Em Blackjack:

Hand	P(Win)	P(Push)	P(Lose)
17	.2875	.1344	.5781
18	.4219	.1397	.4384
19	.5616	.1368	.3016

The new house edges are 14.69% on 17, 1.65% on 18, and 2.08% on 19. At this point, playing a standard hand becomes preferable to making any of the three fixed bets. ∎

Example 5.36. How about a six-deck game where the dealer hits soft 17, which is far more typical of actual casino conditions? Again, we expect an increase in the HA as more decks are added. Table 5.31 contains the probabilities for the dealer's hand.

TABLE 5.31: Probability distribution for dealer's hand in six-deck blackjack where the dealer hits soft 17 [122]

Hand	Probability
17	.1335
18	.1412
19	.1357
20	.1815
21	.1224
Bust	.2858

The corresponding probability distribution for Pick 'Em Blackjack is

Hand	P(Win)	P(Push)	P(Lose)
17	.2858	.1335	.5807
18	.4193	.1412	.4395
19	.5605	.1357	.3038

The new house advantages are 15.20% on 17, 2.02% on 18, and 2.36% on 19. Our initial intuition is confirmed: these bets are worse—though not a lot worse—in a six-deck shoe. ∎

Spanish 21 Super Bonus

Spanish 21 is a blackjack variation played with one or more 48-card decks from which all of the 10s, though none of the face cards, have been removed. Since this starts the player out at a 2% disadvantage relative to standard blackjack, a typical Spanish 21 game includes player bonuses for certain hands. At the Muckleshoot Casino in Auburn, Washington, the 8-deck Spanish 21 game includes a free "Super Bonus", which is paid to any player holding a

21 consisting of three suited 7s when the dealer's upcard is any 7. This bet requires no additional wager; the payoff is a flat $1000 if the player's bet is $5-24 and $5000 if the bet is $25 or more. An "envy bonus" of $50 is paid to every other player at the table if another player wins this bonus.

The probability of winning the Super Bonus is

$$\underbrace{\frac{4 \cdot \binom{8}{3}}{\binom{384}{3}}}_{\text{P(3 suited 7s)}} \cdot \underbrace{\frac{29}{381}}_{\text{P(Dealer 7)}} = \frac{6496}{3,567,625,504} \approx 1.8209 \times 10^{-6},$$

or approximately one chance in 549,188. Since this is a free bet, the expected value is positive. Assuming a $25 wager, the expected return on the Super Bonus is less than 1¢, so if you can convince the player sitting next to you to give you a penny in exchange for your Super Bonus payoff if it hits, you'll come out ahead on the exchange.

The rules for the Super Bonus bet state that the bonus is forfeited if the player splits a pair or doubles down. This may be an incentive—albeit a very weak one—to violate Spanish 21 basic strategy, which calls for a player to split 7s (suited or not) against a dealer 7. Doubling down on a 14 would be foolish against any dealer upcard, so that rule is not a concern.

WINSURE

Along with WINGO and WINSUIT, *WINSURE* was a side bet, for blackjack, proposed in 1990 by *WIN Magazine* writer Stanley Roberts. This wager was a variation on the standard blackjack insurance bet that is offered when the dealer's upcard was a 10 or face card, instead of an ace. Like the insurance bet, this bet won if the dealer turned over a natural 21, winning against any player without a natural but offering a chance for players to hedge their main bet against this event. Like insurance, WINSURE bets were limited by the size of the player's initial wager: a WINSURE bet consisted of 10% of the primary wager, rounded up if necessary to avoid dealing with nonexistent fractional chips in small values [103]. A $10 bet, for example, could be WINSUREd for $1; a $15 bet might require a $2 WINSURE bet if the casino did not stock 50¢ chips. This practice of rounding bets or payoffs to available chip values is called *breakage*.

The clever twist with WINSURE came when the dealer turned over an ace and the player's bet won. If the player did not also have a natural, the WINSURE bet was resolved by simply returning his or her initial wager, paying off WINSURE at up to 10–1 odds. This was easier for dealers, who had no need to dip into their chip tray to pay the bet. A player with a winning WINSURE bet protecting a dealt natural (called a "SURE WIN") received an additional payoff equal to his or her initial wager; effectively an even-money payoff on the main bet, which is in line with the "even money" option for conventional insurance at a table playing 3–2 on naturals.

Example 5.37. In the simplest case, the player is able to wager exactly 10% of his or her main bet on a WINSURE bet. Assume that the game is dealt from the top of a fresh 6-deck shoe and that the player holds no aces. The probability of a dealer ace is then $\frac{24}{309}$. The expectation of the WINSURE wager is

$$E = (10) \cdot \frac{24}{309} + (-1) \cdot \frac{285}{309} = -\frac{45}{309} \approx -.1456,$$

giving the house a 14.56% advantage.

If the player must bet more than 10% to avoid breakage issues, the HA rises, since a winning WINSURE bet simply saves the original wager. Suppose the player's bet is $21, as might occur when the player is using a Fibonacci progression system (page 86) where successive losing bets increase in accordance with the Fibonacci sequence: 1, 1, 2, 3, 5, 8, 13, 21, 34, and so forth. A casino that stocks $2.50 chips or uses half dollar coins would require a $2.50 WINSURE bet, which results in an expectation of

$$E = (21) \cdot \frac{24}{309} + (-2.50) \cdot \frac{285}{309} = -\frac{208.5}{309} \approx -.6748.$$

The corresponding HA would be 26.99%.

Without the $2.50 chips or half dollars, which might be the case in a casino paying only 6–5 on naturals, a $21 blackjack wager would require a $3 WINSURE bet, dropping the payoff odds to 7–1 and raising the HA to 37.86%. ∎

In a single-deck game, again assuming no player aces, these HAs fall to 10.20%, 23.27%, and 34.69% respectively. Once again, we see that increasing the number of decks in play makes the game better for the casino.

Like the blackjack insurance bet, WINSURE might be susceptible to card counters, provided that they are using a separate count to keep track of aces.

Example 5.38. A standard blackjack insurance bet is a good bet for the player if the ratio of non-10s to 10s remaining to be dealt is less than 2:1. For WINSURE, if the ratio of non-aces to aces (which starts at 12:1 in a full deck or shoe) in the undealt portion of the shoe is 10:1, we have

$$E = (10) \cdot \frac{1}{11} + (-1) \cdot \frac{10}{11} = 0,$$

and so WINSURE becomes a fair bet. If that ratio is less than 10:1, indicating a considerable relative surplus of aces, the player is wise to make the WINSURE bet—provided that his or her main wager can be protected by exactly 10% of its value. A wager which requires a WINSURE bet exceeding 10% of its value requires a smaller ratio of non-aces to aces before WINSURE is a wise bet. ∎

A version of WINSURE called "\$uper In\$urance" popped up briefly at the Mohegan Sun Casino in Uncasville, Connecticut. The bet paid 10–1 and carried a HA near 23% when blackjack was dealt from an 8-deck shoe [139]. Super Insurance bets were capped at $\frac{1}{5}$ of the player's main wager instead of $\frac{1}{10}$.

Super 4 Progressive Blackjack

An unusual twist to the *Super 4 Progressive Blackjack* side bet is that it paid off if the dealer was dealt a natural, with a small additional prize if the dealer's upcard was an ace without a 10 in the hole. Super 4 payoffs were based on the 4-card poker hand formed by the dealer's natural and the player's initial hand. The top two prizes were paid off from a progressive jackpot that started at \$50,000, and the \$5 bet paid off in accordance with Table 5.32. All payoffs

TABLE 5.32: Super 4 Progressive Blackjack pay table. All payoffs except the last assume a dealer natural.

4-Card Hand	Payoff
Royal flush (AKQJ): $\diamondsuit$	100% of jackpot pool
Royal flush: $\spadesuit$, $\heartsuit$, or $\clubsuit$	10% of jackpot
3 of a kind	\$750
Straight	\$400
Flush	\$300
Two pairs	\$200
All same color	\$100
Pair	\$50
Any other hand	\$25
Dealer ace up	\$10

included refunding the player's bet, making the "All Same Color" payoff odds 20 for 1, or 19 to 1. Paying off a wager at "x for 1" rather than "x to 1" means that the player's initial wager is returned as part of the payoff; this means that payoffs can be advertised as larger without being larger.

This wager might be thought of as a different form of insurance. It stands as another example of a side bet whose goals oppose the goals o fthe main game.

The probability of winning this bet is the probability of the dealer's upcard being an ace plus the probability of the dealer receiving a natural with the 10-count card as upcard. With a fresh single deck in play, this probability is

$$\frac{1}{13} + \frac{1}{2} \cdot \left(\frac{4}{52} \cdot \frac{16}{51} + \frac{16}{52} \cdot \frac{4}{51} \right) = \frac{1}{13} + \frac{16}{663} \approx .1011.$$

If a six-deck shoe is in use, this probability drops slightly, to

$$\frac{1}{13} + \frac{24}{312} \cdot \frac{96}{311} \approx .1007.$$

Example 5.39. Find the probability that a royal flush in diamonds is dealt, triggering the progressive jackpot.

The winning probability is the number of ways to pick the ace, king, queen, and jack of diamonds out of the shoe and have the ace land in the dealer's hand. In a four-card draw including the $A\Diamond$, the dealer gets the ace half the time. For a full single deck, the probability is

$$\frac{1}{2} \cdot \frac{1}{\binom{52}{4}} = \frac{1}{541,450}.$$

∎

This is just slightly higher than the probability of drawing a 5-card royal flush at poker. It's also the probability of seeing a 4-card diamond royal flush dealt with the ace going to the player—in which case the Super 4 wager loses.

Destiny 21

The majority of blackjack side bets are resolved before hands are played out, and many are simply based on the players' and dealer's initial cards, with no opportunity for player choices that might influence the side bet. *Destiny 21*, an optional wager which is based on the number of hits taken by the player's hand, was designed to involve the player in the side bet throughout the hand. This may lead to circumstances where a player goes against basic strategy in playing out a hand in order to increase the payoff from the side bet.

A sample pay table for Destiny 21 is shown in Table 5.33.

TABLE 5.33: Destiny 21 pay table [154]

Number of Hits	Payoff
0	Lose
1	1–1
2	2–1
3	4–1
4	5–1
5	15–1
6	50–1
7	500–1
8+	1000–1

If the player is dealt an initial hand totaling hard 17–21, he won't hit it, and so a Destiny 21 side bet loses. If that hand is a natural, the casino effectively pays out a smaller net amount. Additionally, if the dealer holds a natural, players have no option to draw more cards, and so there is a second profit-taking opportunity for the house.

There are a number of optional rules that may be incorporated into Destiny 21 [154]:

- One variation provides that the player loses this side bet if his or her hand busts.

- Alternately, the casino may agree that a player who busts wins the last amount he or she would have won from the hand just prior to busting.

  This version might inspire players to prioritize the side bet over the main bet, whereas the previous variation might cause players to prefer protecting the main bet over pursuing an increased Destiny 21 payoff.

- The casino may offer this bet on the dealer's hand, where the fixed rules for dealer play eliminate the tension between a main bet and the Destiny 21 bet.

- Players may be permitted to make a Destiny 21 bet on another player's hand.

- The Destiny 21 bet may be declared an automatic push if the player splits, doubles down, or surrenders, or may proceed regardless.

Example 5.40. Suppose that the player holds $3\heartsuit\ 2\spadesuit$ against a dealer $6\clubsuit$. Naturally, he hits, drawing the $4\spadesuit$ and guaranteeing at least an even-money return on the Destiny 21 bet. If he hits again and draws the $A\diamondsuit$, the hand may be valued at 10 or 20. Without the side bet, this hand should be called 20 and the player should stand. The better decision for the Destiny 21 bet is to hit again without risk of busting, guaranteeing at least a 4–1 return.

If the 5th card (3rd hit) is the $3\clubsuit$, additional tension between the main bet and the side bet is created. Basic strategy says to stand on 13 against a 6, but in a single-deck game, the probability of busting when drawing a 6th card s only $\frac{20}{47}$, less than 50%. Assuming a \$1 Destiny 21 bet and a rule that a busted hand pays the Destiny 21 bet at its last winning odds, we have an expected value of

$$E = (5) \cdot \frac{27}{47} + (4) \cdot \frac{20}{47} \approx \$4.57,$$

so taking the hit boosts the value of the side bet by over 14%. ∎

If the 6th card is also low, this question recurs, with the goal of the main bet and the goal of the side bet in conflict. That the two bets have opposing optimal strategies is a boon to the casino, as the player must choose between competing goals in making choices. Moreover, these competing goals provide a form of casino insurance against card counting, since the main bet is favored when there is a surplus of high cards and Destiny 21 is favored when low cards are more abundant.

One way for casinos to limit their risk from players is to cap the Destiny 21 bet at some fixed percentage of the main bet. Olaf Vancura, inventor of Destiny 21, recommends a maximum side bet wager of 20% of the player's main bet.

The probability distribution for the length of a player's blackjack hand, assuming basic strategy and independent of card counting, is shown in Table 5.34. These are completed non-busting hands; the probabilities are derived

TABLE 5.34: Distribution of completed blackjack hand lengths

Hand length	Probability
2	.5281
3	.3791
4	.0983
5	.0181
6	.0039
7	.0002

from 1 billion simulated hands. We see immediately that hands with 6 or more hits (8 or more cards) are extremely unlikely, and thus that the top 3 payoffs in Table 5.33 are nearly unattainable.

We can use these data to estimate the HA on Destiny 21.

$$E = (1) \cdot .3791 + (2) \cdot .0983 + (4) \cdot .0181 + (5) \cdot .0039 + (15) \cdot .0002 - .5281$$
$$\approx .1425.$$

The advantage, 14.25%, lies with the player. While it may be surprising to see a bet with a player edge, the casino can afford this risk when pursuing the side bet means increasing the risk of losing the main bet, which is at least 5 times bigger than the Destiny 21 wager.

5.7 Exercises

Solutions begin on page 330.

5.1. In a single-deck game, find the probability of a player receiving a pair of 8s or aces, two hands that should be split regardless of the dealer upcard, off the top of the shoe.

5.2. In a one-player electronic blackjack game where every round is dealt from a freshly-shuffled 8-deck shoe, find the probability that the player and dealer are both dealt naturals.

5.3. Using the infinite deck approximation, find the probability that the dealer is dealt a pair of 10s and is beaten by a player natural.

5.4. At a casino paying only 6–5 on naturals, show that accepting an even-money payoff (the equivalent of insurance if naturals pay 3–2) on a natural dealt from the top of a single deck is the better choice for the player.

5.5. A bartop electronic blackjack game manufactured by IGT is dealt from a double deck, reshuffled before each hand. The game uses Las Vegas Strip rules, including the dealer standing on all 17s, with the following changes:

- A player hand of 6 cards under 21 is an automatic win. (This is probably because the screen can only display 6 player cards.)

- Doubling down and splitting pairs are not offered.

- Player naturals pay even money.

Find the house advantage of this game.

5.6. You have bet $10 on a single-deck blackjack game. You have previously seen 12 cards from the deck: the dealer's ace now showing, four 10-count cards, and seven cards ranking 9 or lower. You have a natural 21, whose cards are counted above among the 12 you have seen. Is a $5 insurance bet a good idea? Explain your answer by computing the expected value of the total bet including insurance, bearing in mind that if the insurance bet wins, your original bet pushes, and if the insurance bet loses, your original bet pays off at 3–2.

Variations

5.7. In a four-deck game of Pick 'Em Blackjack where the dealer hits soft 17, Table 5.35 is the probability distribution for the dealer's final hand.

TABLE 5.35: PDF for the dealer's hand in 4-deck Pick 'Em Blackjack [122]

Hand	Probability
17	.1335
18	.1412
19	.1357
20	.1815
21	.1224
Bust	.2858

Confirm that the HA for each of the three fixed-hand bets is between the values for single-deck and six-deck games where the dealer hits soft 17.

5.8. An early single-deck video blackjack machine manufactured by IGT, "Giant Jackpot 21", offered a progressive jackpot. This jackpot did not require a separate bet, but was only available to players wagering max coins; the game accommodated the jackpot and increased the house advantage by barring pair-splitting and restricting doubling down to hands of 10 or 11. The progressive amount was paid to a player whose hand contained the ace through 5 of the same suit. A sixth card could be included so long as the hand was not busted.

a. A 1985 pamphlet describing electronic blackjack machines recommended pursuing this jackpot by drawing cards as long as winning it was a possibility [25]. This may require rejecting basic strategy recommendations. Give an example of a hand where drawing in pursuit of the progressive jackpot calls for going against basic strategy.

b. Find the probability of a 5-card blackjack hand consisting of a suited ace through 5.

c. Find the probability of a 6-card hand that wins this jackpot.

d. Find the probability of winning this jackpot.

5.9. In 1992, the Frontier Hotel & Gambling Hall in Las Vegas launched a free blackjack promotion. Every time a player was dealt a natural with a $5 or greater wager, she received a metal token. Two of these tokens are shown in Figure 5.3. Once she collected 10 tokens, they could be redeemed for a free

FIGURE 5.3: Frontier Hotel blackjack tokens. Ten tokens could be redeemed for a free weekday night at the hotel. Fifteen were worth a weekend night.

night's stay at the hotel from Sunday through Thursday. Fifteen tokens were worth a free weekend night [31].

The Frontier used Las Vegas Strip rules and permitted doubling after a split. Assume that the game was dealt from a 4-deck shoe. Since the token was awarded on a natural regardless of the size of the wager, assume a $5 bet. Suppose that a room at the Frontier cost $35 during the week and $45 on the weekend. What was the player advantage with the token promotion in force, if the player was playing for

a. A weeknight?

b. A weekend night?

5.10. *2urbo Blackjack* is the California variation at the Hollywood Park Casino in Los Angeles. This 6- to 8-deck joker-free game has a target of 21.99. The top-ranked hand is a pair of 9♠s, which is valued at 21.99 and pays 3–1 [64]. A 9♠ dealt in any other hand retains its value of 9, and hands over 21.99 bust. Find the probability that you are dealt a 21.99 from the top of an 8-deck shoe.

Side Bets

5.11. In 1993, the Las Vegas Hilton (now the Westgate) introduced a particularly player-friendly blackjack side bet, the "50/50 Split" wager. This side bet allowed the player to start a new hand after the deal, and after seeing the dealer's upcard, by standing on his or her original two-card hand and making a second wager of the same amount. The game lasted only 3½ days, and the Hilton lost an estimated $230,000 in that time [60, 133].

Advantage players were responsible for the death of 50/50 Split, as they quickly devised a strategy that gave players a 2% advantage without card counting. Can you find a mathematically sound way to take advantage of this offer?

5.12. Lucky Ladies played against Table 5.24 is not suitable for single-deck blackjack since the top two prizes are impossible to win. Another pay table, Table 5.36, is available for casinos wishing to offer this side bet at single-deck tables.

TABLE 5.36: Lucky Ladies single-deck pay table

Hand	Payoff
Pair of queens with dealer natural	250–1
Pair of queens with no dealer natural	25–1
Ranked 20 (TT, JJ, or KK)	9–1
Suited 20	6–1
Unsuited 20	3–1

Show that the probability of a ranked 20 is smaller than the probability of a suited 20. Note that in a single-deck game, a hand of 20 that is both ranked and suited is impossible.

5.13. How much does the top payoff in single-deck Lucky Ladies contribute to the expected value of the bet when made on the first hand of the deck?

5.14. Suppose that the Over/Under side bet were revised so that both Over and Under bets pushed rather than losing on a dealt hand of 13. What would the new HAs be?

5.15. For a \$21 main blackjack bet, what ratio of non-aces to aces makes the WINSURE bet a fair wager? Assume no aces in the player's hand and a casino using \$2.50 chips.

5.16. *Top of the Deck* is a blackjack side bet that is available only immediately after the shuffle. This bet pays off if the player is dealt a natural at the beginning of a new deck or shoe. Find the probability of winning this bet in a single deck, 2-deck, 4-deck, and 6-deck game, and in the infinite deck approximation.

5.17. The pay table for 21 + 3 offers a 9–1 payoff on all winning outcomes regardless of their relative scarcity. This is convenient for dealers, who simply pay off all winners at the same rate. An electronic 8-deck blackjack game produced by Unity Technologies includes a variation of 21 + 3 called *Straight 2 Flush*. This game pays off on the same 3-card combinations formed in the player's first 2 cards and the dealer's upcard as 21 + 3, but offers different payoffs for different hands, as shown in Table 5.37.

TABLE 5.37: Straight 2 Flush pay table

Hand	Payoff
Straight flush	10–1
Straight	10–1
Flush	8–1
3 of a kind	7–1

Find the house advantage on a Straight 2 Flush bet. Is it a better bet for the player than 21 + 3?

Answers to Selected Exercises

Chapter 2

Exercises begin on page 94.

2.1. The expectation would be

$$E = (36) \cdot \frac{1}{39} + (-1) \cdot \frac{38}{39} = -\frac{2}{39} \approx .0513,$$

and so the HA would be 5.13%—lower than for a \$1 bet on a standard American roulette wheel.

2.2a. $E(z) = -\dfrac{z}{36 + z}$.

2.2b. At $z = 12$, the HA is exactly 25%. (Playing keno is approximately as bad as playing roulette on a wheel with a 000000000000 pocket!)

2.3.

Digit	Winning numbers	Proposed payoff	HA
0	5	5–1	7.89%
1,2	13	7–4	5.92%
3	10	5–2	7.89%
4–6	4	8–1	5.26%
7–9	3	11–1	5.26%

For the digits 1 and 2, paying 9–5 gives a HA of 4.21%.

2.4. The standard payoff remains $\dfrac{24 - n}{n}$ for a bet on n numbers while the probability of winning on this wheel drops from $\dfrac{n}{25}$ to $\dfrac{n}{26}$. The common HA is then 7.69%.

2.5. Suppose that the bet pays off at x to 1. The HA of a \$1 bet would be

$$-\left[(x) \cdot \frac{5}{25} + (-1) \cdot \frac{20}{25} \right] = \frac{20 - 5x}{25}.$$

For integer values of x, the maximum payoff for which this is positive is 3. If the Luxor's pay table paid this bet at 3–1, the HA would be 20%. Paying the bet at 7–2 (3½–1) would result in a 10% HA. At 4–1, this would be a fair bet.

2.6. $\dfrac{48 - n}{n}$ to 1. 4.00%.

2.7. 2.04%.

2.8. 6.25%.

2.9a. 2.633×10^{-5}.

2.9b. .0203.

2.10a. 1.37%.

2.10b. 1.37%.

2.11.

Numbers covered	Proposed payoff	HA
1	15–1	11.11%
2	7–1	11.11%
3	4–1	16.67%
4	3–1	11,11%
8	1–1	11.11%

2.12. 4.72%.

2.13.

Exercise	Win probability	HA
a.	$\dfrac{12}{37}$	2.70%
b.	$\dfrac{7}{37}$	2.70%
c.	$\dfrac{8}{37}$	2.70%
d.	$\dfrac{17}{37}$	2.70%
e.	$\dfrac{5}{37}$	2.70%

2.14. The 3-number bet pays 32–1, not the $32\frac{1}{3}$–1 that the formula delivers. Its HA is 5.71%.

2.15a. 5.45%.

2.15b. 12.72%.

2.15c. 7.27%.

2.15e. 4.29%.

2.16. .6977.

2.17. −$.1966.

2.18a. $\dfrac{52 - n}{n}$ to 1; 5.45%.

2.18b. 7.27%.

2.18c. A 16–1 payoff gives a house edge of 8.93%. At 17–1, the HA is 3.57%.

2.19b. .0018.

2.20. 8.89%.

2.21. There are 10 red odd numbers and 10 black even numbers. A Newar bet in either configuration gives a HA of –16.22%—which gives the gambler a huge edge and explains why these bets are not on offer.

2.22. 414–1.

2.23. The HA is 7.13%, very slightly less than the HA with the original pay table.

2.24. The probability of no repeated numbers is simply the probability that all 16 numbers are different:

$$P(\text{No repeats}) = \frac{38}{38} \cdot \frac{37}{38} \cdot \frac{36}{38} \cdot \ldots \cdot \frac{23}{38} = \frac{_{38}P_{16}}{38^{16}} \approx .0246.$$

This is a pretty good indication that the roulette results board does very little to identify so-called "hot" numbers—the vast majority of the time, there will be at least one number that appears to be "hot" in the sense of appearing more than once in the last 16 spins.

2.25. Tracking the wheel as you read through Caro's system will reveal that *all* 38 of the numbers are excluded from betting. If you follow this system, you will make no roulette bets, therefore the house advantage is 0%.

2.26. Net loss: –$1. The next bet will be for $3.

Chapter 3

Exercises begin on page 184.

3.1. .0384.

3.2. .0514.

3.4. 9.09%.

3.5a. 4–5 odds.

3.5b. 16.67%.

3.5c. 8.33%.

3.5d. $\dfrac{4}{18} \approx .2222.$

3.6. 2.27%.

3.8. 8–1 on 4 and 10; 10–1 on 6 and 8.

3.9. $p_3(k,m) = \dfrac{K}{K+M}.$

3.10. Yes. $P(9) + P(11) = \dfrac{4}{36} + \dfrac{2}{36} = \dfrac{6}{36} = P(7).$

3.11. 18.69%.

3.12. Since the highest sum with these 2 dice is 9, any wager that gives the gambler a number greater than 9 and the hustler a number at most 9 cannot win for the gambler. One plausible-sounding bet is "Even money if he rolls a 10 before a 4," since the sums of 4 and 10 are equally likely on 2 standard d6.

3.13. 9.09%.

3.15. 42.86%.

3.16. The probability of a player win is .5269; the player holds a 5.38% advantage.

3.17. 0.677%.

3.18. 0.83%.

3.19. 16 × .

3.20. 8.33%.

3.21. Using the analog of standard craps with no barred numbers, Pass wins on an initial 6 or 9 and loses on an initial 2, 3, or 10. Don't Pass wins on an initial 2, 3, or 10 and loses on an initial 6 or 9. Possible point numbers are 4, 5, 7, and 8. These rules give

$$P(\text{Pass wins}) = \frac{461}{900} > .5.$$

$$P(\text{Don't Pass wins}) = \frac{439}{900}.$$

To give the casino an edge, some come-out roll or rolls must be barred on the Pass line wager. An easy way to restore the edge is to bar a come-out 3–3 on the Pass line, while all other sums of 6 win for the Pass line. The resulting house edges are 5.56% on Pass and 2.44% on Don't Pass. (Paying 11–10 on a winning Pass line bet would reduce the HA to .83%, which might make this more palatable to gamblers.)

3.22. A bet paying x to 1 has a HA of $\dfrac{201 - 15x}{216}$; paying at 13–1 gives a 2.78% HA. 12–1 payoff odds raise the HA to 9.72%.

3.23. 10–1.

3.24a. $\frac{1}{7}$.

3.24b. $\frac{1}{10}$.

3.24c. $\frac{5}{14}$.

3.24d. $\frac{1}{4}$.

3.24e. $\frac{1}{4}$.

3.24f. $\frac{1}{10}$.

3.26. $\frac{1}{630}$.

3.27. $\frac{6}{7}$.

3.28. .021096.

3.30. 11.11%.

3.31a. .1127.

3.31b. 9.88%.

3.31c. 8.80%

3.32a. $P(k) = \left(\dfrac{8}{36}\right)^k \cdot \dfrac{28}{36}$; $4 \leqslant k \leqslant 6$. $P(7) = \left(\dfrac{8}{36}\right)^7$.

3.32b. $27,083.

3.33. 5.56%.

3.34. 16.67%.

3.35. 28.13%.

Chapter 4

Exercises begin on page 232.

4.1. Player: 1.23%. Banker: 1.06%.

4.2. Player: .02836%. Banker: .0337%. Tie: .275%.

4.3. $P(\text{Tie}) = \frac{1}{13}$. Note that the 10, jack, queen, and king are not equal in this game.

4.4. 3.85%.

4.5. 30.77%.

4.6. $P(3 \text{ Immortals}) = \dfrac{\binom{90}{3}}{\binom{390}{3}} \approx .0120.$

4.7. The infinite deck approximation gives .0123 for this probability, which is only 2.63% higher than the value computed in Exercise 4.6.

4.9. $P(n) = \dfrac{\binom{5n}{4}}{\binom{65n}{4}} = \dfrac{(5n-1)(5n-2)(5n-3)}{13(65n-1)(65n-2)(65n-3)}.$

As $n \to \infty, P(n) \to \dfrac{1}{28,561}.$

4.10. 12.28% on Player; 15.24% on Banker.

4.11. 1.88%.

4.12a. $(.0955)^3 \approx 8.710 \times 10^{-4} \approx \dfrac{1}{1148.1}.$

4.12b. 12.81%.

4.12c. 1101–1.

4.13a. .0225.

4.13b. 7.61%.

4.14a. .0345.

4.14b. 10.19%.

4.15a. 4.439×10^{-5}.

4.15b. 1.280×10^{-4}.

4.15c. 1.335×10^{-4}.

4.16 .0018. This is much higher than $\dfrac{1}{6000}$, indicating a significant player edge.

4.17. .0073.

4.18a. $\dfrac{1}{2197}.$

4.18b. $\dfrac{6}{2197}.$

4.18c. $\dfrac{1}{2197}.$

4.18d. $\dfrac{21}{2197}.$

4.19a. .0049.

4.19b. .0269.

4.19c. .0128.

4.20. The best bets for both Banker and Player are on 0. The house advantage on the Banker bet is 7.66%; on Player, the HA is 7.83%.

4.21. 6.63%.

4.22. P(Suited KQ) $\approx$ .0030. P(Other suited hand) $\approx$ .2452.

4.23. A bet on a tie at 3 has the highest HA: 6.95%.

4.24. Rabbit: $+10$. Tiger: $+2$. Monkey: $+10$. Zoo: $+12$. All of the bets except Tiger are favored.

4.25. The count is -8, so the banker bet is favored over Player, though neither bet has an edge over the house.

Chapter 5

Exercises begin on page 320.

5.1. .0090.

5.2. .0089 $\approx \dfrac{1}{113}$.

5.3. .0022.

5.4. The return if you accept even money is $5. The expected value of a $5 wager without insurance is $4.65. If a $2.50 insurance bet is taken, the expected value of the combined bet is $3.84. Taking the even-money payoff is the best option.

5.5. 4.89%.

5.6. If you make the insurance bet, then your expected win is $10, win or lose. If you forego insurance, the probability that the dealer has a natural is $\frac{12}{40}$, and the expected value of your total bet is then $10.50. Your expectation is higher if you do not make the insurance bet.

5.8a. One such hand—an extreme one—would be A235$\heartsuit$. Pursuing the progressive jackpot would call for drawing to this soft 21 hand in the hope of turning it into a hard 15 with the 4$\heartsuit$. No version of basic strategy directs a player to hit a soft 21 in a standard blackjack game.

If you're counting cards, especially at a game where naturals pay only 6–5, doubling down on a natural 21 against a highly positive deck, if allowed, may be advisable, as would doubling down on a natural in tournament blackjack play in an effort to win more money near the end of a tournament round, but these situations are outside the reach of basic strategy.

5.8b. $\dfrac{1}{649,740}$.

5.8c. 2.924×10^{-5}.

5.8d. $\dfrac{1}{32487} \approx 3.078 \times 10^{-5}$.

5.9a. 2.45%.

5.9b. 1.97%.

5.10. 3.244×10^{-4}.

5.11. To make the best use of this side bet, start a second hand whenever you hold 12–16 and the dealer's upcard is a 2–6. Basic strategy dictates that you only risk busting against a 2–6 with a 12 against a dealer 2 or 3, so you would be standing much of the time anyway.

Standing on low totals against a 2–6 is done in the hopes that the dealer will bust, which means that this strategy will give you a second hand against a relatively weak dealer hand.

5.12.

$$P(\text{Ranked }20) = \frac{3 \cdot \binom{4}{2}}{\binom{52}{2}} = \frac{18}{1326}.$$

$$P(\text{Suited }20) = \frac{4 \cdot \binom{4}{2} + 4}{\binom{52}{2}} = \frac{28}{1326}.$$

There are 10 more ways to draw a suited 20 (including suited A/9) than a ranked 20, so suited 20 has a greater probability.

5.13. The probability of a pair of queens with a dealer natural is

$$\frac{4}{52} \cdot \frac{3}{51} \cdot \frac{4}{50} \cdot \frac{10}{49} \approx 7.388 \times 10^{-5}.$$

Multiplying by the $250 payoff gives a contribution of 1.85¢.

5.14. The HA on Under is 1.78%. For Over, the *player* has the same 1.78% advantage.

5.15. 9.2:1.

5.16. In an n-deck game, we have

$$P(\text{Natural}) = 2 \cdot \frac{4n}{52n} \cdot \frac{16n}{52n-1} = \frac{128}{2704n-52}.$$

5.17. The HA is 6.18% for this 8-deck game, so 21 + 3 offers a better gamble.

Number of decks	P(Natural)
1	.0483
2	.0478
4	.0476
6	.0475

In the infinite deck approximation, the probability of a natural approaches $\frac{8}{169} \approx .0473$.

References

[1] "An Adept", *A Grand Exposé of the Science of Gambling*. Frederic A. Brady, Publisher, New York, 1860.

[2] Alphabet Roulette Wagering. Online at http://alphabetroulette.com/index.php/products/alphabetroulette-wheel-version/wager. Accessed 24 June 2014.

[3] Baldwin, Roger R., Wilbert E. Cantey, Herbert Maisel, and James P. McDermott, The Optimum Strategy In Blackjack. *Journal of the American Statistical Association* **51**, September 1956, pp. 429–439.

[4] Barlowe, Katie, Blackjack Too Mathematically Challenging? New Format Simplifies Game, Casino.org, 15 August 2017. Online at https://www.casino.org/news/blackjack-too-mathematically-challenging-new-format-simplifies-game/. Accessed 17 August 2020.

[5] Benjamin, Arthur T., Michael Lauzon, and Christopher Moore, Why the Player Never Wins in the Long Run at LA Blackjack. *The UMAP Journal* **20** #2, 1999, pp. 127–138.

[6] Bennett, Barbara J., *Randomness*. Harvard University Press, Cambridge, MA, 1998.

[7] Boardwalker, Not So Progressive 21...(yawn). *Blackjack Forum* **XIII** #3, September 1993, p. 42.

[8] Bollman, Mark, *Basic Gambling Mathematics: The Numbers Behind The Neon*. CRC Press/Taylor & Francis, Boca Raton, FL, 2014.

[9] Bollman, Mark, *Mathematics of Casino Carnival Games*. CRC Press/Taylor & Francis, Boca Raton, FL, 2020.

[10] The Bone Man, New Craps Fire Bet Emerging in Las Vegas. Online at http://www.nextshooter.com/firebet. Accessed 23 June 2012.

[11] Bonus Baccarat Zero Mondays. Online at http://www.grandpoker.eu/Info/BaccaratZero.html. Accessed 231 July 2020.

[12] Brahms, John, Blackjack in Asia. *Blackjack Forum* **IX** #3, September 1989, pp. 9–13.

[13] Brokopp, John, Gaming Company Makes Roulette More Colorful. Online at http://brokopp.casinocitytimes.com/article/gaming-company-makes-roulette-more-colorful-58988. Accessed 21 May 2012.

[14] Buro, Chris, Atlantic City Update. *Blackjack Forum* **XVI** #4, Winter 1996, pp. 49–52.

[15] Card Deck and Method of Playing Card Games With Same, United States Patent Application Publications, # US 2014/0138915 A1. Online at http://www.google.com/patents/US20140138915. Accessed 2 May 2015.

[16] Carlson, Bryce, Why Casino Craps Can't Be Beaten. *Las Vegas Advisor*, 26 October 2020. Online at https://www.lasvegasadvisor.com/gambling-with-an-edge/why-casino-craps-cant-be-beaten/. Accessed 4 November 2020.

[17] Caro, Mike. Games You Can Beat and Games You Can't. Online at http://www.gamblingtimes.com/writers/mcaro/mcaro_winter2001.html. Accessed 15 July 2015.

[18] Cash Card Games, LLC. Online at http://www.cashcardgames.com. Accesse 12 June 2020.

[19] Cash, Les, More Indian Reservation Problems. *Blackjack Forum* **XVII** #4, Winter 1997, pp. 30–32.

[20] Catlin, Donald E., A Really Hard Hardway Bet, in *Finding the Edge*, Olaf Vancura, Judy A. Cornelius, and William R. Eadington, editors. Institute for the Study of Gambling and Commercial Gaming, Reno, NV, 2000, pp. 297–302.

[21] Ceglasvegas1 / CC BY-SA (https://creativecommons.org/licenses/by-sa/4.0), File:Dealer school las vegas3.png. Online at https://commons.wikimedia.org/wiki/File:Dealer_school_las_vegas3.png. Accessed 27 July 2020.

[22] Chambliss, Carlson R., The Eighth International Gambling Conference. *Blackjack Forum* **X** #4, December 1990, pp. 11–14.

[23] The Chickasaw Nation Division of Commerce Game Rules for Card Craps. Online at http://www.winstarworldcasino.com/wp-content/uploads/2015/07/Card-Based-Craps-Official-Rules.pdf. Accessed 14 February 2018.

[24] Choctaw Casinos & Resorts, Craps. Online at https://www.choctawcasinos.com/choctaw-durant/game/table-games/craps/. Accessed 14 February 2018.

[25] Cohen, R. Carl, *Beating The Blackjack Slot Machines*, R. Carl Cohen, Publisher, Philadelphia, 1985.

[26] Craps Game With A Repeated Number Based Wagering Area, United States Patent Application Publications, # US 2014/0138911 A1, 22 May 2014. Online at https://patents.google.com/patent/US20140138911. Accessed 7 April 2018.

[27] Craps layout.png. Online at http://en.wikipedia.org/wiki/File: Crapslayout.png. Accessed 15 May 2013.

[28] Curtis, Anthony, Las Vegas Advisor. *Blackjack Forum* **IX** #4, December 1989, pp. 29–31.

[29] Curtis, Anthony, Las Vegas Advisor. *Blackjack Forum* **XI** #1, March 1991, pp. 22–24, 29.

[30] Curtis, Anthony, Las Vegas Advisor. *Blackjack Forum* **XI** #4, December 1991, pp. 32–34.

[31] Curtis, Anthony, Las Vegas Advisor. *Blackjack Forum* **XII** #2, June 1992, pp. 35–37

[32] Curtis, Anthony, Las Vegas Advisor. *Blackjack Forum* **XII** #4, December 1992, pp. 27–28, 33.

[33] Curtis, Anthony, The Gambling Conference, or, How I Finally Found Positive Expectation in Atlantic City. *Blackjack Forum* **V** #1, March 1985, pp. 9–18.

[34] Dace, Mark, Around the World: Peru. *Blackjack Forum* **XXI** #2, Sumemr 2001, p. 70.

[35] Devlin, Keith, *The Unfinished Game*. Basic Books, New York, 2008.

[36] Dice–The Soul of Honesty. *Casino & Sports* **22**, 1982, pp. 26–55

[37] Double Bonus Spin Roulette. Online at http://www.roulettediary.com/double-bonus-spins-roulette.html. Accessed 16 March 2015.

[38] Double Bonus Spin Roulette—Player's Suite, IGT Interactive (United States). Online at http://www.igt.com/us-en/interactive/game-page.aspx?type_id=51. Accessed 7 March 2015.

[39] Empire Global Gaming, Inc. Launches a New 50 Card Deck. Online at http://empireglobalgaminginc.com/empire-global-gaming-inc-launches-a-new-50-card-deck/. Accessed 3 May 2015.

[40] Epstein, Richard A., *The Theory of Gambling and Statistical Logic*. Oxford, Academic Press, 1977.

[41] Ethier, Stewart, *The Doctrine of Chances: Probabilistic Aspects of Gambling*. Berlin, Springer-Verlag, 2010.

[42] Famous Betting Systems: Fibonacci Series. *Casino & Sports* **1**, pp. 6–41, 1977.

[43] Foster, R.F., *Hoyle's Games: Autograph Edition*. New York, A.L. Burt Company, 1926.

[44] File: European Roulette Wheel.png. Online at https://commons.wikimedia.org/wiki/File:European_Roulette_wheel.png. Accessed 12 September 2015.

[45] Film8ker, American roulette table layout.gif. Online at http://commons.wikimedia.org/wiki/File%3AAmerican_roulette_table_layout.gif. Accessed 16 May 2013.

[46] Film8ker, American roulette wheel layout.gif. Online at http://commons.wikimedia.org/wiki/File:American_roulette_wheel_layout.gif. Accessed 26 July 2013.

[47] Flexedge Gaming, LLC., Supreme Baccarat. Online at https://oag.ca.gov/sites/all/files/agweb/pdfs/gambling/Aviator.pdf. Accessed 14 August 2014.

[48] Flip Card Blackjack. Online at http://www.bigbetgaming.com/flip-card.html. Accessed 1 May 2015.

[49] Flip Card Black Jack side bet—Rules & Procedures. Online at http://www.inag11.com/#!flip-card-betting-spot/cysh. Accessed 30 July 2015.

[50] Flush Roulette®. Online at http://www.newtablegames.com/flush-roulettereg.html. Accessed 21 February 2018.

[51] Friedman, Stacey, Hard Pass Craps Wager, United States Patent # US 8,118,309 B1, 21 February 2012. Online at https://patents.google.com/patent/US8118309. Accessed 4 May 2020.

[52] Gambler's Book Club staff, $100,000 Club. *Systems & Methods* **5**, 1975, pp. 59–60.

[53] Gambling Commission (Great Britain), Rules of casino games in Great Britain, August 2007. Online at https://docplayer.net/3281127-Contents-1-introduction-3-2-general-rules-5-3-roulette-7-4-blackjack-16-5-three-card-poker-25-6-   punto-banco-29-7-casino-stud-poker-38-8.html#show_full_text. Accessed 5 May 2020.

[54] Garry, Around the World: Russia. *Blackjack Forum* **XXI** #1, Spring 2001, pp. 67–70.

[55] Great 8 Baccarat—Rules. Online at http://www.great8baccarat.com/#!rules/cuxg. Accessed 4 September 2015.

[56] Griffin, Peter, *The Theory of Blackjack*, 6th edition. Huntington Press, Las Vegas, 1999.

[57] Griffin, Tina, Letter to Ashford Kneitel, Ashford Gaming LLC, 24 April 2014. Online at https://fortress.wa.gov/wsgc/etransfer/ OnlineServices/activities/ViewGameRule.cshtml?gid=9. Accessed 2 March 2019.

[58] Group of eight glass astragali (knucklebones). Online at https://commons.wikimedia.org/wiki/File:Group_of_eight_glass_ astragali_(knucklebones)_MET_DP121111.jpg. Accessed 29 August 2020.

[59] Hadsall, Joe, No dice: Casino invents version of craps played with cards. *The Joplin Globe*, Joplin, MO, 16 April 2010.

[60] Hannum, Robert C., A Primer on the Mathematics of Blackjack, in *The Oxford Handbook of the Economics of Gambling*, Williams, Leighton Vaughan and Donald S. Siegel, editors. Oxford University Press, Oxford, 2013, pp. 370–386.

[61] Hannum, Robert C., and Anthony N. Cabot, *Practical Casino Math*, 2nd edition. Institute for the Study of Gambling and Commercial Gaming, Reno, NV, 2005.

[62] Hawaiian Gardens Casino, Hawaiian Blackjack, in *Gaming Activity Report—As Of December 28, 2004*. Online at https://oag.ca.gov/sites/all/files/agweb/pdfs/gambling/hawaiian-gardens.pdf. Accessed 28 May 2020.

[63] Holloway, Louis G, Trave;ling gamblers: Stick to Nevada. *Systems & Methods* **10**, 1976, p. 43.

[64] Hollywood Park Casino, 2urbo Blackjack. Online at https://oag.ca.gov/sites/all/files/agweb/pdfs/gambling/ hollywood_park.pdf. Accessed 18 June 2020.

[65] Hunter, Samuel C., Craps game with progressive jackpot. United States Patent Application Publication, # US 2004/0130094 A1, 8 July 2004. Online at https://patents.google.com/patent/US20040130094. Accessed 7 May 2020.

[66] Jacobs, Kim, California No Bust Blackjack: A Variation of Blackjack Created for Card Rooms. *Cardroom Insider*, 7 October 2015. Online at https://cardroominsider.wordpress.com/2015/10/07/california-blackjack-blackjack-variation-created-for-card-rooms/. Accessed 24 May 2020.

[67] Jacobson, Eliot, *Advanced Advantage Play*. Blue Point Books, Santa Barbara, CA, 2015.

[68] Jacobson, Eliot, Baccarat Combinatorial Analysis The Easy Way. Online at https://www.888casino.com/blog/baccarat-tips/baccarat-combinatorial-analysis-the-easy-way. Accessed 15 May 2018.

[69] Jacobson, Eliot, Card Counting the Natural 9 and Natural 8 Baccarat Side Bets. 11 August 2013. Online at https://www.888casino.com/blog/side-bets/card-counting-the-natural-9-and-natural-8-baccarat-side-bets. Accessed 5 August 2020.

[70] Krigman, Alan, Variations on Craps: Are they Fixes for What Ain't Broke?. *Casino City Times*, 8 March 2006. Online at http://krigman.casinocitytimes.com/article/variations-on-craps-are-they-fixes-for-what-aint-broke-25070. Accessed 7 May 2020.

[71] The Latest News From The Great 8 Team. Online at http://www.great8baccarat.com/#!latest-news/cjp8. Accessed 5 September 2015.

[72] Lightning Launch Roulette—Scientific Games. *Casino Journal*, 3 May 2018. Online at https://www.casinojournal.com/articles/92082-lightning-launch-roulette—scientific-games. Accessed 11 June 2018.

[73] Limey Craps. *Systems & Methods* **14**, p. 37, 1976.

[74] Lowery, Jerrery Roy, Craps game with jackpot. United States Patent Application Publication, # US 2008/0128990 A1, 5 June 2008. Online at https://patents.google.com/patent/US20080128990. Accessed 7 May 2020.

[75] Lubin, Dan, *The Essentials of Casino Game Design*. Huntington Press, Las Vegas, 2016.

[76] Lucky 13s Blackjack. Online at http://lucky13s.com.au/. Accessed 6 May 2014.

[77] Lucky 13s Blackjack Rack Card. Online at http://lucky13s.com.au/ETG-Brochure.pdf. Accessed 6 May 2014.

[78] Malmuth, Mason, *Blackjack Essays*. Two Plus Two Publishing, Las Vegas, 2000.

[79] Marina Bay Sands Casino, Singapore Table Games. Online at https://www.sandscasino.com/singapore/games/table-games.html. Accessed 21 June 2020.

[80] May, John, Card-counting at Baccarat. *Casino City Times*, 9 September 1999. Online at http://may.casinocitytimes.com/article/card-counting-at-baccarat-1149. Accessed 30 July 2020.

[81] Meltzer, Marc, New Blackjack Variant: Jack Jack. Online at https://www.blackjackonline.com/857/new-blackjack-variant-jack-jack/. Accessed 7 April 2018.

[82] Mini Roulette by NetEnt. Online at http://www.casinogamespro.com/play-casino-games/mini-roulette-netent. Accessed 9 May 2018.

[83] Nehrt, Philip, *Winning Gambling Strategies*. AuthorHouse, Bloomington, IN, 2012.

[84] Nelson, Rex, California Aces. *WIN Magazine* **14** #1, June 1992, pp. 36–37, 57.

[85] Nevada OKs craps-style table game. *Gaming Today*, 22 April 2009. Online at https://www.gamingtoday.com/casino_games/table_games/article/21067-Nevada_OKs_craps_style_table_game. Accessed 28 Auugust 2020.

[86] New betting option livens up craps, *Gaming Today*, 14 October 2008. Online at https://www.gamingtoday.com/casino_games/table_games/article/18459-New_betting_option_livens_up_craps. Accessed 16 March 2018.

[87] Norseman, The Ol', Blackjack in Denmark. *Blackjack Forum* **XII** #2, June 1992, p. 34.

[88] Ohio Lottery, The Lucky One. Online at https://www.ohiolottery.com/Games/The-Lucky-One#1. Accessed 17 April 2018.

[89] Paulsen, Dennis, Just How Much of an Edge Does 'Crap-less' Craps Offer?. *Casino & Sports* **16**, 1981, pp. 42–43.

[90] Paulsen, Dennis, A Theoretical & Computer Analysis of Vegas World's Innovative 'No Lose Free Roll' Craps. *Casino & Sports* **26**, pp. 71–78.

[91] Pentagon Craps. *WIN Magazine*, **13**, #12, May 1992, pp. 8–9.

[92] Pontoon Pandemonium (MBS) Game Rules, Version 4, 4 November 2016. Online at https://www.cra.gov.sg/docs/default-source/game-rule-documents/mbs/pontoon-pandemonium-mbs-game-rules-version-4.pdf. Accessed 21 June 2018.

[93] Prime Time (Roulette side bet). Online at
https://gaming.nv.gov/Modules/ShowDocument.aspx?documentid=8997.
Accessed 12 Ocotber 2018.

[94] Progressive Roulette, United States Patent # US 6,776,714 B2. Online
at http://www.google.com/patents/US6776714. Accessed 17 July 2015.

[95] Rainbow Roulette side bet—Procedures. Online at
http://www.inag11.com/#!rainbow-roulette-sidebet/c1zfr. Accessed 1
May 2015.

[96] Random Gambling Playing Pieces and Layout and Game Table for Use
With the Same, United States Patent # US 4,900,034, 13 February 1990.
Online at https://patents.google.com/patent/US4900034. Accessed 19
February 2018.

[97] Resorts World Genting Gaming Guide: Three Pictures. Online at
https://www.rwgenting.com/casino/Gaming-Guide/Three-Pictures/.
Acessed 27 July 2018.

[98] Richardson, Joey, Is Baccarat Card Counting Worth the Effort?
Gamblingsites.net, 15 August 2018. Online at
https://www.gamblingsites.net/blog/is-baccarat-card-counting-worth-
the-effort/. Accessed 31 July 2020.

[99] Richfield, Rebecca, How the Free Ride Really Died. *Blackjack Forum*
XVI #4, Winter 1996, pp. 30–32.

[100] "Riverboat Roulette" to be rebranded as "Rainbow Bet Roulette".
Online   at   http://doubleluckgaming.com/riverboat-roulette-to-be-
rebranded-as-rainbow-bet-roulette/. Accessed 1 May 2015.

[101] Riverwind Casino, How to Play Card-Based Roulette at Riverwind
Casino. Online at https://www.youtube.com/watch?v=MTO3-yncZ_4.
Accessed 28 February 2018.

[102] Roberts, Stanley *et al. The Gambling Times Guide to Blackjack.* Gambling Times Inc., Fort Lee, NJ, 2000.

[103] Roberts, Stanley, Roberts' Rules. *WIN Magazine* **11**, #11, June 1990,
pp. 10–11.

[104] Roberts, Stanley, Roberts' Rules. *WIN Magazine* **11**, #12, July 1990,
pp. 10–11, 68.

[105] Roberts, Stanley, Roberts' Rules. *WIN Magazine* **12**, #1, August 1990,
pp. 10–11, 62.

[106] Roulette Dinner, *Casino & Sports* **2**, p. 30, 1978.

[107] Roulette Evolution™. Online at http://www.igt.com/us-en/games/game-page.aspx?type_id=5075&showtab=1. Accessed 12 September 2015.

[108] Roulette Royale—Progressive Jackpot. Online at http://www.32red.com/games/roulette-royal---progressive-jackpot.html. Accessed 23 August 2015.

[109] Roulette with Multiple Color and Number Variations for American and Euro Tables. Online at http://empireglobalgaminginc.com/roulette/. Accessed 2 May 2015.

[110] Royal Roulette Pty Ltd, *Royal Roulette*. Online at http://www.royalroulette.com. Accessed 28 May 2012.

[111] Sands Casino, Singapore Table Games. Online at https://www.sandscasino.com/singapore/games/table-games.html. Accessed 15 August 2020.

[112] Scarne, John, *Scarne On Card Games*. Dover Publications, Inc., Mineola, NY, 2004 reprint of 1965 revision.

[113] Scarne, John, *Scarne On Dice*, 8th revised edition. Crown Publishers, Inc., New York, 1980.

[114] Scarne, John, *Scarne's Complete Guide to Gambling*. Simon and Schuster, New York, 1961.

[115] Schlesinger, Don, The Gospel According To Don. *Blackjack Forum* **XI** #3, September 1991, pp. 19–21.

[116] Schlesinger, Don, *Blackjack Attack*, 2nd edition. RGE Publishing, Oakland, CA, 2000.

[117] Scoblete, Frank, *I Am A Dice Controller: Inside The World Of Advantage-Play Craps*. Triumph Books, Chicago, 2015.

[118] 2nd Official Atlantic City Craps Layout. *Casino & Sports* **4**, pp. 32–33.

[119] Shackleford, Michael, Baccarat—The Number of Combinations of Each Possible Hand. Online at https://wizardofodds.com/games/baccarat/appendix/1/. Accessed 1 August 2020.

[120] Shackleford, Michael, Bonus Craps. Online at https://wizardofodds.com/games/craps/side-bets/bonus-craps/. Accessed 21 May 2020.

[121] Shackleford, Michael, Craps Side Bets. Online at http://wizardofodds.com/games/craps/appendix/5/#firebet. Accessed 23 June 2012.

[122] Shackleford, Michael, Dealer probabilities in blackjack. Online at http://wizardofodds.com/games/blackjack/appendix/2b/. Accessed 6 April 2015.

[123] Shackleford, Michael, Gambling in the United Kingdom. Online at https://wizardofodds.com/blog/gambling-united-kingdom/. Accessed 16 February 2018.

[124] Shackleford, Michael, Lucky Ladies. Online at https://wizardofodds.com/games/blackjack/side-bets/lucky-ladies/. Accessed 17 August 2020.

[125] Shackleford, Michael, Raving Table Games Show 2014. Online at http://wizardofodds.com/blog/raving-2014/. Accessed 6 April 2015.

[126] Shackleford, Michael, Roul 8: Roulette Side Bet. Online at https://wizardofodds.com/games/roulette/side-bets/roul-8/, 26 July 2017. Accessed 12 June 2018.

[127] Shackleford, Michael, Touchdown Roulette. Online at https://wizardofodds.com/games/touchdown-roulette/. Accessed 13 August 2020.

[128] Smallwood, Jake, Blackjack in New York. *Blackjack Forum* **IX** #2, June 1989, pp. 16–17.

[129] Smart, R.J., *Playing Roulette as a Business*. Carol Publishing Group, Secaucus, NJ, 1996.

[130] Smith, John, *No Limit: The Rise and Fall of Bob Stupak and Las Vegas' Stratosphere Tower*. Huntington Press, Las Vegas, 1997.

[131] Snyder, Arnold, *The Big Book of Blackjack*. Cardoza Publishing, New York, 2006.

[132] Snyder, Arnold, Inside the Over/Under. *Blackjack Forum* **IX**, #3, September 1989, p. 5.

[133] Snyder, Arnold, Anthony Curtis, and The Flower, LV Hilton Throws A Party!. *Blackjack Forum* **XIII** #2, June 1993, pp. 70–71.

[134] Snyder, Arnold, Response to "Letter from Utah" (letter to the editor). *Blckjack Forum* **VII**, #1, March 1987, pp. 47–48.

[135] Snyder, Arnold, Surprise Party at the Klondike Hotel. *Blackjack Forum* **XVI** #3, Fall 1996, pp. 9–17.

[136] Snyder, Arnold, There's Gold in Them Thar Hills!. *Blackjack Forum* **XX**, #4, Winter 2000, pp. 7–14.

[137] Snyder, Arnold, The Unbalanced Over/Under. *Blackjack Forum* **XI**, #1, March 1991, pp. 7–11.

[138] Soares, John, *Loaded Dice: The True Story of a Casino Cheat.* Taylor Publishing Company, Dallas, 1985.

[139] Socks, Red, Around the States: Connecticut. *Blackjack Forum* **XVII**, #4, Winter 1997, pp. 61–62.

[140] StateImpact Oklahoma, Tribal Gaming: Net Funds From Gambling Fun. Online at https://stateimpact.npr.org/oklahoma/tag/tribal-gaming/. Accessed 14 February 2018.

[141] Super 31 Blackjack. Online at http://super31bj.com/. Accessed 14 June 2015.

[142] Three Way Roulette, Ltd., 55 Card Version. Online at http://threewayroulette.com/55-card-version/. Accessed 23 July 2018.

[143] Three Way Roulette, Ltd., About Three Way Roulette. Online at http://threewayroulette.com/about/. Accessed 23 July 2018.

[144] Thorp, Edward O., *Beat the Dealer: A Winning Strategy for the Game of Twenty-One.* Blaisdell Publishing Co., New York, 1962.

[145] Thorp, Edward O., *Beat the Dealer: A Winning Strategy for the Game of Twenty-One*, second edition. Vintage Books, New York, 1966.

[146] Thorp, Edward O., The "Free Ride" Optimal Strategy. *Blackjack Forum* **XVI** #4, 1996, pp. 11–18.

[147] Thorp, Edward O., and William Walden, *A Favorable Side Bet in Nevada Baccarat.* Journal of the American Statistical Association **61**, #314, Part 1 (June 1966), p. 313–28.

[148] Tilton, Nathaniel, *The Blackjack Life.* Huntington Press, Las Vegas, 2012.

[149] Today's Gambing Myth: The Monte Carlo Fallacy. Online at https://www.caesars.com/casino-gaming-blog/latest-posts/table-games/roulette/gambling-myth-monte-carlo-fallacy#.X8lEJrP_rb0. Accessed 3 December 2020.

[150] Touchdown Roulette—Interblock USA, *Casino Journal*, 13 June 2013. Online at https://www.casinojournal.com/articles/88313-touchdown-rouletteinterblock-usa. Accessed 13 August 2013.

[151] Turn & Burn Craps. Online at http://www.turnandburncraps.net. Accessed 23 July 2015.

[152] UK Casino Table Games: Lucky Ball Roulette—Side Bet. Online at http://www.ukcasinotablegames.info/luckyballroulette.html. Accessed 16 February 2018.

[153] Urbanphotos, Own work, CC BY-SA 3.0. Online at https://commons.wikimedia.org/w/index.php?curid=6530969. Accessed 27 June 2020.

[154] Vancura, Olaf, Method of playing a casino blackjack side wager. United States Patent #5,673,917, 7 October 1997. Online at https://patents.google.com/patent/US5673917. Accessed 25 May 2020.

[155] Vegas Aces, How to Play Muggsy's Corner. Online at https://www.youtube.com/watch?v=tFa5lUxmMzE. Accesses 23 October 2020.

[156] Villano, Matt, California-style card game. *SFGate*, 8 February 2007. Online at https://www.sfgate.com/entertainment/gaming/article/ California-style-card-game-Blackjack-1873-2618922.php. Accessed 24 May 2020.

[157] Walker, Bert, Review of *The Guaranteed Roulette System By Howard P. Johnson*, by Wolfgang Bartschelly. *Casino & Sports* **23**, 1983, pp. 48–49.

[158] Wechsberg, Joseph, Blackjack Pete. *Collier's*, 25 July 1953, pp. 34–37.

[159] Win Systems launches Chinese roulette. Online at https://calvinayre.com/2018/02/01/press-releases/win-systems-launches-revolutionary-chinese-roulette/. Accessed 2 Februray 2018.

[160] Wong, Stanford, *Basic Blackjack*. Pi Yee Press, Las Vegas, 1995.

[161] Wong, Stanford, *Blackjack Secrets*. Pi Yee Press, Las Vegas, 1993.

[162] Wong, Stanford, *Professional Blackjack*. William Morrow & Co., New York, 1981.

[163] Wong, Stanford, *Wong On Dice*. Pi Yee Press, Las Vegas, 2005.

[164] Zender, Bill, Banking Games in California: The Last Gold Rush. *Casino Enterprise Management*, July 2010, pp. 94–95.

Index

Aarhus, Denmark, 263
Aberdeen Proving Grounds, 247
Aces Up Gaming, 171, 178
Action button, 287, 298
Addition Rule, 74, 76
All-sevens set, 139
All/Tall/Small, 64, 175–178
Alphabetic Roulette, 40–41, 95
American roulette, 27, 33, 39, 43, 49, 70, 73, 83, 94, 102, 143
Any Craps, 179
Any Pair, 146, 187
Any Seven bet, 114, 126, 139, 189
Artichoke Joe's Casino, 287
Aruba, 57
Astragalus, 105, 184
Atlantic City, NJ, 77, 189, 190, 239, 244, 250, 307
Atlantis Casino, 273
Au-Yeung, Stephen, 81
Auburn, WA, 314
Australia, 280
Average, 18
Aviator Casino, 230

Baccarat, 8, 191–238, 313
Baccarat Bonus, 223–224
Baccarat World, 64, 221–222
Backline bet, 216
Backline betting, 287
Bahamas, 111
Bai Cao Monkey 9, 235
Baldwin, Roger, 247
Barlowe, Katie, 333
Barona Casino, 160
Bartschelly, Wolfgang, 90

Basic strategy, 246–249, 259, 260, 277, 279, 305, 322, 330
Basket bet, 34, 57, 62, 91, 99
Battle Creek, MI, 203
Bay 101 Casino, 226, 229
Beat the Dealer, 258, 264
Bell Gardens, CA, 290
Benjamin, Arthur, 333
Bennett, Barbara J., 333
Bereuter, Bernard, 146
Bicycle Hotel & Casino, 290, 298
Big 6, 146, 184, 185
Big 8, 146, 184, 185
Big Bet Gaming, 310
Big Red bet, 112, 115, 152, 280
Big Six wheel, 55
Bingo, 14
Binion's Horseshoe, 245
Binomial distribution, 20
Binomial experiment, 20
Binomial formula, 21
Binomial random variable, 21
Biribi, 32–33
Black Jackpot, 273–276
Blackjack, 8, 9, 17, 198, 208, 209, 239–324
Blackjack Elite, 284
Blackjack II, 239
Blackjack Ranch, 293
Bond. James Bond, 191
Bonus Baccarat Zero, 232
Bonus Box bet, 40
Borgata Casino, 77, 78
Brahms, John, 333
Breakage, 315
Brokopp, John, 334
Buro, Chris, 334

Bust pair, 282–284
Busting, 247
Buy bet, 144, 158

Cabazon, CA, 234
Cabot, Anthony N., 337
Caesars Tahoe, 308
Caesarscasino.com, 47
California Aces, 284, 290–293
California Baccarat, 234
California Blackjack, 288–290, 293,
    294
California Grand Casino, 300
Cambodia, 232
Cancellation system, 89–90
Cannery Casino, 186
Cantey, Wilbert, 247
Capitol Casino, 230, 285, 286
Card counting, 257–266, 279, 306
Card craps, 155–167
Card eating, 266
Card-Based Craps, 158
Card-Based Roulette, 55
Carlson, Bryce, 334
Caro, Mike, 102, 327, 334
Ca$h Card, 100–101
Cash, Les, 334
Catlin, Donald, 183, 334
Center for Gaming Innovation, 311
Chambliss, Carlson R., 334
China Bear, 234
Chinese roulette (game), 49, 55
Chinese roulette (system), 84–85
Choctaw Casino, 158
Cincinnati, OH, 271
Cohen, R. Carl, 335
Coin tossing, 19
Collection fee, 214, 229, 285, 289
Color Match 21, 279
Colors bet, 64, 67–68
Columbus, OH, 143
Combination, 14
Come bet, 145
Commerce Casino, 215, 229
Complement, 3

Complement Rule, 3, 149
Compton, CA, 236
Conditional probability, 9
Continental Casino, 252
Continuous shuffling machine, 159,
    243
Contra Costa, CA, 300
Crapless Craps, 142–143, 187
Craps, 105–190, 208, 230, 242
Craps Free Craps, 143–144
Crooked dice, 124–135
Crown Casino, 205, 272
Crown Pontoon, 272–273
Crystal Casino, 250
Crystal Park Casino, 236
Curaçao, 208
Curtis, Anthony, 335

D Casino, 40
Dace, Mark, 335
Dai Bacc, 204
David, Florence Nightingale, 105
Delano, CA, 230
Deluxe Devil Dice, 154–155, 188
Des Plaines, IL, 45
Destiny 21, 318–320
Devil Dice, 153–155
Devlin, Keith, 335
Dice, 1, 17, 105–106
Different Doubles, 173–175
Digits bet, 39–40, 94
Disjoint sets, 2
Don't Pass bet, 107–109, 142, 184
Double Bonus Spin Roulette, 49–53,
    98
Double D, 174–175
Double down, 241, 250, 264
Double Exposure, 267–270
Double Exposure II, 270
Double hardway, 156
Double pitch, 138
Downstream Casino, 157, 160
Dragon 7, 234
Dragon Tiger, 232–233

Early surrender, 242, 254

Easy Jack, 271–272
El Condado Casino, 252
El Rancho Casino, 252
Empire Global Gaming, Inc., 213, 277
Empirical Rule, 24–25, 141
En prison, 30–31
Engel, Jacob, 149
Envy bonus, 315
Epstein, Richard, 149, 335
European roulette, 27, 68, 75, 80, 81, 155, 280
Even-money option, 241
Even/Odd bets, 185
Event, 2
Expectation, defined, 18
Experiment, 1–2
Experimental probability, 4–5
Experto 21, 266–267
EZ Baccarat, 204, 234

Fa Fa Fabulous 4 Baccarat, 233, 234
Fab 4, 262
Fabulous 4, 223
Fabulous 4 Baccarat, 222–223, 233, 234
Factorial, 12
Fair game, 19
Fast Action Roulette, 41–45, 96
Favored 5, 84
Fermat, Pierre de, 1
Fibonacci progression, ix, 86–89, 316
Field bet (Blackjack), 301–302
Field bet (Craps), 64, 111, 116, 121, 132, 152, 155, 181, 209, 301
Field bet (Scarney Baccarat), 208
Field bet (Turn & Burn), 145
Fielder's Choice, 171–172, 178
50-card deck, 213, 277–281
50/50 Split, 323, 330–331
Fire Bet, 172–174
FireKeepers Casino, 203
Firelake Grand Casino, 55, 160
First Addition Rule, 6
Fitzgerald's Casino, 40

Five Color Blackjack, 277–280
Flex Action, 230–231
Flip Card Blackjack, 310–312
Florida Lottery, 20, 21, 25
Flush Roulette, 81–82
Fortune 7 Baccarat, 236
Foster, R.F, 336
Four Queens Casino, 32, 112, 171, 309
Free odds, 109–110
Free Ride, 254–256
Friedman, Stacey, 336
Frome, Leonard, 183
Frontier Hotel, 322
Fundamental Counting Principle, 11, 15

Gambler's Fallacy, 8, 82
Gambling Times, 102
Gardena, CA, 216
Gardens Casino, 223, 295
General Multiplication Rule, 10
Geyserville, CA, 145
Giant Jackpot 21, 321
Global Gaming Expo Asia, 280
Gold Coast Casino, 186
Golden Frog Baccarat, 229, 234
Golden Nugget Casino (London), 65
Golden Palace, 250
Golden ratio, 86
Grand Casino (Brussels), 80
Grand Monkey Bonus, 234
Grand Series, 99
Great 8 Baccarat, 211–213
Great Britain, 68
Griffin, Peter, 246, 337
Griffin, Tina, 337
Grossman, Howard, 244
Guaranteed Roulette System, 90

Hadsall, Joe, 337
Hannum, Robert C., 337
Hard double, 250
Hard Hardway bet, 183
Hard Pass, 169–170

Hardway odds, 186
Hardway set, 136, 137, 187
Harrah's Casino (Las Vegas), 185
Harrah's Marina, 239
Harrah's Resort Southern California,
  160
Hart's Palace, 250
Hawaiian Blackjack, 295–298
Hawaiian Gardens Casino, 295
Hawaiian Gardens, CA, 223
Hazard, 116–119
Henderson, NV, 211
Hi-Opt I, 262–264
Hide-Away Casino, 246, 253
High Dice, 186
High horn, 113
High-Low count, 258–262, 279, 309
Holloway, Louis G., 337
Hollywood Casino, 143
Hollywood Park Casino, 323
Hop bet, 115, 146
Horn bet, 114
Hot Action Blackjack, 285, 300
House advantage, defined, 19
Hunter, Samuel C., 337
Hustler Casino, 216
Hustler's hardway, 120

IGT, 49, 98, 321
Illustrious 18, 260
INAG International, 55
Independent events, 7–8
Index numbers, 260
Indian Gaming Regulatory Act, 62,
  213
Infinite deck approximation, 9, 156,
  193, 212, 222, 250–252, 257,
  271, 272, 275, 284, 301, 308
Infinite deck assumption, 260
Inside work, 124–128
Instant 18, 312
Insurance, 241, 250, 260, 315
Interblock USA, 77

Jack Casino, 271

Jack Jack, 271
Jackpot, 189
Jackpot Streets Double Six, 81
Jacobs, Kim, 337
Jacobson, Eliot, 220, 338
Jersey City, NJ, 156
Jeu 0, 98
Joker, 2, 55, 57, 60, 95, 99, 160, 251,
  285, 290, 293
Joker Seven, 62–63

Kallabat, Nicholas, 100
Keno, 94
Kewadin Casino, 251
Key-card locating, 265
Klondike Casino, 254–256
Kock, Aron, 310
Koi 8, 234
Krigman, Alan, 338
Kum-kum betting, 286

LA Blackjack, 284, 285, 293–296
Labouchère system, 89
Lammer, 196, 254
Las Vegas, 31, 32, 40, 67, 95, 110,
  142, 185, 186, 197, 245, 252,
  254, 267, 309, 322
Las Vegas Hilton, 323
Las Vegas Resort Hotel Association,
  264
Las Vegas Strip, 251, 255
Las Vegas Strip Rules, 243, 321
Late surrender, 242
Lauzon, Michael, 333
Law of Large Numbers, 5, 8
Lay bet, 114, 126, 158
Lightning Launch Roulette, 53–54
Lightning Roulette, 54
Lima, Peru, 252
London, 112
Los Angeles, CA, 214, 215, 295, 323
Lottery, 183
Lotto Store at Primm, 214
Low Dice, 186
Lowery, Jeffery Roy, 338

Lucky 13s Blackjack, 280–284
Lucky Ball, 76–77
Lucky Ladies, 64, 304–305, 323
Lucky One, The, 46
Lucky Ruby Border Casino, 232
Luxor Casino, 31, 95

M Resort, 211
Macau, 206, 253
Maisel, Herbert, 247
Majestic Match, 237
Malaysia, 205
Malmuth, Mason, 338
Marina Bay Sands Casino, 41, 43, 204, 207, 222, 233
Martingale, ix, 35, 85–90
May, John, 339
McDermott, James, 247
Mean, 18
Melbourne, Australia, 205
Michigan Lottery, 19
Midway bet, 190
Mini Roulette, 47
Mini-baccarat, 191
Mississippi, 68
Missouts, 127–128
Mohegan Sun Casino, 317
Monkey 9, 235
Monkey Baccarat, 234
Montbleu Casino, 308
Moore, Christopher, 333
Morongo Casino, 234, 235
Moscow, 250
Muckleshoot Casino, 314
Muggsy's Corner, 186
Multiplication Rule, 8, 74, 80
Mutually exclusive events, 6, 7
Mystery Card Roulette, 55

Natural, 252, 258, 305–307
Natural 8, 217–219
Natural 9, 217–219
Nehrt, Philip, 339
Neighbors bet, 99
Nelson, Rex, 339

Nepal Baccarat, 204, 205, 230
Nevada, 265
New Games Lab, 272
New Jersey, 156, 184, 265
New York, 250
New York City, 105
New York-New York Casino, 31
Newar bets, 65, 102
9 Bonus, 235
No Lose Free Roll, 67, 181–183
No-Bust 21st Century Blackjack Second Chance, 298–299
Norman, OK, 55
North Las Vegas, NV, 186
Northern Nevada Rules, 244

Ocean's Eleven Casino, 215, 216
Oceanside, CA, 215
Ohio Lottery, 46
101 Casino, 158
100 To 1 Roulette, 47–49, 99
One Up, 236
Orchid Casino, 57
Orleans Casino, 67, 244
Orphelins, 98
Outside work, 128–135
Over/Under, 308–309
Ox 8, 234

Pair splitting, 244, 264
Pala Casino, 60, 161, 162, 165, 189
Pala, CA, 60
Panda 8, 234
Partage, 31, 92
Party Craps, 158
Pascal, Blaise, 1
Pass bet, 107–109, 145, 184
Paulsen, Dennis, 339
Pechanga Casino, 160
Pentagon Craps, 146–150
Permutations, 12
Petaluma, CA, 158
Phase One, 84
Pick 'Em Blackjack, 312–314, 321
Place bet, 114, 133, 137, 144, 185, 187, 230

Planet Hollywood Casino, 31
Play Craps, 159–160
Playing cards, 3
Plaza Casino, 142, 192
Point Right Back, 170–171
Poker, 17
Polish Roulette, 35–36
Pontoon Pandemonium, 272, 273
Powerball, 8, 14, 183
Precious Pair, 222, 233
Prial, 62
Prime number, 66, 190
Prime Time (Roulette), 66–67, 102, 340
Primm, NV, 214
Pro-Aggressive Roulette, 70–72
Probability distribution, 17, 50
Probability distribution function, 17
Probability, defined, 3
Progressive 21, 307–308
Progressive jackpot, 64
Protection bet, 282–284
Pure 21.5 Blackjack, 284
Put bet, 143, 187
Python, 176, 177

Quadratic formula, 159

Rainbow Roulette, 73–75
Random variable, 16–18
Red/Black, 309–310
Reno, NV, 214, 273
Repeater, 178–181
Resorts International Casino, 189
Resorts World Genting Casino, 205
Rhythmic rolling, 136–141
Richardson, Joey, 340
Richer Roulette, 55–57
Rincon Craps, 160
Rio Casino, 185
River Rock Casino, 145
Riverboat Roulette, 36–39, 73
Rivers Casino, 45
Riverwind Casino, 55
Riviera Casino, 110, 131, 132, 191

Roberts, Stanley, 67, 217, 219, 315, 340
Roul 8, 80–81
Rouleno, 13
Roulette, 8, 19, 27–103, 242
Roulette 73, 96
Roulette Dinner, 92–94
Roulette Evolution, 98–99
Roulette Rage, 68–69
Roulette Royale, 75–76
Royal Caribbean, 280
Royal Roulette, 95–96
Royal Scandinavian Casino, 263
Running count, 259, 265

Sacramento, CA, 230
Sahara Casino, 197
Sample space, 2
San Bruno, CA, 287
San Diego, CA, 159
San Jose, CA, 226
Sands Casino, 251
Sands Roulette, ix, 31, 42, 43, 94
Santa Barbara, CA, 293
Sault Sainte Marie, MI, 251
Scarne, John, 207–211, 341
Scarney Baccarat, 207–211
Schlesinger, Don, 341
Scientific Games, 53
Scoblete, Frank, 341
Second Addition Rule, 6
September Surrender, 254
7 Point 7, 169
7 Up Baccarat, 207
Shackleford, Michael, 178, 341–342
Shawnee, OK, 55
Shuffle Master, 310
Shuffle tracking, 265
Sic bo, 154
Side bet, 260
Siegel, Donald S., 337
Silver City Casino, 254
Silver Sevens Casino, 252
Singapore, 204, 207
Skip straight, 154

Smallwood, Jake, 342
Smart, R.J., 342
Smith, John, 342
Snyder, Arnold, 342–343
Soares, John, 131, 343
Spanish 21, 314–315
Spanish deck, 272
Sparc, 149
Spider Craps, 149–153, 178
Spinette, 55
Splitting pairs, 241, 247, 253
Standard deviation, 22–25
Stasi, Perry, 172
Stateline, NV, 308
Stiff hand, 261
Straight 2 Flush, 324
Strat Casino, 143, 266
Stream, Matthew, 271
Street Craps, 119–131, 133, 187
Stupak, Bob, 35, 142, 181, 267
Subset, 11
Super 31, 305–307
Super 4 Progressive Blackjack, 317–318
Super 62 Roulette, 39–40, 96
Super 64 Roulette, 96
Super Bonus bet, 314–315
Super Fun Blackjack, 244–245
Super Green Bet, 61
$uper In$urance, 317
Supreme Baccarat, 230
Surrender, 241, 244, 253, 262

Theoretical probability, 4–5
Thorp, Edward O., 217, 255, 256, 258, 264, 267, 343
Three Faces, 206
3 Lucky Immortals, 233
Three Pictures, 205–206
Three-Card Baccarat, 205–206
3V set, 138
3-Way Seven, 189
Tie bet, 313
Tiers du cylindre, 98
Tiger 6 hand, 211–212

Tilton, Nathaniel, 343
Top of the Deck, 324
Tops, 128, 130
Total Shot, 230
Touchdown Roulette, 77–79
Triple Flop Roulette, 57–60, 99–100
Tropicana Casino (Atlantic City), 250
Tropicana Casino (Las Vegas), 40, 255
True count, 259–260, 262, 265
2urbo Blackjack, 323
Turn & Burn Craps, 151
Turn & Burn Craps, 145–146, 187
Turn bet, 146
21st Century Baccarat, 215
21 (movie), 258, 265
21+3, 302–304, 324
2–1 Baccarat, 205

Ultimate Counting System, 264
Uncasville, CT, 317
Under and Over 7, 185, 186
Union Plaza Casino, 192
United Kingdom, 280
Unity Technologies, 236–237, 324
University of Nevada, Las Vegas, 310
Upcard, 259, 309

Valley Center, CA, 55
Valley View Casino, 55
Vancura, Olaf, 320, 344
Vegas World Casino, 35, 181, 183, 266, 267
Venetian Casino, 31, 42, 94, 251
Venus, 184
Viejas Casino, 159
Vietnam, 232
Vigorish, 157
Villano, Matt, 344
Voisins du Zero, 99

Walden, William, 217, 343
Walker, Bert, 344
Watching for 0s, 83
Wechsberg, Joseph, 344

West Wendover, NV, 246
Westgate Casino, 323
Wheel clocking, 98
White Out, 37
Williams, Leighton Vaughan, 337
WIN Magazine, 168, 219, 315
Win Systems, 49
WINGO, 64, 67
Winstar World Casino, 158

WINSUIT, 64, 219–221
WINSURE, 315–317, 324
Wisted, Roger, 288, 290, 293
Wong, Stanford, 255, 256, 344
World bet, 114

Zender, Bill, 344
Zero 50/50 protection, 49
ZooBac, 226–229, 238
Zweikartenspiel, 267

Printed in the United States
by Baker & Taylor Publisher Services